LPWAN Technologies for IoT and M2M Applications

LPWAN Technologies for IoT and M2M Applications

Edited by

Bharat S. Chaudhari

School of Electronics and Communication Engineering,
MIT World Peace University, Pune, India

Marco Zennaro

T/ICT4D Laboratory, The Abdus Salam International
Centre for Theoretical Physics, Trieste, Italy

ACADEMIC PRESS

An imprint of Elsevier

ELSEVIER

Academic Press is an imprint of Elsevier
125 London Wall, London EC2Y 5AS, United Kingdom
525 B Street, Suite 1650, San Diego, CA 92101, United States
50 Hampshire Street, 5th Floor, Cambridge, MA 02139, United States
The Boulevard, Langford Lane, Kidlington, Oxford OX5 1GB, United Kingdom

Notices
Knowledge and best practice in this field are constantly changing. As new research and experience broaden our understanding, changes in research methods, professional practices, or medical treatment may become necessary.

Practitioners and researchers must always rely on their own experience and knowledge in evaluating and using any information, methods, compounds, or experiments described herein. In using such information or methods they should be mindful of their own safety and the safety of others, including parties for whom they have a professional responsibility.

To the fullest extent of the law, neither the Publisher nor the authors, contributors, or editors, assume any liability for any injury and/or damage to persons or property as a matter of products liability, negligence or otherwise, or from any use or operation of any methods, products, instructions, or ideas contained in the material herein.

British Library Cataloguing-in-Publication Data
A catalogue record for this book is available from the British Library

Library of Congress Cataloging-in-Publication Data
A catalog record for this book is available from the Library of Congress

ISBN: 978-0-12-818880-4

For Information on all Academic Press publications
visit our website at https://www.elsevier.com/books-and-journals

Publisher: Mara Conner
Acquisitions Editor: Tim Pitts
Editorial Project Manager: Gabriela D. Capille
Production Project Manager: Anitha Sivaraj
Cover Designer: Greg Harris

Typeset by MPS Limited, Chennai, India

Working together
to grow libraries in
developing countries

www.elsevier.com • www.bookaid.org

Contents

8. **TV white spaces for low-power wide-area networks**   **167**

ANJALI ASKHEDKAR, BHARAT S. CHAUDHARI, MARCO ZENNARO AND
ERMANNO PIETROSEMOLI

9. **Performance of LoRa technology: link-level and cell-level performance**   **181**

DANIELE CROCE, MICHELE GUCCIARDO, GIUSEPPE SANTAROMITA,
STEFANO MANGIONE AND ILENIA TINNIRELLO

List of contributors

Ijaz Ahmad   University of Oulu, Oulu, Finland; VTT Technical Research Centre of Finland, Espoo, Finland

Zeinab E. Ahmed   Department of Computer Engineering, University of Gezira, Sudan

Raluca Maria Aileni   Politehnica University of Bucharest, Faculty of Electronics, Telecommunications and Information Technology, Bucharest, Romania

Elmustafa Sayed Ali   Department of Electrical and Electronics Engineering, Red Sea University, Port Sudan, Sudan

Bharadwaj Amrutur   Robert Bosch Centre for Cyber Physical Systems, IISc, Bengaluru, Karnataka

S.V.R. Anand   Robert Bosch Centre for Cyber Physical Systems, IISc, Bengaluru, Karnataka

G. Araniti   DIIES Department, Mediterranean University of Reggio Calabria, Reggio Calabria, Italy

Paventhan Arumugam   ERNET India, India

Mukunth Arunachalam   Robert Bosch Centre for Cyber Physical Systems, IISc, Bengaluru, Karnataka

Anjali Askhedkar   School of Electronics and Communication Engineering, MIT World Peace University, Pune, India

Hari Krishna Atluri   ERNET India, India

Suresh Borkar   Department of Electrical and Computer Engineering, Illinois Institute of Technology, Chicago, IL, United States

Suresh R. Borkar   Department of Electrical and Computer Engineering (ECE), Illinois Institute of Technology, Chicago, IL, United States

Ahcène Bounceur   University of Brest, LabSTICC, Brest, France

Guillermo Cañada   Technical University of Madrid, Universidad Politecnica de Madrid, Madrid, Spain

Bharat S. Chaudhari   School of Electronics and Communication Engineering, MIT World Peace University, Pune, India

Laurent Clavier   IMT Lille, Lille, France

Daniele Croce   Engineering Department, University of Palermo, Palermo, Italy

Guillermo del Campo   Technical University of Madrid, Universidad Politecnica de Madrid, Madrid, Spain

Pierre Dufour   Actility, Lannion, France

Muhammad Ehsan   University of Pau, LIUPPA, Pau, France

Christophe Fourtet   Co-Founder and Chief Science Officer, Sigfox, Labège, France

Radek Fujdiak   Brno University of Technology, Brno, Czech Republic; Technical University of Ostrava, Ostrava, Czech Republic; VSB—Technical University of Ostrava, Ostrava, Czech Republic

Sheetal N. Ghorpade   RMD Sinhgad School of Engineering, Pune, India

Igor Gomez   Technical University of Madrid, Universidad Politecnica de Madrid, Madrid, Spain

Michele Gucciardo   Engineering Department, University of Palermo, Palermo, Italy

Rohit Gupta   Actility, Lannion, France

Mona Bakri Hassan   Department of Electronics Engineering, Sudan University of Science and Technology (SUST), Khartoum, Sudan

Olivier Hersent   Actility, Lannion, France

Derek Hunt   LoRa Alliance®, Fremont, CA, United States

A. Iera   DIIES Department, Mediterranean University of Reggio Calabria, Reggio Calabria, Italy

Thorsten Kramp   Formerly Semtech (International) AG, Rapperswil-Jona, Switzerland

Chetan Kumar S   Aikaan Labs Pvt. Ltd, Bengaluru, Karnataka

Rajagopal Maheswar   School of Electrical & Electronics Engineering (SEEE), VIT Bhopal University, Bhopal

Lukas Malina   Brno University of Technology, Brno, Czech Republic

Stefano Mangione   Engineering Department, University of Palermo, Palermo, Italy

Pavel Masek   Brno University of Technology, Brno, Czech Republic

Konstantin Mikhaylov   Brno University of Technology, Brno, Czech Republic; University of Oulu, Oulu, Finland

Petr Mlynek   Brno University of Technology, Brno, Czech Republic

Rania A. Mokhtar   Department of Electronics Engineering, Sudan University of Science and Technology (SUST), Khartoum, Sudan; Department of Computer Engineering, Taif University, Alhawiya, Taif, South Africa

A. Molinaro   DIIES Department, Mediterranean University of Reggio Calabria, Reggio Calabria, Italy

Amitava Mukherjee   Department of Computer Science and Engineering, Adamas University, Barasat, Kolkata, India

Hoang-Sy Nguyen   VSB—Technical University of Ostrava, Ostrava, Czech Republic; Binh Duong University, Thu Dau Mot City, Vietnam

Tuan Nguyen Gia   Turku Intelligent Embedded and Robotic Systems (TIERS) Group, University of Turku, Turku, Finland

Umber Noreen   University of Brest, LabSTICC, Brest, France

Sever Pasca   Politehnica University of Bucharest, Faculty of Electronics, Telecommunications and Information Technology, Bucharest, Romania

Congduc Pham   University of Pau, LIUPPA, Pau, France

Ermanno Pietrosemoli   T/ICT4D Laboratory, The Abdus Salam International Centre for Theoretical Physics, Trieste, Italy

Luca Piovano   Technical University of Madrid, Universidad Politecnica de Madrid, Madrid, Spain

S. Pizzi   DIIES Department, Mediterranean University of Reggio Calabria, Reggio Calabria, Italy

Benoît Ponsard   Director of Standardization, Sigfox, Labège, France

Pawani Porambage   University of Oulu, Oulu, Finland

Ari Pouttu   University of Oulu, Oulu, Finland

Jorge Peña Queralta   Turku Intelligent Embedded and Robotic Systems (TIERS) Group, University of Turku, Turku, Finland

Rakshit Ramesh   Robert Bosch Centre for Cyber Physical Systems, IISc, Bengaluru, Karnataka

Rashid A. Saeed   Department of Electronics Engineering, Sudan University of Science and Technology (SUST), Khartoum, Sudan; Department of Computer Engineering, Taif University, Alhawiya, Taif, South Africa

K. Samuylov   Peoples' Friendship University of Russia (RUDN University), Moscow, Russia

Asuncion Santamaria   Technical University of Madrid, Universidad Politecnica de Madrid, Madrid, Spain

Giuseppe Santaromita   Engineering Department, University of Palermo, Palermo, Italy

Nicolas Sornin   Semtech (International) AG, Rapperswil-Jona, Switzerland

Ramez Soss   Actility, Lannion, France

Martin Stusek   Brno University of Technology, Brno, Czech Republic

George Suciu   Politehnica University of Bucharest, Faculty of Electronics, Telecommunications and Information Technology, Bucharest, Romania

Ilenia Tinnirello   Engineering Department, University of Palermo, Palermo, Italy

Carlos Alberto Valderrama Sukuyama   Department of Electronics and Microelectronics, Faculty of Engineering, University of Mons, Mons, Belgium

O. Vikhrova   DIIES Department, Mediterranean University of Reggio Calabria, Reggio Calabria, Italy

Miroslav Voznak   Technical University of Ostrava, Ostrava, Czech Republic; VSB—Technical University of Ostrava, Ostrava, Czech Republic

Tomi Westerlund   Turku Intelligent Embedded and Robotic Systems (TIERS) Group, University of Turku, Turku, Finland

Alper Yegin   Actility, Lannion, France

Marco Zennaro   T/ICT4D Laboratory, The Abdus Salam International Centre for Theoretical Physics, Trieste, Italy

About the editors

Bharat S. Chaudhari received the ME degree in electronics and telecommunication engineering and the PhD degree from Jadavpur University, Kolkata, India in 1993 and 2000, respectively. After being a Full Professor in electronics and telecommunication engineering with the Pune Institute of Computer Technology and the Dean of the International Institute of Information Technology, Pune, India, he joined MIT World Peace University (then MIT Pune) Pune, as a Professor, in 2014. With more than 30 years of experience in teaching and research, he has authored more than 72 research articles in the fields of wireless, telecommunication, and optical networks, and edited a number of conference proceedings and a book. A recipient of a young scientist research grant from the Department of Science and Technology, Government of India, he has successfully executed several research projects and assignments. He has accomplished a number of international research collaborations and delivered technical talks at various conferences and platforms in different countries. His current research interests include low-power wide-area networks, Internet of things, wireless sensors networks, and optical networks. He has been the Simons Associate of the International Centre for Theoretical Physics (ICTP), Trieste, Italy, since 2015. He is a fellow of the IETE and IE (I), the Founder Chair of the IEEE Pune Section (R-10), and a Senior Member of IEEE. He is the Program Evaluator of Engineering Accreditation Commission (EAC) of ABET, United States, for accreditations of computer and communications engineering programs.

Marco Zennaro received the MSc degree in electronic engineering from the University of Trieste, Trieste, Italy, and the PhD degree from the KTH Royal Institute of Technology, Stockholm, Sweden. He is a Research Engineer with the Abdus Salam International Centre for Theoretical Physics, Trieste, where he coordinates the Telecommunications/ICT4D Laboratory, Wireless Group. He is a Visiting Professor with the Kobe Institute of Computing (KIC), Japan. His research interests include ICT4D and the use of ICT for development. In particular, he investigates the use of IoT in developing countries. He has given lectures on wireless technologies in more than 30 countries.

Preface

Low-power wide-area network (LPWAN) is a promising solution for long-range and low-power Internet of things (IoT) and machine-to-machine communication applications. The LPWANs are resource-constrained networks and have critical requirements for long battery life, extended coverage, high scalability, and low device and deployment costs. There are several design and deployment challenges such as media access control, spectrum management, link optimization and adaptability, energy harvesting, duty cycle restrictions, coexistence and interference, interoperability and heterogeneity, security and privacy, and others. This book is intended to provide a one-stop solution for study of LPWAN technologies as it covers a broad range of topics and multidisciplinary aspects of LPWAN and IoT. Primarily, the book focuses on design requirements and constraints; channel access; spectrum management; coexistence and interference issues; energy efficiency; technology candidates; use cases of different applications in smart city, health care, and transportation systems; security issues; hardware/software platforms; challenges; and future directions. This book will be helpful to the students, academicians, researchers, industry professionals, and practitioners to understand LPWAN technologies, in designing the networks, for research, in implementing and deploying IoT applications. The book is organized in 18 chapters, as described below:

Chapter 1, Introduction to low-power wide-area networks, presents a general introduction to LPWANs, innovative applications/services and their requirements, wireless access, and LPWAN application characteristics.

Chapter 2, Design considerations and network architectures for low-power wide-area networks, discusses the different key design considerations of LPWANs, networks, and topological aspects to give an overall architectural and design framework. It also briefly describes the major LPWAN technology solutions available as a segue to the upcoming chapters.

Chapter 3, LoRaWAN protocol: specifications, security, and capabilities, covers the LoRa and long-range wide-area network (LoRaWAN) protocol. It focuses on the technical specifications, regional parameters, activation and roaming, network-based and multitechnology geolocation, security, and capabilities.

Chapter 4, Radio channel access challenges in LoRa low-power wide-area networks, presents the review LoRA physical layer, orthogonality properties, network scalability, interferences and mitigation techniques, channel access mechanism, and reliability of clear channel assessment.

Chapter 5, An introduction to Sigfox radio system, introduces with the ultra-narrow band Sigfox technology along with its benefits, communication rules, coding, frame structure, interfaces, and unique features.

Chapter 6, NB-IoT: concepts, applications, and deployment challenges, discusses the narrowband IoT technology. It covers fundamentals concepts, benefits, characteristics, architectures, standards, working principles, frame structure, and applications.

Chapter 7, Long-term evolution for machine-type communication, presents long-term evolution (LTE) for machine-type communication. The chapter focuses on the LTE foundation, major applications, architecture, and operational aspects; interrelationships from LTE and LTE-M; future coexistence between LTE-M and 5G networks; and a summary of LTE-M evolution along with selected use cases.

Chapter 8, TV white spaces for low-power wide-area networks, covers the study on TV white spaces for LPWAN applications. The chapter presents the needs and advantages of TV white spaces, architectures, and protocols, using TVWS for LPWAN applications, future challenges, and deployment opportunities.

Chapter 9, Performance of LoRa technology: link-level and cell-level performance, presents numerical and experimental studies for link-level and cell-level performance of LoRa in the presence of interference. It covers the impact of interspreading factor, interference and fading, and scalability of the networks.

Chapter 10, Energy optimization in low-power wide area networks by using heuristic techniques, describes the review of various metaheuristics optimization techniques used for energy optimization in wireless sensor networks, along with analysis, evaluation, and applicability to LPWANs.

Chapter 11, Energy harvesting—enabled relaying networks, deals with energy harvesting—enabled relaying networks. It discusses the issues related to cooperative communication techniques, concerned factors, impairing of wireless relaying networks, and solutions.

Chapter 12, Energy-efficient paging in cellular Internet of things networks, discusses various solutions for paging aimed at improving the energy efficiency of IoT applications in cellular-based radio access technologies. It describes the basic power-saving solutions, paging strategies, and their applications, and open issues related to paging and radio wake-up schedules.

Chapter 13, Guidelines and criteria for selecting the optimal low-power wide-area network technology, presents the guidelines and criteria for selecting optimal LPWAN technology. It covers different aspects that affect the decision-making process, ranging from technical parameters to implementation to functional issues. It also covers the properties of LPWANs and comparison, along with some examples of technology selection use cases.

Chapter 14, Internet of wearable low-power wide-area network devices for health self-monitoring, describes different aspects concerning LPWAN-based wearable devices for remote health monitoring. The chapter deals with efficient algorithms for data processing and high-level optimization for minimal power consumption and enhanced data accuracy.

Chapter 15, LoRaWAN for smart cities: experimental study in a campus deployment, describes the use cases of LPWAN-based applications for smart cities, the experience in deploying interoperable LoRaWANs, management aspects in a campus environment, the impact of dense foliage, and other parameters on optimal network deployment.

Chapter 16, Exploiting LoRa, edge, and fog computing for traffic monitoring in smart cities, presents a hybrid edge-fog-cloud computing architecture for monitoring environmental parameters and traffic flow in a city. It also discusses a lightweight image processing algorithm to estimate traffic density.

Chapter 17, Security in low-power wide-area networks: state-of-the-art and development toward the 5G, is focused on the security aspects in LPWANs and the way toward 5G. It covers potential security threats, features, interrelations, interfaces, and significant features related to LPWANs. The chapter also discusses possible security issues and gaps for LPWAN and 5G integration.

Chapter 18, Hardware and software platforms for low-power wide-area networks, presents the study of various LPWAN hardware and software platforms available in the market for research and deployment. It also describes various open-source tools available for simulations and research.

Acknowledgment

The editors would like to acknowledge the interest and help of all the people directly and indirectly involved during the preparation of this book. The editors are sincerely thankful to all the chapter authors for their reader-friendly contributions of a very new and emerging technology. Our gratitude goes to all the reviewers for sparing their time and expertise in reviewing the book chapters thoroughly and helping in quality improvement. Editors would like to offer special thanks to Dr. Suresh Borkar, Illinois Institute of Technology, Chicago, United States, and Dr. Laurent Clavier, IMT Lille Douai, France, for their constructive inputs. Without support from authors and reviewers, this book would not have become a reality. We are grateful to the entire Elsevier team, especially Mr. Tim Pitts, Senior Acquisitions Editor, Ms. Gabriela Capille, Editorial Project Manager, and Ms. Anitha Sivaraj, Project Manager, for their untiring efforts in bringing out a quality publication. The editors are also grateful to leadership and colleagues at their respective serving organizations: MIT World Peace University, Pune, India, and International Centre for Theoretical Physics, Trieste, Italy, for encouraging and providing all necessary support for this project. Last but not least, the editors are indebted to their family members and well-wishers for their continuous support and understanding.

1

Introduction to low-power wide-area networks

Bharat S. Chaudhari[1], Marco Zennaro[2]

[1]SCHOOL OF ELECTRONICS AND COMMUNICATION ENGINEERING, MIT WORLD PEACE UNIVERSITY, PUNE, INDIA [2]T/ICT4D LABORATORY, THE ABDUS SALAM INTERNATIONAL CENTRE FOR THEORETICAL PHYSICS, TRIESTE, ITALY

1.1 Introduction

With the emergence of the Internet of things (IoT) and machine-to-machine (M2M) communications, massive growth in the sensor node deployment is expected soon. According to the forecast by Ericsson [1], around 29 billion devices will be connected to the Internet by 2022. These connected IoT devices include connected cars, machines, meters, sensors, point-of-sale terminals, consumer electronics products, wearables, and others. IoT survey reported on the Forbes website [2] forecasts more than 75 billion IoT device connections by 2025. HIS Markit [3] forecasted that the number of connected IoT devices would grow to 125 billion in 2030. The exponential growth in IoT is impacting virtually all stages of industry and nearly all market areas. It is redefining the ways to design, manage, and maintain the networks, data, clouds, and connections.

With highly anticipated developments in the fields of artificial intelligence, machine learning, data analytics, and blockchain technologies, there is immense potential to exponentially grow the deployments and its applications in almost all the sectors of society, profession, and industry. Such progression allows any things such as sensors, vehicles, robots, machines, or any such objects to connect to the Internet. It enables them to send the sensed data and parameters to the remote centralized device or server, which provides intelligence for making an appropriate decision or actuating action.

In general, IoT applications require energy-efficient and low-complexity nodes for a variety of uses that are to be deployed on scalable networks. Currently, wireless technologies such as IEEE 802.11 wireless local area networks (WLAN), IEEE 802.15.1 Bluetooth, IEEE 802.15.3 ZigBee, low-rate wireless personal area networks (LR-WPAN), and others are being used for sensing applications in the short-range environments. In contrast, wireless cellular technologies such as 2G, 3G, 4G, and 5G can be extended to long-range applications. Primarily, WLAN and Bluetooth were designed for high-speed data communication, whereas ZigBee and LR-WPAN were designed for wireless sensing applications in the local

LPWAN Technologies for IoT and M2M Applications. DOI: https://doi.org/10.1016/B978-0-12-818880-4.00001-6

environments and are used for low data—rate application for communication distances ranging from a few meters to a few hundred meters, depending on the line of sight, obstacles in the path, interference, transmit power, etc. Wireless cellular networks such as 2G, 3G, and 4G are designed for voice and data communication, not primarily for wireless sensing applications. Although these technologies are used for sensing for one or other ways in some of the applications, their performance in terms of performance metrics used in the wireless sensor networks may not be acceptable.

Hence, to support such requirements, a new paradigm of IoT, called low-power wide-area networks (LPWAN) is evolved. The LPWAN is a class of wireless IoT communication standards and solutions with characteristics such as large coverage areas, low transmission data rates with small packet data sizes, and long battery life operation [4]. The LPWAN technologies are being deployed and have shown enormous potential for the vast range of applications in IoT and M2M, especially in constrained environments.

1.2 Intelligent applications and services

The growing popularity of IoT use cases in domains that rely on connectivity spanning large areas and able to handle a massive number of connections is driving the demand for massive IoT technologies. With the advancement in the field of miniaturized electronics, communication, computing, sensing, actuating, and battery technologies, it is possible to design low--power, long-range networking technologies with many years of battery life and tens of kilometers coverage. These technologies have to be Internet-compatible so that data, device, and network management can be undertaken through cloud-based platforms. The most critical requirements of wireless IoT/M2M devices are low power consumption with extended transmission range, support to massive number of devices, the capability to handle RF interference, low cost, easy deployment, and robust security for the both, applications and network level. LPWAN technologies are promising and can be deployed for a broad range of smart and intelligent applications, including environment monitoring, smart cities, smart utilities, agriculture, health care, industrial automation, asset tracking, logistics and transportation, and many more as given in Table 1—1.

1.2.1 Application requirements

Various applications have varying requirements. Coverage, capacity, cost, and low-power operation are of course the primary drivers for all LPWAN applications. However, any LPWAN solution may entail significant tradeoffs between different requirements, for example, coverage versus cost. In addition, some applications are comparatively homogeneous, for example, meters, whereas others have a plethora of heterogeneous devices with varying expectations. In addition, selected applications require other capabilities, for example, interworking with other technologies, voice support, among others. Hence a specific LPWAN solution may be customized to a narrow set of applications, whereas another solution may be designed to cover a range of applications and attributes.

Table 1–1   Applications of LPWANs.

Field	Major applications
Smart cities	SMART CITY	Smart parking, structural health of the buildings, bridges and historical monuments, air quality measurement, sound noise level measurement, traffic congestion and traffic light control, road toll control, smart lighting, trash collection optimization, waste management, utility meters, fire detection, elevator monitoring and control, manhole cover monitoring, construction equipment and labor health monitoring, environment and public safety
Smart environment		Water quality, air pollution, temperature, forest fire, landslide, animal tracking, snow level monitoring, and earthquake early detection
Smart water		Water quality, water leakage, river flood monitoring, swimming pool management, and chemical leakage
Smart metering		Smart electricity meters, gas meters, water flow meters, gas pipeline monitoring, and warehouse monitoring

(Continued)

Table 1–1   (Continued)

Field		Major applications
Smart grid and energy	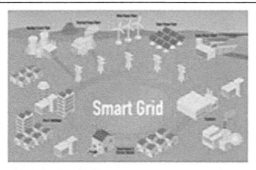	Network control, load balancing, remote monitoring and measurement, transformer health monitoring, and windmills/solar power installation monitoring
Security and emergencies		Perimeter access control, liquid presence detection, radiation levels, and explosive and hazardous gases
Retail		Supply chain control, intelligent shopping applications, smart shelves, and smart product management
Automotives and logistics	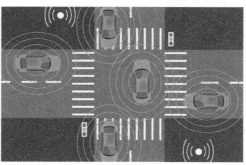	Insurance, security and tracking, lease, rental, share car management, quality of shipment conditions, item location, storage incompatibility detection, fleet tracking, smart trains, and mobility as a service

(Continued)

Table 1–1   (Continued)

Field		Major applications
Industrial automation and smart manufacturing	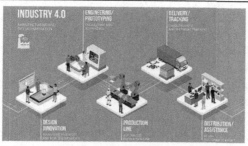	M2M applications, robotics, indoor air quality, temperature monitoring, production line monitoring, ozone presence, indoor location, vehicle auto-diagnosis, machine health monitoring, preventive maintenance, energy management, machine/equipment as a service, and factory as a service
Smart agriculture and farming		Temperature, humidity, alkalinity measurement, wine quality enhancing, smart greenhouses, agricultural automation and robotics, meteorological station network, compost, hydroponics, offspring care, livestock monitoring and tracking, and toxic gas levels
Smart homes/buildings and real estate	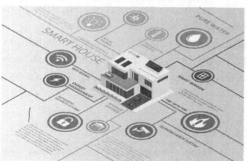	Energy and water use, temperature, humidity, fire/smoke detection, remote control of appliances, intrusion detection systems, art, goods preservation, and space as a service
eHealth, life sciences, and wearables		Patient health and parameters, connected medical environments, health care wearable, patients surveillance, ultraviolet radiation monitoring, telemedicine, fall detection, assisted living, medical fridges, sportsmen care, tracking chronic diseases, tracking mosquito and other such insects population and growth

Table 1−2 provides a mapping of selected applications to the corresponding emphasis that needs to be placed by the intended LPWAN solution similar to what has been done in Ref. [5]. In addition to the primary categories of coverage, capacity, cost, and low-power operation, another requirement area added is "additional specific" one. This covers the additional features mentioned that may be needed for a specific application. The relative scales for applicability of the requirement to the application are High (H), Medium (M), and Low (L). Table 1−2 provides the context for an LPWAN solution for carrying out architectural and design decision driven by which application or set of applications the technology is being targeted to. Some selected examples of the categorization in Table 1−2 are highlighted below.

Coverage is of fundamental value to almost all LPWAN applications and hence it is identified to be of high relevance to them. However, typical manufacturing environment may entail localized operations. In such a case, tradeoffs may be carried out to focus on types and number of devices to be supported and the intense coverage requirement may be compromised. Low-power operation is driven primarily by availability of electric power supply, for example, agricultural applications. In such situations, various sensors are in far out and sometimes difficult to reach locations and hence batteries lasting 10 + years without recharging are needed. Low-power operation is considered to be of high significance in such applications. In others, for example, retail, electric power may be readily available and low-power operation may be considered of low priority. In many instances, an application with massive number of devices requires very low-cost devices, for example, smart metering, whereas others such as smart homes may be able to absorb reasonable cost. This is hence captured as of low relevance for such applications.

For design considerations to be addressed in Chapter 2, Design challenges and network architectures for low-power wide-area networks, further granularity is needed to these requirements categories. The major characteristics corresponding to these requirements are summarized in Fig. 1−1 and elaborated on in the next section.

Table 1–2 Mapping of applications with their requirements.

Applications	Coverage	Capacity	Cost	Low power	Additional specific
Smart cities	H	H	H	M	H
Smart environment	M	H	H	H	M
Smart water	H	M	M	M	L
Smart metering	H	H	H	M	L
Smart grid and energy	H	H	M	M	M
Security and emergencies	H	L	M	H	H
Retail	H	H	H	L	M
Automotives and logistics	H	H	M	L	H
Industrial automation and smart manufacturing	L	H	H	L	L
Smart agriculture and farming	H	H	M	H	L
Smart homes/building and real estate	H	M	L	L	L
eHealth, life sciences, and wearables	H	H	M	H	H

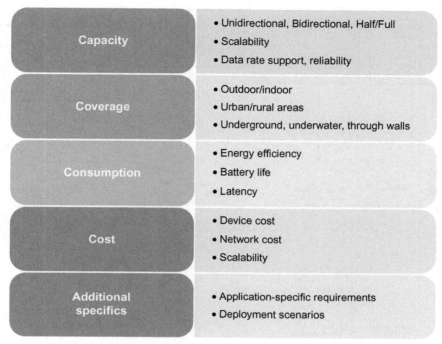

FIGURE 1–1 Application requirement priorities and characteristics.

1.3 Wireless access

IoT and LPWAN provide the basic foundational system for many applications. It plays a critical role in fulfilling the agile and dynamic requirements of applications and services and provides the framework for offering effective and efficient solutions. For communications and interconnections of such applications, a range of proprietary and standards-based solutions are available. The networks span different geographic ranges, as shown in Fig. 1−2.

Wireless proximity networks based on radio frequency identification and near-field communication are the near-me area network-type communication networks for the devices in close proximity. WPANs are used to convey information over short distances among the group of participant devices with little or no infrastructure. These networks can be connected to cloud platforms through a centralized device or server. Most of the WPANs are designed for low data−rate, power-efficient, short distance, and inexpensive solutions. The prominent WPAN technologies include IEEE 80.15.4 low-rate WPANs, ZigBee, WirelessHART, ISA100.11a, 6LoWPAN, Wibree, Bluetooth low energy, INSTEON, Wavenis, Z-Wave, ANT + , Enocean, and CSRMesh. WLANs are primarily designed for high-speed data exchange between the devices with campus-wide coverage limited to a few hundred meters. WLAN technologies include the different flavors of IEEE 802.11 standard. Wireless neighborhood

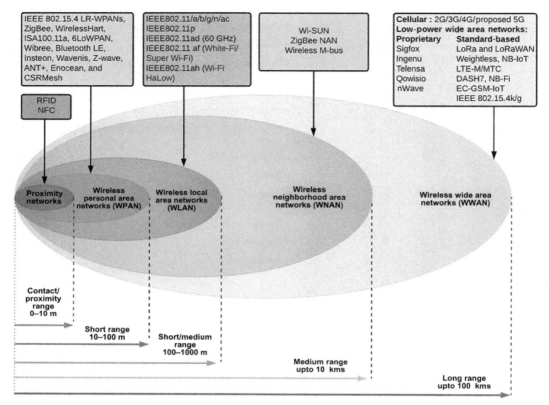

FIGURE 1-2 Wireless access geographic coverage.

area network (WNAN) has evolved in a new architectural system element for broadband wireless local distribution applications, which comprise of service area smaller than metropolitan but larger than local area networks. It can be used for residential, campus, street-level environments for utility and smart grid applications. The technologies for WNAN are Wi-SUN, ZigBee NAN, and Wireless M-bus.

WWANs are designed to cater to larger areas compared to LANs and WNANs. They have different requirements for different applications in terms of coverage, power efficiency, data rates, scalabilities, resource reuse, and others. WWANs can be broadly classified into cellular and LPWANs. Cellular networks such as 3G and 4G are primarily designed to transfer data at high rate for a few to tens of kilometers. These networks support mobility and hence provide extended coverage beyond the range of a single cell via handover mechanisms. LPWANs are the wireless communication technologies designed to allow long-range communications with low power consumption, low-cost interface, and a relatively low bit rate for IoT and M2M applications. Most of the intelligent applications will require some combinations of the above wireless access solutions.

There is a major segment of IoT-based applications which span long distance and are sensitive to both cost and power consumption. Such emerging networks are classified as LPWAN. It is estimated that one-fourth of overall IoT/M2M devices are to be connected to the Internet using either proprietary or standard LPWAN technologies. LPWAN-based applications are expected to be one-third of all IoT applications. Technologies other than LPWAN typically focus on achieving higher data rates, lower latency, and higher reliability. LPWAN solutions typically involve a massive number of end devices, send small-sized infrequent messages, and are tolerant of reasonably long end-to-end delays. Reliability requirements are varied depending upon the application. LPWAN technologies complement and sometimes supersede the conventional cellular and short-range wireless technologies in performance for various emerging applications [6].

1.4 Low-power wide-area network application characteristics

The extensive range of LPWAN applications requires interconnection and communications between a diverse set of devices. These devices span coverage ranging from very short to remote distances, from stationary to moving positions, from the battery based low power to commercial power-based connections, and a range of friendly to hostile environments. A significant share of low-power wide-area solutions typically send small-sized messages infrequently, are delay-tolerant, do not need high data rates, and require low power consumption and low cost.

IoT applications can be categorized as per the coverage needs and performance requirements in terms of transmission rates, delay, power consumption, etc. The coverage requirements for different applications are highly localized, for example, indoor stationary deployments. For the applications involving device mobility such as asset tracking requires global service coverage [7]. LPWAN applications are categorized as Massive IoT applications in contrast to critical IoT applications that require ultralow latency and ultra-high reliability. The characteristics and requirements for the applications that characterize LPWAN solutions are indicated below. The crucial characteristics include handling of M2M traffic, massive capacity, energy-efficient, and low-power operations, extended coverage, security, and interworking.

1.4.1 Coverage

1.4.1.1 Traffic characteristics
The inherent communication mechanism of LPWAN networks is traffic generated by distributed sensors. In addition to the possible presence of traffic created by smartphones or other devices, the LPWAN traffic itself can vary in a wide range of attributes such as the number of messages, message size, and reliability requirements. LPWAN technologies have diverse categories of applications with varying requirements. Some of the applications are delay-tolerant (e.g., smart metering); while applications such as fire detection, nuclear radiation detection, and home security require prioritized and immediate transmission. In some

applications, a priority message scheduling may be required for event-triggered transmissions. With the massive number of active devices, there is a possibility of a service level agreement (SLA) requirement of each application that might not be satisfied. Mechanisms need to be supported for the coexistence of different traffic types, the required quality of service (QoS), and SLA. In LPWAN applications, provision may need to be made for handling multiple classes of the end devices based on their communication needs in uplink or downlink. In some applications, device mobility support is needed, requiring being connected anywhere and ensuring seamless service on the move.

1.4.1.2 Coverage
The range of operations requires both long-range and short-range communications. Typically, LPWAN needs to provide long-range communication up to 10−40 km in rural/desert zones and 1−5 km in urban zones with +20 dB gain over the legacy cellular networks [6,8]. Indoor hard to reach locations such as underground locations and basements, and also the coverage which results in signal propagation through buildings and walls is needed, especially for the application involved in monitoring and collecting data. Coverage needs to be consistent with expectations on adaptable data rates and managed data error rates. Use of the sub-GHz band helps most of the LPWANs to achieve robust and reliable communication with a lower power budget as the lower frequencies of the sub-GHz band have better propagation characteristics as compared to 2.4 GHz band. Additionally, the slow modulation techniques used for LPWAN put more energy for each bit and hence increase the coverage. Slow modulation also helps the receivers in demodulating the signal correctly.

1.4.1.3 Location identification
The location identification for devices is a crucial requirement. Location accuracy plays critical role in applications such as logistics and livestock monitoring. It varies from a few centimeters to meters. Monitoring and security for sensing unusual events such as changed device location and facilitating the proper level of authentication need to be supported. Location identifications can be achieved by GPS, GPS-like systems, or by running smart algorithms with the help of network infrastructure.

1.4.1.4 Security and privacy
The security requirements for LPWAN devices are particularly stringent because of the massive number, vulnerabilities, and simplicity of the devices. The essential attributes of authorization, authentication, trust, confidentiality, data security, and nonrepudiation need to be supported. The security support should be able to handle malicious code attacks (such as worms), handle hacking into LPWAN devices and system, and manage eavesdropping, sniffing attacks, and denial-of-service attacks [9]. It is also important to protect the device identity and its location privacy from the public. Additionally, it should also support security for the forward and backward transmission as required in various applications.

1.4.2 Capacity

1.4.2.1 Capacity and scalability

One of the essential requirements for LPWAN is to support a massive number of simultaneously connected devices with the low data rate. Many applications require support for 100,000 + devices in a scalable manner. Scalability refers to the ability for seamlessly growing from a network of the small number of heterogeneous devices to massive numbers of devices, new devices, applications, and functions without compromising the quality and provision of existing services [9]. As LPWAN end-devices have low computational and power capabilities, network devices such as gateways and access stations can also play a vital role in enhancing scalability. Employing multichannel and multiantenna based on different diversity techniques can also significantly improve the scalability of LPWAN networks. However, it is to be ensured that such features do not compromise other performance metrics. A better solution could be a tradeoff to support the optimized performance and the requirements of the application. Secondly, the environment requires the transmission of data over confined and often shared radio resources. Such a large number of devices also results in high densification [6]. In such a case, there is always possibility of bottleneck at media access, large interference, and hence substantial degradation of performance of the network.

1.4.3 Cost

1.4.3.1 Cost-effectiveness

LPWAN applications are particularly sensitive to the device and operational cost. In addition to the standard requirements of low deployment and operating costs for the network, the large number of devices involved puts major constraints on cost, operational expenses, and an imperative of low power consumption. Software upgradability without changing hardware is a key attribute that needs to be supported. Besides, it becomes imperative to support scalability, easy installation and maintenance, and cost-effective functionality.

1.4.4 Low-power operations

1.4.4.1 Energy-efficient operations and low-power sources

In several applications of LPWANs, the environment and the constraints do not allow recharging of batteries. The battery is expected to last over 10 years without charging for AA or coin cell batteries. If the battery loses power and even the replacement of the battery is possible, it may not be doable in short periods. The cost of battery sources needs to be low. The LPWAN should be operated with strict and very low duty cycle limit so that node lifetime can be enhanced. Hence, ultralow-power operation is a crucial requirement for battery-powered IoT/M2M devices.

1.4.4.2 Reduced hardware complexity

In order to handle the large number, low cost, and long-range coverage, the design of small-sized and low-complex devices becomes an essential requirement. The reduced hardware complexity structure enables the reduction of the power consumption in battery-powered devices, without sacrificing too much performance. The devices generally are expected to possess low processing capabilities. Simple network architecture and protocols need to be supported by the hardware. From a technology point of view, in order to achieve the required adaptability of the LPWAN devices, radio transceivers need to be flexible and software-reconfigurable devices.

1.4.5 Additional specific requirements

1.4.5.1 Range of solution options
To allow flexibility and choices for the customer, operation support in both licensed and unlicensed bands is desired. Unlicensed spectrum may be derived from the industrial, scientific, and medical band. In many instances, customers prefer solutions that are upgradable from existing wireless access systems. There are demands for both custom proprietary and standards-based solutions. Applications require configurability between different topologies, including star, mesh, and tree.

1.4.5.2 Operations, interrelationships, and interworkings
The network should be able to handle heterogeneous devices. These large numbers of devices may share the same radio resources causing intra- and internetwork and technology interference resulting in degradation of network performance. Hence, LPWAN devices should possess the ability to connect and operate in varied LPWAN technology environments with interference tolerance, handling, and mitigation capabilities. The network should be able to enable connectivity of devices irrespective of hardware infrastructure and application programming interface. It is required to have seamless end-to-end interoperability between different network technologies. It requires standardization and gateway with adaptability protocols between various communications technologies. Full end-to-end application integration is expected.

1.5 Summarized objectives and expectations for low-power wide-area network

The LPWANs are resource-constrained networks and have critical requirements for long battery life, extended coverage, high scalability capabilities, low device cost, and low deployment cost. There are several challenges such as network virtualization, software-defined radio, further simplifying the media access control, dynamic spectrum management, use of TV white spaces, link optimization and adaptability, energy harvesting, duty cycle restrictions, scalability, localization, coexistence and interference mitigation, mobility, higher data rate and packet size, QoS support, interoperability and heterogeneity, security and privacy, congestion control, fulfilling the SLAs, integration with data analytics, use of artificial intelligence and machine learning techniques for performance improvement, development of testbeds and associated tools. These are covered in Chapter 2, Design challenges and network architectures for low-power wide-area networks along with how these characteristics and requirements translate into architectural and design considerations. An extensive amount of research is necessary to realize all these challenges and to expand the application landscape of LPWANs further and to compete with other cellular technologies.

References

[1] Internet of things forecast. <https://www.ericsson.com/en/mobility-report/internet-of-things-outlook>.

[2] Return on IoT: dealing with the IoT skills gap. <https://www.forbes.com/sites/danielnewman/2019/07/30/return-on-iot-dealing-with-the-iot-skills-gap/#27017efb7091>.

[3] The Internet of things: a movement, not a market. <https://ihsmarkit.com/Info/1017/internet-of-things.html>.

[4] Low-power wide area network (LPWAN) overview. <https://tools.ietf.org/pdf/rfc8376.pdf>.

[5] Frost and Sullivan, growing convergence of LPWAN & IoT technologies. <https://rfdesignuk.com/uploads/9/4/6/0/94609530/murata_lpwan_study.pdf>.

[6] U. Raza, P. Kulkarni, M. Sooriyabandara, Low power wide area networks: an overview, IEEE Commun. Surv. Tutor. 19 (2) (2017) 855−873.

[7] <https://www.ericsson.com/4ada75/assets/local/publications/white-papers/wp_iot.pdf>.

[8] K. Mekki, E. Bajic, F. Chaxel, F. Meyer, A comparative study of LPWAN technologies for large-scale IoT deployment, ICT Express 5 (1) (2019) 1−7.

[9] G.A. Akpakwu, B.J. Silva, G.P. Hancke, A.M. Abu-Mahfouz, A survey on 5G networks for the Internet of things: communication technologies and challenges, IEEE Access 6 (2017) 3619−3647.

2

Design considerations and network architectures for low-power wide-area networks

Bharat Chaudhari[1], Suresh Borkar[2]

[1]SCHOOL OF ELECTRONICS AND COMMUNICATION ENGINEERING, MIT WORLD PEACE UNIVERSITY, PUNE, INDIA [2]DEPARTMENT OF ELECTRICAL AND COMPUTER ENGINEERING, ILLINOIS INSTITUTE OF TECHNOLOGY, CHICAGO, IL, UNITED STATES

2.1 Introduction

Chapter 1, Introduction to low-power wide-area networks, discussed intelligent applications and their characteristics and requirements to be addressed by low-power wide-area network (LPWAN) technologies. This chapter maps the characteristics and attributes to the corresponding set of architectural facilities and design constructs. The intent is to provide an overall architectural and design framework in the context of which a given LPWAN technology can be evaluated with respect to its coverage of the LPWAN requirements. At the end of the chapter, major LPWAN technology solutions available in the marketplace are also mentioned and summarized as a *segue* to the upcoming chapters which cover the major solutions.

As stated in Chapter 1, Introduction to low-power wide-area networks, the primary emphasis for LPWAN technologies is on low-cost devices, small-sized data with low bandwidth, long battery life, large number of devices, and extended coverage. The key expectations are device costs in the $3−$7 range, packet sizes from 10 to 1000 bytes at uplink speeds of up to 200 kbps, battery life of 10 + years, support for 100k + devices, and coverage from 2 to 1000 km [1]. These characteristics are expanded in many cases to consider new applications. Hence, in addition to these core LPWAN characteristics, there are applications that may require larger data sizes and varying bandwidth, may tolerate somewhat reduced coverage, and may allow some compromise on cost. A typical LPWAN technology is designed using a set of design considerations and architectural constructs derived from these core and additional characteristics. Application of these design constructs results in various tradeoffs especially cost, performance, and hardware and software complexity. A repertoire of these design considerations is discussed next.

LPWAN Technologies for IoT and M2M Applications. DOI: https://doi.org/10.1016/B978-0-12-818880-4.00002-8

2.2 Design considerations for low-power wide-area networks

The characteristics and requirements for LPWAN solutions can be classified into the following major categories: Traffic characteristics, capacity and densification, energy-efficient operations and low-power sources, coverage, location identification, security and privacy, cost-effectiveness, reduced device hardware complexity, range of solution options, and finally operations, interrelationships, and interworking. Unlike traditional voice, data, or video-based applications, machine type communication (MTC)−based applications entail not just comparatively homogeneous types of devices and traffic characteristics but a wide range of technologies designs and architectures.

Table 2−1 summarizes the key design considerations needed to meet these requirements. Such traceability approach of associating requirements with their corresponding design considerations is a powerful and effective tool to ensure that each requirement can be mapped to its corresponding design entities for ascertaining the coverage of the requirement. It may be noted that a design construct may be addressing multiple requirements as well. The design considerations are grouped into two categories—Desirable and Enhancement. Desirable or Expected design constructs are the typical basic capabilities for LPWAN networks. They are generally applicable for homogeneous-type lower end of Internet of things (IoT) applications. Design constructs identified as enhancements are applicable for selective applications, may be needed to enhance specific characteristics, handle heterogeneous traffic environments, or manage varying types of devices in the same LPWAN network. Many of these enhancements primarily impact the access platform and not necessarily the device. The design considerations in the table are elaborated on subsequently.

2.2.1 Traffic characteristics

Support for the varying nature of devices and traffic with differing attributes of bandwidth, latency, size of packets, etc. drive a range of design considerations. Minimally, LPWAN technology needs to provide basic admission and user traffic management depending upon its network access architecture. Some LPWAN technologies may send data without having to undertake any elaborate admission activity and data bearer setup. On the other hand, another technology may decide to go through protocol-based admission control and data bearer set up. Basic admission and user data transmission are particularly relevant to applications where the traffic is homogeneous and limited and all devices are of same type, for example, meters only. However, for heterogeneous devices and coexistence of varying types of devices in the same LPWAN network, these capabilities need to be enhanced. Admission control mechanisms may need to have the capabilities of handling congestion, selectively blocking off admission for new traffic requests, and be able to handle admission on a priority basis. In some applications, multiple levels of device traffic priority management may be needed with different attributes of guaranteed and nonguaranteed bit rates for device data bearers.

Table 2–1 Design considerations summary for LPWAN networks.

Characteristics/requirements	Design considerations	Desired	Enhancement
Traffic	Admission and user traffic management	X	
	Selective admission management		X
	Priority structure for traffic management		X
	Mobility		X
	Control plane to carry user traffic for mobility applications		X
Capacity and densification	Globally unique identifiers	X	
	Software upgrade capabilities	X	
	Reliable and energy-efficient communication	X	
	Interference management	X	
	Multichannel operation	X	
	Software-defined radio (SDR)	X	
	Efficient multiaccess techniques		X
	Access, congestion, and overload		X
	Utilize diversity in channel, time, space, and hardware		X
	Adaptable and high link quality		X
	Transmission power control		X
	Dynamic state and context management		X
Energy-efficient operations and low-power sources	Energy-saving modes for devices	X	
	Media access control (MAC) support	X	
	Lightweight MAC	X	
	Off-load complex activities from devices to access stations	X	
	Utilization of simple waveforms for transceivers	X	
Coverage	Repeated transmissions	X	
	Power boosting	X	
	Interference management	X	
	Higher sensitivity of antennas and transceivers		X
	Adaptive modulation rate		X
Location identification	Cell ID in cellular systems	X	
	Time difference of arrival	X	
	Satellite-based global positioning system		X
Security and privacy	Basic mechanisms for authentication, security, and privacy	X	
	Over-the-air updates	X	
	Authentication by personalization	X	
	Automating posture management		
Cost-effectiveness	Software update capability	X	
	Reliable and energy-efficient communication	X	
	Simple MAC protocols and techniques	X	
Reduced device hardware	Flexible and software reconfigurability	X	
	Low-complexity transceiver structures	X	
Range of solution options	Support multiple physical layers		X
	Hardware, software, and protocols alignments		X
	Multimode and multifrequency devices		X
Operations, interrelationships, and interworkings	IPv6 and constrained application protocol	X	
	Internet Engineering Task Force IP network stack		X
	Service-level agreements		X
	Application, identity, security, and service management models		X

For applications requiring mobility, use of control plane to carry selected data traffic may be invoked in order to handle overload traffic conditions. For minimizing the processing load on the devices, several capabilities, for example, support for handovers, need to be primarily implemented in the access station instead of in the device.

2.2.2 Capacity and densification

From capacity viewpoint, the LPWAN technology needs to support and provision a large number of devices in a scalable manner. A large number of devices if concentrated in a comparatively small geographic area result in densification issues.

Globally unique identifiers and software upgrade capabilities are needed so that each device is uniquely addressable and scalability can be carried out without requiring physical access to or hardware changes in the device. To maximize scalability and network capacity, individual links need to be optimized for high link quality and reliable and energy-efficient communication needs to be supported. For applications with device heterogeneity, standardized protocols are required. Large number of devices and dense deployments of base stations also cause high levels of interference between signals from different devices. The basic approach to reduce interference is to use frequency reuse as part of network planning. In some instances, communication may need to be made resilient to interference by using adaptive modulation schemes, multiple channels, and doing redundant transmissions [2].

Efficient exploitation of diversity in channel, time, space, and hardware may be used. In some specialized applications, parallelization in transmission may need to be implemented by taking advantage of techniques like multichannels and multiple-input multiple-output (MIMO) configurations. As technology is advancing, software-defined radio (SDR) can be a good option for optimizing access and performance in an efficient manner. For selected applications where cost considerations and energy efficiency are important, such robust techniques are optional.

In situations with a large number of devices and densification, strong and efficient multi-access techniques are needed due to limited air resource issues and potential for cross-technology interference. A possible multiaccess approach can be nonorthogonal multiple access (NOMA) [3]. It may be noted that NOMA is also being considered for 5G access networks. NOMA can be applied to support diverse quality of service (QoS) requirements for differing types of end devices. Using NOMA, devices can achieve multiple access by exploiting power and code domains [3]. All the devices can utilize resources simultaneously. However, this can create interdevice interference. To mitigate it, multidevice detection techniques, for example, joint decoding or successive interference cancellation can be used to retrieve the device's signals at the receiver.

Multiple and adaptable link-level configurations may need to be supported with tradeoffs between different performance metrics such as bandwidth, latency, error rate, etc. This implies a need for adaptive techniques that can readjust its parameters for better performance based on monitor link quality. Adaptive modulation schemes, selection of better channels to reach distances reliably, or adaptive transmission power control are some of the techniques that can be utilized.

There is a significant need to develop novel densification approaches especially in many proprietary solutions. Such technologies may not be able to utilize synchronization technique and well-coordinated radio resource management. The dynamic state and context information needs to be stored for the devices in the access station and the core entities. Managing the state information of massive number of connected devices is also an issue that needs to be addressed.

2.2.3 Energy-efficient operations and low-power sources

Energy-efficient operations involve decreasing complexity and enabling efficient utilization of resources. The following are some key design considerations for extending the lifetime of devices to 10 + years of operation.

For energy-efficient designs, use of multiple energy-saving low-power modes, for example, sleep mode, is a desirable approach. Significant number of LPWAN applications involves sending of frame sizes of the order of tens of bytes transmitted a few times per day at ultra-low speeds [4]. The intent is to operate the nodes with extremely low duty cycle so that the nodes will be in sleep mode most of the time. Selected high power-consuming elements such as data transceivers may be turned off when not required. This reduces the amount of consumed energy. Only when data are to be transmitted or received, the transceiver is to be turned on. In case data are to be transmitted by the access station, the access station may wait until the time of sleep has expired or it may send explicit wake-up signals.

Media access control (MAC) is particularly influential in improving resource utilization efficiency and extending battery life. Efficient resource utilization can be done via scheduling optimization. Efficient scheduling involves assignment of transmission resources to selected devices with higher levels of channel quality. Another approach for extending battery life is by harvesting energy from other sources. Natural sources, for example, wind or solar power can be used. However, these sources may be unpredictable and unreliable. Capacitors may be utilized to store energy from these types of alternate sources. For consistent availability of energy, power may be derived from radio-frequency (RF) signals which transfer control and data information. There is tradeoff between lifecycle of a battery without these enhancements versus the resulting availability of additional power from the alternate sources after you take into account the overhead of these enhancements.

Lightweight MAC protocols and techniques to off-load complexity from end devices are some typical techniques that may be considered for energy efficiency. Use of simple media access techniques is an example of such an approach. In order to match the range and data rate requirements, flexible and inexpensive hardware designs are needed for implementing multiple physical layers [3]. Each of these layers can offer complementary solutions to corresponding applications.

Generally from an energy consumption perspective, communication operations consume more energy than processing operations in end devices. Power utilization in the devices can be reduced by using less complex or less frequent communication mechanisms as long as the corresponding additional processing power is still manageable. As an example, to reduce communication overhead, sending of a large number of unformatted data can be replaced by consolidating them into one formatted package prior to their transmission.

With the increasing number of connected LPWAN devices, additional design considerations that can be explored include the use of channel diversity, opportunistic efficient spectrum utilization, and adaptive transmission. The use of MIMO at the LPWANs access stations can also significantly increase the diversity gain or the data rates. Using space division multiplexing, the access station can boost the number of LPWAN devices that it can support.

Efficient spectrum utilization pertains to optimized usage of the overall spectrum resources in time, frequency, bandwidth, and spatial dimensions. In many instances, spectrum is not fully utilized. The access station or the device can sense and identify the gaps in spectrum usage and improve efficiency of spectrum utilization [3]. Such an approach can also reduce cross-interference.

2.2.4 Coverage

Coverage is a critical attribute for LPWAN-based applications. There are several aspects of coverage—reaching larger geographical area, areas around obstructions, and areas inside buildings. Techniques to overcome issues associated with degradation of signal include repeated transmission, boosting of transmitter power, higher sensitivity of antennas and transceivers, and decreasing the modulation rate which increases the probability of successful detection. It may be noted that similar to the scenarios for capacity, weak signals in enhanced coverage also result in additional interference. The techniques for interference management mentioned in handling capacity are also applicable to such situations.

Repeated transmissions entail sending the same data a defined number of times with the same or with different error-detecting codes. The receiver processes multiple copies of the packets received and extracts the data information. Power boosting implies increasing the transmission power to provide higher coverage. To handle low levels of signals and increase coverage, antenna configuration may be enhanced and receiver sensitivity can be increased. It may be noted that lower the modulation rate, the higher is the probability of detecting bits correctly. However, lower modulation rate also decreases the spectrum efficiency (bps/Hz), thus reducing the data bandwidth. This may be acceptable for many applications.

2.2.5 Localization

Determining the location of the devices is an important requirement for typical IoT applications. Devices in cellular network—based infrastructure have a built-in location method based on Cell ID. This capability is cost- and battery-effective but is usually only as accurate as the cell site's footprint. Time differences of arrival designs locate devices using the available infrastructure including gateways. Because gateways and cells are able to collect and transmit timing information, the devices can be located by comparing the times the signals arrive at multiple gateways or cells, sometimes called the triangulation scheme. It is one of the lowest cost solutions. Satellite-based location methods also work nicely for use cases that are not particularly battery-sensitive.

2.2.6 Security and privacy

Security threats arise especially due to heterogeneity of and physical accessibility to devices and the openness of the systems connected to the Internet through LPWAN wireless air interface [5]. Risk of vulnerability to cyberattacks is particularly relevant to IoT LPWAN MTC environments due to the massive number of devices involved. To avoid intrusion and hacking, encryption of both the application payload and the network admission request needs to be considered.

Due to cost and energy considerations, it becomes necessary for LPWAN networks to settle for simpler communication protocols for authentication, security, and privacy. Over the air as an alternative to authentication by personalization is a key facility to assure that end devices are not exposed to any security risks over prolonged duration [6]. This may be comparatively expensive in proprietary LPWAN networks operating in unlicensed bands.

In many networks, strong encryption and authentication schemes such as advanced encryption suite are used for confidential data transport, Diffie-Hellman utilized for key exchange and management, and Rivest-Shamir-Adleman are applied to authenticate digital signatures and key transport [2]. These are based on cryptographic suites with robust protocols. Similarly, for 3rd Generation Partnership Project (3GPP) cellular-based systems, subscription identity module—based authentication technique provides comparatively robust protection.

Automating posture management and use of software updates to patch vulnerabilities may be needed to isolate devices from potential attackers [4]. In order to provide protection and to isolate the device and its application, an overlay network may need to be explored. This provides security since the unique local addresses used in the overlay network do not allow access from outside the overlay.

2.2.7 Reduced device hardware complexity

Reduction in hardware complexity has multiple advantages including reduction of cost, device size reduction, and reduced power consumption.

One key area of effecting hardware simplicity is application of software reconfigurability to radio transreceivers [2]. Radios need to be able to operate over multiple-frequency bands, support varying types of waveforms, as well as various air interface technologies depending upon LPWAN-related and other interfaces. They need to be reconfigured based on software updates, without hardware changes. Moreover, SDR is considered one of the key technologies that enable the use of optimized spectrum access. The adoption of integrated circuit technologies and low-complexity transceiver structures, such as the direct-conversion radio (DCR) architecture [2], results in manufacturing efficiency. DCR neither requires external intermediate frequency filters nor image rejection filter. Use of such simplification runs the risks of RF imperfections, for example, in-phase and quadrature imbalance, phase noise, and amplifier nonlinearities. Several digital processing techniques that may cause energy consumption need to be applied to minimize the negative effects. These approaches usually require high-complexity processing, which is energy-demanding. Therefore there are trade-offs in implementing the simplification techniques.

2.2.8 Range of solutions options

Options and flexibility of using different LPWAN technologies create competition, reduce cost, and provide meaningful LPWAN network options to the user. For this, the hardware, software, and protocols supported in the device need to be able to interface with multiple technologies. The device needs to support multiple physical layers and align with the corresponding proto-cols associated with selected LPWAN technologies. Such adaptability also extends to availability of multiple options in terms of bandwidth and latency consistent with the application attributes. Multimode and multifrequency operations need to be supported on the devices. Multimode operations can interface with multiple technologies whereas multifrequency oper-ation refers to capability of operations on multiple frequencies on the same technology.

2.2.9 Operations, interrelationships, and interworking

Proprietary and standards-based LPWAN technologies have different strengths and weaknesses in terms of operations, coexistence, and interworking with each other. There are different chal-lenges for the technologies operating in unlicensed and licensed bands [7]. For unlicensed bands, flexibility and faster time-to-market are major advantages. However, the transceivers become more complex and interference needs to be managed. Proprietary technologies also need to adhere to regional regulations on the use of shared spectrum. These may put limits on their operations, for example, transmitted RF power [2]. For technologies operating in licensed bands, standardized solutions, QoS management facilities, and more robust control of the net-work and devices are major advantages. But cost, complexities associated with acquiring and managing the allocated spectrum, and delays in time-to-market are issues to be handled. In either case, use of sub-GHz frequencies achieves wider coverage, better penetration into the building and underground installation. Use of sub-GHz frequencies also leads to lower interfer-ence and increased sensitivity, making these technologies more energy-efficient.

Traditionally, proprietary technologies optimize their costs by using customized imple-mentations and lean and thin proprietary protocols. As competitiveness, portable services, availability of options to the users of multiple technologies, and utilization of common inter-faces become more important, customized and simple protocols need to be replaced by standardized ones. Typically, communication based on Internet protocol (IP) between net-work server applications and the end devices becomes the interface of choice. This allows portable and more generic implementations based on IPv6 and constrained application pro-tocol (CoAP) [8]. To incorporate simplicity, diverse radio technologies may need to exploit compressed form of IPv6 and CoAP. LPWAN device implementation requires small frame size due to their very low data rates. So traditional compression techniques that rely on octet or two available to signal both IP and CoAP may not be appropriate. The higher-level Internet Engineering Task Force (IETF) stack structure can also be used. This involves trade-offs with cost and simplicity as well.

Cross-technology interference also influences the performance of LPWAN technologies. Therefore, service-level agreements need to be put in place especially for unlicensed band operations.

Interoperability between heterogeneous technologies is crucial to the long-term viability of LPWAN technologies. The issues mentioned in the previous paragraph also need to be addressed when it comes to interoperability. Again, standardized design structures and protocols become an imperative. In addition, the emergence of new communication technologies introduces many integration challenges including alignment of common practices, service descriptions, standards, and discovery mechanisms. This creates incompatibilities and does not allow interworking. Common application, identity, security, and service management models need to be utilized by any LPWAN technology desiring to work with others. For the technologies, this has impact primarily on the physical and MAC layers in addition to work related to adherence to IP and IETF standards for higher layers.

2.3 Internet of things/low-power wide-area network layer model

IoT is considered as the third wave of the Internet after static web pages and social networking-based web. It uses IP to connect different types of sensors and objects worldwide [9]. In the last few years, different layered architectures have been proposed by different researchers. However, there is no consensus on a single globally agreed architecture for IoT. A generic four-layer model for IoT is shown in Fig. 2−1.

Like Transmission Control Protocol/IP protocol stack, the bottom layer, sometimes called as sensing and identification layer, is a physical layer that is mainly responsible for integrating hardware such as sensors, objects, actuators, etc. This layer is called as IoT perception layer. The IoT devices include remote sensor nodes, information collection devices, smart meters, smart devices, and intelligent electronic devices. This layer collects information from IoT devices and transmits the collected data to a network infrastructure layer. It also has capabilities to connect to different access stations and core entities of the networks. It performs modulation/demodulation, power control, transmission and reception of the signals. For LPWAN networks, some of the technologies have proprietary physical layer, for example, LoRa and Sigfox. Above the physical layer, there is network infrastructure layer for providing networking support and data transfer over wired and wireless networks. There can be one or two such layers, for example, data link and network, depending on the type of networks. For star topology architectures, data link layer comprising MAC and logical link control (LLC) sublayers is sufficient. However, whenever data are to be sent to a server on the cloud, network layer with a capability of routing the packets on the Internet is required based on IP. For LPWAN technologies, generally MAC/LLC is defined for single-hop networks based on star topology. For communications on the Internet, there is a wide range of implementations. Some systems may use all the principal layers of open system interconnection (OSI) reference model—transport, session, presentation, and application whereas others may use selected layers, for example, transport and application.

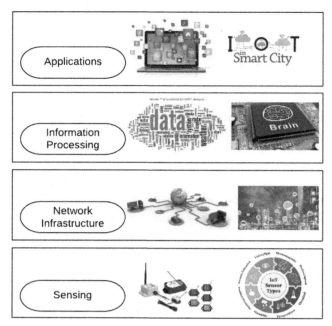

FIGURE 2–1 IoT four-layer model.

For IoT applications, the network layer can be wired or wireless. Depending on the type of IoT devices, appropriate communication network is used. As an example, ZigBee is used by sensor nodes in order to transmit the collected data wirelessly for very short distances. The remote communication network layer can also be wired or wireless. For wired connections, optical networking may be used. For wireless connectivity, use can be made of 2G, 3G, long-term evolution (LTE), or LPWANs.

The next layer above the network infrastructure layer is defined as information or service processing layer. It is responsible for managing the services as per the customer needs. Primary responsibilities include information analytics, security control, process modeling, and device control.

The application layer has integrated applications and provides interaction methods for users and applications. In several cases, support sublayers are added for special needs such as edge/fog computing and cloud computing [10]. Similar layers can be added and used on LPWAN gateway and server side.

2.4 Low-power wide-area network topologies and architecture

This section summarizes the topologies applicable to LPWAN networks and the architectural framework associated with them. Proprietary and standards-based LPWAN solutions use a set of topologies geared toward the applications they serve and the set of requirements they want to focus on. From topology viewpoint, LPWAN networks can be classified into two major groups—star and mesh. A selected technology can be configured into either of these groups if it is equipped with the designs needed for the topology and if deployment facilities exist to support it. Cellular technologies are generally versatile in this respect, and they also support mobility. There is a set of basic entities that form the architecture for LPWAN networks. Typical architectures are covered subsequently in this section.

2.4.1 Low-power wide-area network topologies

Prominent topologies are star and mesh. Generally, star or star-on-star topology is preferred for LPWAN over mesh network for preserving battery power and increasing the communication range. LPWAN's long-range connectivity allows such single-hop networks access to a large number of nodes, thus reducing the cost. From coverage viewpoint, traditional wireless sensor technologies such as ZigBee, Bluetooth, and Wi-Fi are not designed for wide coverage and hence are not directly applicable as LPWAN technologies.

The simplest form of wireless network topology is a point-to-point network in which nodes communicate directly with a central node. It is often used for remote monitoring applications and can be useful in hazardous environments where running wires is difficult or dangerous. Such LPWAN technologies support a star topology, as shown in Fig. 2−2A. A star network consists of one gateway node to which all other nodes connect. Nodes can only communicate with each other via the gateway. Node messages are relayed to a central server via gateways. Each end node transmits the messages to one or multiple gateways. The gateway forwards the messages to the network server where redundancy, errors, and security checks are performed. Star networks are fast and reliable because of their single-hop feature. Faulty nodes can also be easily identified and isolated. But, if the gateway fails, all the nodes connected to it become unreachable. Since the end node sends messages to multiple gateways, there is no need for gateway-to-gateway communication. This simplifies the design as compared to networks where the end nodes are mobile.

Mesh topology network consists of a gateway node, sensor nodes, and sensor-cum-routing nodes connected, as shown in Fig. 2−2B.

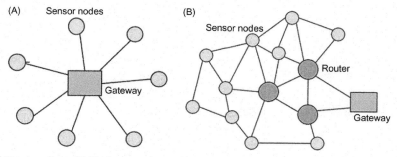

FIGURE 2–2 (A) Star topology. (B) Mesh network topology.

All the nodes can connect directly to each other in a full mesh topology. In a partial mesh topology, some nodes are connected to some of the others, but others are only connected to those with which they exchange the most messages with.

Mesh networks have several advantages such as availability of multiple routes for reachability, simultaneous up/downstream transmissions, easy scalability of the network, and capability of self-healing. These networks have some disadvantages including complexity due to redundant nodes, added latency because of multihop communication, and increase in cost. Redundancy of nodes also compromises the energy efficiency of the network.

2.4.2 Major architectures in low-power wide-area network technologies

The basic LPWAN architecture requires wireless access and connectivity to the Internet and cloud.

Based on the varying range of desirable and enhancement design constructs summarized in Table 2−1 and varied levels of layer operations proposed in Fig. 2−1, a specific LPWAN technology may require a selected set of architectural entities as shown in Fig. 2−3.

The basic function of an LPWAN device is to collect data and respond to inputs from the LPWAN network. Collected data are sent on a specific radio link to the wireless access station and on to the IoT network. The access station provides the radio link for device management and device traffic exchange. It maintains the integrity of the radio link by handling acceptable bit error rates, admissions, security, etc. The access station interfaces with the gateway/concentrator, in some instances as also called a core. There are different implementations depending upon proprietary or standards-based LPWAN technology being considered. The core is responsible for handling control and user plane traffic. It provides a conduit for information exchange between the access station and the IoT network and translation between the protocols supported by the access station on one side and by the network on the other. Depending upon the technology, a concentrator may provide edge computing

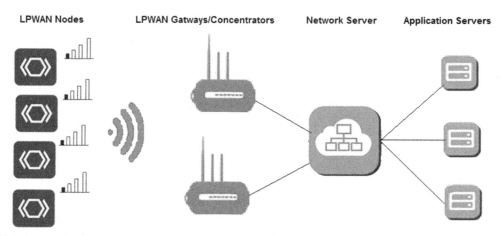

FIGURE 2–3 Typical LPWAN network entities.

and storage facilities to off-load the cloud. Because of its relative proximity to the end devices, this is particularly applicable to cases where device requires real-time support with low latency. For selected LPWAN technologies, the core may provide robust admission, priority treatment, and mobility support if applicable. LPWAN server is responsible for provisioning, registering, and operations for the LPWAN entities. It may also share or augment core functionalities such as routing of traffic, security, and priority handling with the core. The application servers and the cloud support the LPWAN network in carrying out the management of the database that contains the messages received from all the connected objects. It may use big data analytics to analyze and act upon the data.

2.4.3 Mixed hybrid architectures

The basic architecture mentioned in Section 2.4.2 provides direct device connectivity to the access station associated with the LPWAN technology. There are other access configurations also being considered. Two such prominent ones are discussed in the following.

Fig. 2—4 shows architecture wherein the primary connectivity to a device is provided by different access technologies such as ZigBee, Wi-Fi, etc. The corresponding gateway interfaces with the access point of an LPWAN network. This is particularly relevant for cellular LPWAN networks.

Another architecture that uses multiple LPWAN networks for providing a range of interfaces to end devices [11] is shown in Fig. 2—5. It shows an example where devices have

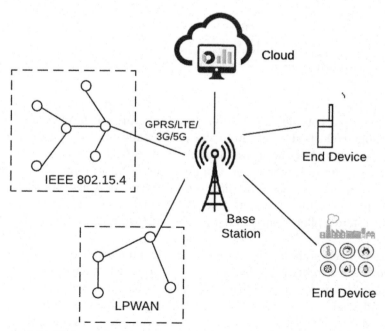

FIGURE 2–4 A generalized cellular-type architecture.

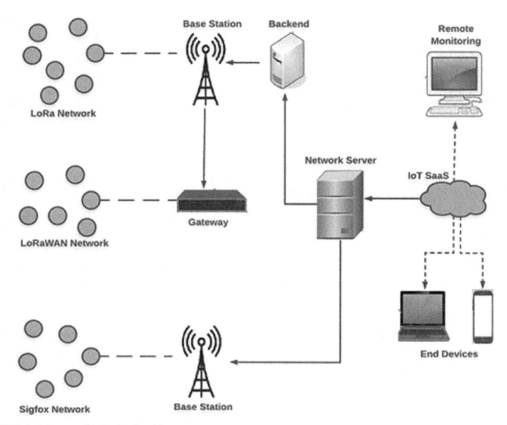

FIGURE 2–5 Proposed IoT mixed architecture.

access to both LoRa and Sigfox systems. Such multitechnology resident hybrid networks are particularly relevant for complex applications requiring different LPWAN technologies. Each LPWAN technology collects the data from the devices and nodes located in their coverage areas. The corresponding base or access stations are deployed consistent with their coverage area. The LPWAN gateway or nodes can communicate the user traffic data to the core network and the cloud. In such mixed architecture, the associated network server or core network entities perform the functions of device management such as registration, authentication, resource allocation, and data traffic management to the devices connected to their network.

New approaches and architectures based on artificial intelligence (AI) to create cognitive-LPWAN solutions are also being studied [12]. These approaches have led to a powerful cognitive computing ability to support advanced communication, handle heterogeneous IoT applications, and enable software-defined networking. Cognitive LPWANs support coexistence and interworking of a mix of a variety of LPWA technologies and provide users more efficient and convenient intelligent services. It has several applications including smart cities, green IoT, general heterogeneous networks, as well as AI applications such as smart home, health monitoring, automatic driving, and emotional interaction.

2.5 Introduction to proprietary and standards-based solutions

As mentioned in Chapter 1, Introduction to low-power wide-area networks, the major LPWAN solutions can be classified into two categories—proprietary and standards-based. In the following, the major LPWAN technologies are introduced.

2.5.1 Proprietary technologies

2.5.1.1 Sigfox

Sigfox [13] is a proprietary ultra-narrowband LPWAN technology that uses a slow modulation rate to achieve longer range. Initially, Sigfox supported only unidirectional uplink communication, that is, between the sensor devices to the aggregator with differential binary phase-shift keying (DBPSK) modulation. The later releases support bidirectional communication where Gaussian frequency-shift keying (GFSK) modulation is used for downlink. This ultra-narrowband feature of Sigfox allows the receiver to only listen in a tiny slice of the spectrum so that the effect of noise can be mitigated. Like LoRa, Sigfox also uses unlicensed industrial, scientific, and medical (ISM) bands. Sigfox has inexpensive sensor devices and cognitive SDR-based access stations to manage the network and for Internet connectivity. Sigfox supports very low data rate compared to other LPWAN technologies. To provide reliability, Sigfox transmits the message multiple times, resulting in high energy consumption. One of the main differences between Sigfox and LoRa is business distinction. Sigfox is deployed by network operators and the users need to pay the subscription charges, whereas LoRa can be deployed as an own independent network with no subscription charges. Sigfox gateway can handle up to a million connected objects, with a coverage area of 30−50 km in rural areas and 3−10 km in urban areas [14].

2.5.1.2 Ingenu

Ingenu is based on the proprietary random phase multiple access technique with more flexible spectrum regulations, allowing higher throughput and capacity [15]. The solution is proprietary in the sense that the company is the sole developer and manufacturer of the hardware. It uses direct sequence spread spectrum technique with a peak data rate of up to 80 kbps. Ingenu operates in the 2.4 GHz band, which gives it a shorter range than Sigfox and LoRa, and also encounters more propagation loss from obstructions, such as water or packed earth. 2.4 GHz band is widely used by many other personal and local area network technologies such as Wi-Fi, Bluetooth, and ZigBee, making it more congested, and hence results in an increase in the interference level. It offers low-power, low-cost, robust, and bidirectional communication. To add reliability to the transmission, it provides acknowledged transmission. As it has higher data transmission rates, the power consumption is more than Sigfox and LoRa. Ingenu was originally designed and focused on applications in utility, oil, and gas sectors. Nowadays, it is being proposed for a diverse range of applications such as smart city, agriculture, asset tracking, fleet management, smart grids, among others.

2.5.1.3 Telensa

Telensa is ultra-narrowband LPWAN technology. It works in 868 and 915 MHz unlicensed ISM bands. It has bidirectional communication capabilities, and hence it can be used for monitoring as well as for control. It has central management system (CMS) called Telensa PLANet, which is used for end-to-end operations adopted for an intelligent street lighting system. It consists of wireless nodes connecting individual lights in a dedicated network [16]. CMS reduces overall energy consumption and maintenance costs through its automatic fault detection system. These sensor nodes on the street light poles can be used for gathering the data of various parameters such as pollution, noise level, temperature, humidity, and radiation level, as required in the smart city applications. One Telensa base station can control 5000 nodes with low power for around 2 km in urban and 4 km in rural areas. It supports integration with support services such as asset management, metering, and billing systems. Telensa is presently available in more than 30 countries worldwide.

2.5.1.4 Qowisio

Qowisio is an ultra-narrowband, dual-mode technology for LPWAN applications. It is compatible with LoRa and provides technological choices and flexibility to the end-users [17]. It offers connectivity as a service to the end-users by providing the end devices and deploying the network infrastructure, developing customized applications, and hosting them at a back-end cloud. Qowisio has a full range of intelligent devices, supporting different applications such as asset management, perimeter control, motion detection, lighting, environment parameter monitoring, energy and power monitoring, tracking, and several others.

2.5.1.5 Nwave

Nwave LPWAN is primarily a solution developed by nWave [18] for smart parking systems. This ultra-narrowband technology is also based on sub-1 GHz unlicensed ISM band operation. It claims long range and high node density as compared to Sigfox and LoRa at the cost of higher power consumption. It works in a star topology and supports mobility of the nodes. nWave end node transmits power from 25 to 100 mW, thus covering longer distance for a data rate of up to 100 bps. It also claims coverage of up to 7 km and 8 years for inbuilt node battery. It has its own real-time data collection and management software system for monitoring and control.

2.5.2 Standards-based technologies

2.5.2.1 LoRa and LoRaWAN

LoRa is a physical layer technology that works in unlicensed sub-GHz ISM band and is based on chirped spread spectrum (CSS) technique [7]. CSS is a wideband linear frequency modulation in which carrier frequency varies for the defined extent of time. LoRa works on pure ALOHA principles and supports different ISM frequencies, namely 868 (Europe), 915 (North America), and 433 MHz (Asia). It is basically single-hop technology, which relays the messages received from LoRa sensor nodes to the central server via gateways. The data transmission rate supported by LoRa varies from 300 bps to 50 kbps, depending on spreading factor (SF) and channel bandwidth settings. LoRa transmissions with different SFs are quasiorthogonal [10] and allow multiple transmissions with different SFs simultaneously. To support LoRa on Internet, LoRa alliance has developed long-range wide-area network (LoRaWAN) [8], which includes the network and upper layer functionalities. LoRaWAN provides three classes of end devices to address the different requirements of a wide range of IoT applications, for example, latency requirements. LoRa is one of the best candidates for long-distance and low-power transmissions.

2.5.2.2 Weightless

Weightless Special Interest Group (Weightless-SIG) proposed Weightless, an open standard offering LPWAN connectivity. There are three versions of weightless: Weightless-W, Weightless-N, and Weightless-P. Weightless-W is designed to operate in TV white space (TVWS) spectrum (470−790 MHz band), and it has better signal propagation as compare to Weightless-N and P. It supports a wide range of spreading codes and modulation techniques such as DBPSK and quadrature amplitude modulation. The data packet size can be up to 10 bytes with throughput ranging from 1 kbps to 10 Mbps, subject to link budget and other settings [19]. To enhance the energy efficiency, the end nodes communicate to the gateway with a narrow spectrum and at low power. Since the use of TVWS is not permitted in many countries, Weightless-SIG has introduced two other variations, namely Weightless-N and Weightless-P. Weightless-N (nWave) is similar to Sigfox and uses slotted ALOHA in the unlicensed band, supporting only unidirectional communication for end devices to the base station. It achieves a communication range of up to 3 km with a maximum data rate of 100 kbps. Weightless-P uses Gaussian minimum shift keying (GMSK) and quadrature phase-shift keying (QPSK) in the unlicensed band and offers bidirectional communication with support for acknowledgments. It achieves a data rate of around 100 kbps and has a comparatively shorter communication range (2 km) and shorter battery lifetime.

2.5.2.3 Narrowband Internet of things

Narrowband IoT (NB-IoT) is a 3GPP Release 13 LPWAN technology offering flexibility of deployment by allowing the use of a small portion of the available spectrum in the LTE band. As a 3GPP technology, NB-IoT can coexist with global system for mobile communications (GSM) and LTE in licensed frequency bands of 700, 800, and 900 MHz. It supports bidirectional communication where orthogonal frequency-division multiple access (OFDMA) is used for downlink, and single-carrier frequency-division multiple access is used for uplink [20]. It connects up to 50k devices per cell and requires a minimum of 180 kHz of bandwidth to establish communication. NB-IoT can also be deployed as a standalone carrier with a spectrum of more than 180 kHz within the LTE physical resource block. NB-IoT is designed by optimizing and reducing the functionalities of LTE so that it can be used for infrequent data transmissions and with low power requirements. The data rate supported is 200 kbps for downlink and 20 kbps for uplink. The maximum payload size for each message is 1600 bytes.

2.5.2.4 LTE-M

Long Term Evolution—Machine Type Communications (LTE-M) is also a 3GPP standard-based technology and operates in the licensed LTE spectrum. It is compatible with LTE networks and provides connection for MTC-type traffic. Also, migration path from legacy 2G and 3G networks is available. It provides extended coverage as compared to LTE networks, coverage for MTC applications similar to 5G Networks, and offers a seamless path toward 5G MTC solution [20]. LTE-M is focused on providing variable data rates and support for both real-time and nonreal-time applications. It supports low-latency applications, as well as deferred traffic applications that can operate with latencies in a few seconds range. It has low power requirements and supports operations ranging from low bandwidth to bandwidth as high as 1 Mbps. Also, it supports devices with a very wide range of message sizes. Since it is derived from LTE as a base, mobility is supported as part of standard LTE functionality but not in extended coverage scenarios. It is software upgradable from LTE. Its capacity is up to 100,000 + devices per base station for applications where devices have very low data throughput requirements.

2.5.2.5 DASH7

DASH7, also known as DASH 7 Alliance Protocol, was developed for wireless sensor and actuator communications and is originated from ISO 18000-7 standard. An extension of active radio-frequency identification technology, DASH7 is a low-power long-range ISM band technology primarily operated at 433 MHz. However, it also supports communication with other bands at 868 and 915 MHz [16]. It uses two-level GFSK modulation with a channel bandwidth of 25 or 200 kHz along with data whitening and forward error correction features. It has a tiny open-source protocol stack, supporting multiyear battery life, low latency, and more flexibility. It is used for low-rate bursty data traffic of up to 167 kbps. DASH7 supports multihop communication and mobility of nodes up to a range of 2 km. The architecture comprises of endpoints, subcontrollers, and gateways. Endpoint nodes follow a strict duty cycle schedule, while subcontrollers collect the data packets from the endpoint nodes with some sleep cycles and low power restrictions. Gateways are continuously active to collect the packets from subcontrollers and endpoints and then send them to server. DASH7 supports tree topology in the presence of subcontrollers or star in the presence of the endpoints.

2.5.2.6 NB-Fi

Narrowband fidelity (NB-Fi) LPWAN technology is designed for the narrowband, low-power, and long-range bidirectional MTC applications [16]. This solution is designed by WAVIoT—an infrastructure as a solution LPWAN provider. It works in 868 and 915 MHz unlicensed ISM bands as well as other sub-GHz unlicensed spectra. NB-Fi is an open, full-stack protocol with all the seven layers of OSI reference model for robust, reliable, and energy-efficient sensor communication. To achieve improved spectral efficiency and performance in the narrow bands, it employs the smart and optimized spectrum utilization algorithms based on SDR technology, neural, and AI techniques. NB-Fi has decentralized architecture allowing base stations to perform significant operations, making it more robust and reliable in the case of network failure. It is a highly scalable solution in which one NB-Fi base station can support up to 2 million sensor nodes. As a sub-1 GHz ISM bands are crowded, the gateways are designed to work with an interference avoidance algorithm. NB-Fi provides coverage of up to 10 km in urban areas and up to 30 km in rural areas.

2.5.2.7 Enhanced coverage—global system for mobile Internet of things

Enhanced coverage—global system for mobile Internet of things (EC-GSM-IoT), developed by 3GPP is one of the promising candidates for low-power, long-range cellular IoT (cIoT) for providing similar coverage and battery life to NB-IoT [21]. It is based on enhanced General Packet Radi Service (GPRS) and designed for scalable, low-complex LPWAN applications. As the majority of current cIoT devices are based on GPRS/Enhanced Data rates for GSM Evolution (EDGE) to connect to the Internet, EC-GSM-IoT provides an easy path to improve energy efficiency and a 20 dB coverage improvement. It is optimized and improved by means of software upgrades to GPRS/EDGE networks and also to support new devices. The traffic of legacy GSM devices and EC-GSM-IoT is multiplexed on the same physical channels without much compromise of the performance of the legacy traffic. The bandwidth of EC-GSM-IoT channel is 200 kHz. Like GSM, it is FDMA + Time Division Multiple Access + Frequency Division Duplex technology, supporting peak data rates of 70 and 240 kbps based on GMSK and 8-Phase Shiforce Keying, respectively. EC-GSM-IoT provides multifold improvement in the coverage for low-rate applications. It also has an ability to reach challenging locations such as deep indoor basements, where many smart meters and parking sensors are installed, or remote areas in which sensors are deployed for agriculture or infrastructure monitoring use cases [22]. The important features of physical layer include new logical channels, repetitions to provide necessary robustness to support up to 164 dB maximum coupling loss, and the use of overlay CDMA to increase cell capacity.

2.5.2.8 IEEE 802.15.4k

This standard is developed by IEEE 802.15.4k Task Group (TG4k) for low-energy critical infrastructure monitoring applications. It works in unlicensed sub-GHz and 2.4 GHz ISM bands with multiple discrete channel bandwidths ranging from 100 kHz to 1 MHz [2]. It uses direct sequence spread spectrum and frequency-shift keying (FSK) at the physical layer. The MAC is based on carrier sense multiple access/collision avoidance (CSMA/CA) and works with three different approaches, namely without priority channel access (PCA), CSMA, and ALOHA with PCA. With PCA, the end devices and base stations can prioritize their traffic in accessing the medium. It supports star topology and has the capability to exchange messages asynchronously. It supports the peak data rate of up to 128 kbps and a maximum coverage of 5 km.

2.5.2.9 IEEE 802.15.4g

IEEE 802.15 WPAN Task Group 4g (TG4g) has proposed this standard to extend the short range of IEEE 802.15.4 base standard for smart utility networks. It defines three physical layers based on FSK, OFDMA, and offset QPSK (OQPSK). Except for operation in one licensed band used in the United States, it is mostly operated in unlicensed sub-GHz and 2.4 GHz ISM band [2]. It gives coverage of several kilometers. The data rates supported are from 40 kbps to 1 Mbps depending on the physical layer and region in which it is operated. It works on the principle of CSMA/CA and supports star, mesh, and other topologies.

References

[1] M. Rouse, LPWAN (low-power wide area network). <https://internetofthingsagenda.techtarget.com/definition/LPWAN-low-power-wide-area-network>.

[2] U. Raza, P. Kulkarni, M. Sooriyabandar, Low power wide area networks: an overview, arXiv:1606.07360v2 [cs.NI], 2017.

[3] S.M. Islam, M. Zeng, O.A. Dobre, K.-S. Kwak, Non-orthogonal multiple access (NOMA): how it meets 5G and beyond, arXiv preprint arXiv:1907.1000, 2019.

[4] P. Thubert, A. Pelov, S. Krishnan, Low-power wide-area networks at the IETF, IEEE Commun. Stand. Mag. 1 (1) (2017) 76−79.

[5] S. Chacko, D. Job, Security mechanisms and vulnerabilities in LPWAN, IOP Conf. Ser. Mater. Sci. Eng. 396 (1) (2018) 012027.

[6] B. Thoen, G. Callebaut, G. Leenders, S. Wielandt, A deployable LPWAN platform for low-cost and energy-constrained IoT applications, Sensors 19 (3) (2019).

[7] K. Mekki, E. Bajic, F. Chaxel, F. Meyer, A comparative study of LPWAN technologies for large-scale IoT deployment, ICT Express 5 (1) (2019) 1−7.

[8] Key technology choices for optimal massive IoT devices, Ericsson Technology Review. <https://www.ericsson.com/en/ericsson-technology-review/archive/2019/key-technology-choices-for-optimal-massive-iot-devices>.

[9] M. Bilal, A review of Internet of things architecture, technologies and analysis smartphone-based attacks against 3D printers, arXiv preprint arXiv:1708.04560, 2017.

[10] P. Sethi, S.R. Sarangi, Internet of things: architectures, protocols, and applications, J. Electr. Comput. Eng. 2017 (2017).

[11] J. Rubio-Aparicio, F. Cerdan-Cartagena, J. Suardiaz-Muro, J. Ybarra-Moreno, Design and implementation of a mixed IoT LPWAN network architecture, Sensors 19 (3) (2019).

[12] M. Chen, Y. Miao, X. Jian, X. Wang, I. Humar, Cognitive-LPWAN: towards intelligent wireless services in hybrid low power wide area networks, IEEE Trans. Green. Commun. Netw. 3 (2) (2018) 409−417.

[13] <www.sigfox.com>.

[14] M. Centenaro, L. Vangelista, A. Zanella, M. Zorzi, Long-range communications in unlicensed bands: the rising stars in the IoT and smart city scenarios, IEEE Wirel. Commun. 23 (2016) 60−67.

[15] Ingenu, RPMA technology for the Internet of Things, 2010.

[16] J. Finnegan, S. Brown, A comparative survey of LPWA networking, arXiv preprint arXiv:1802.04222, 2018.

[17] <www.qowisio.com>.

[18] <www.nwave.com>.

[19] <www.weightless.org>.

[20] G.A. Akpakwu, B.J. Silva, G.P. Hancke, A.M. Abu-Mahfouz, A survey on 5G networks for the Internet of things, communication technologies and challenges, IEEE Access. 6 (2017) 3619−3647.

[21] S. Lippuner, B. Weber, M. Salomon, M. Korb, Q. Huang, EC-GSM-IoT network synchronization with support for large frequency offsets, in: 2018 IEEE Wireless Communications and Networking Conference (WCNC), IEEE, 2018, pp. 1−6.

[22] <https://www.ericsson.com/en/press-releases/2016/2/ericsson-and-orange-in-internet-of-things-trial-with-ec-gsm-iot>.

3

LoRaWAN protocol: specifications, security, and capabilities

Alper Yegin[1], Thorsten Kramp[2], Pierre Dufour[1], Rohit Gupta[1], Ramez Soss[1], Olivier Hersent[1], Derek Hunt[3], Nicolas Sornin[4]

[1]ACTILITY, LANNION, FRANCE [2]FORMERLY SEMTECH (INTERNATIONAL) AG, RAPPERSWIL-JONA, SWITZERLAND [3]LORA ALLIANCE®, FREMONT, CA, UNITED STATES [4]SEMTECH (INTERNATIONAL) AG, RAPPERSWIL-JONA, SWITZERLAND

3.1 Technical overview of LoRaWAN specifications

As the Internet of Things (IoT) began to mature, the need for a long-range, low-power connectivity standard started to impede its growth. Whereas the short-range, low-power device market is already being addressed by technologies such as Bluetooth Low Energy and Zigbee, and the long-range (but not low-power) device market is being addressed by cellular technologies, the long-range, low-power market has been left to proprietary technologies. As the IoT demands low-cost interoperable solutions more than any other communication field, the need for a standard-based open technology emerged. This gave rise to LoRaWAN.[1]

LoRaWAN is an end-to-end wireless system architecture that provides a low-power, long-range, low-cost, secure, and scalable connectivity solution to public operators and private networks for a wide range of IoT use cases. The LoRaWAN architecture uses the spread spectrum modulation-based LoRa physical layer and defines several protocols to create an end-to-end system.

LoRaWAN enables end devices to operate on small batteries for up to 10 years, for which it uses radio gateways with a range of up to 30 miles in rural areas. It is able to penetrate dense urban as well as deep indoor environments. LoRaWAN is based on the 128-bit Advanced Encryption Standard (AES128) to ensure full network security, including mutual end-point authentication, data origin authentication, replay and integrity protection, and privacy. It enables GPS-free location applications thanks to its spread spectrum and fine time-stamping capability. Its use of industrial, scientific, and medical (ISM) radio bands allows high-capacity (millions of messages per gateway), low-cost operation and highly optimized

---

[1] LoRa Alliance is a mark used under license from the LoRa Alliance. LoRaWAN is a mark used under license from the LoRa Alliance. The LoRa Mark is a trademark of Semtech Corporation or its subsidiaries.

LPWAN Technologies for IoT and M2M Applications. DOI: https://doi.org/10.1016/B978-0-12-818880-4.00003-X

ground-up design that is driven specifically by IoT requirements. It also relies on the availability of open standards and an open ecosystem.

The system architecture and protocols are being developed by the LoRa Alliance (LoRa-Alliance.org), an open, nonprofit association with a large and ever-growing ecosystem spanning a wide range of players from chip makers to cloud providers. The LoRa Alliance facilitates the production of interoperability specifications that are publicly available and free of charge, a certification program to support the proliferation of high-quality and interoperable devices, and technology marketing.

The first specification developed by the LoRa Alliance, which also forms the core of the architecture, is the LoRaWAN link-layer specification [1] that describes the layer residing above the LoRa physical layer and below the application layer between the end device and the network. This link layer, which acts as an over-the-air transport, ensures that end devices can send and receive application-layer payloads to and from the network.

The LoRaWAN link-layer specification prior to Version 1.0.2 included the physical-layer parameters that vary across regulatory regions, such as channel frequencies, transmission power, and data rates. Later, the LoRa Alliance separated these specifications into a dedicated document of regional parameters [2] that can evolve in response to regulatory changes and the addition of new regions.

As the LoRaWAN architecture evolved, the LoRaWAN backend interface specifications were subsequently introduced [3]. As networks started to become available around the world, the next logical step was to make these networks collaborate by establishing mutual roaming capabilities. Integrating radio access networks across multiple LoRaWAN networks required an interoperable interface among their core networks. Furthermore, to facilitate device provisioning, LoRaWAN core networks also evolved to externalize the entity that can store the long-term credentials of devices while allowing them to be activated on any network in the world. The backend interface specifications describe the protocols needed for roaming and activation to be performed across separate administrative domains that use platforms from various independent vendors.

One of the key enablers of achieving low-cost devices is the ability to deploy them once and then not having to deal with them for 5–10 years. Software installed on these devices, whether in the application stack or the firmware, requires updates for the obvious reasons of delivering new features and bug fixes. For that reason, it is critical for IoT devices to have an over-the-air firmware update mechanism. LoRaWAN's firmware update over-the-air (FUOTA) feature is based on two specifications, one for downlink fragmentation [4] and the other for multicast management [5]. These specifications define a secure, reliable, and efficient mechanism to transport large files simultaneously to a set of devices. Reliance on simultaneous forward error correction and physical broadcasting has been a key element of achieving these features under the constraints of ISM band regulations. Fragmentation and multicast protocols have been defined at the application layer (i.e., above the link layer) in order for them to be applicable to legacy versions of LoRaWAN link layers.

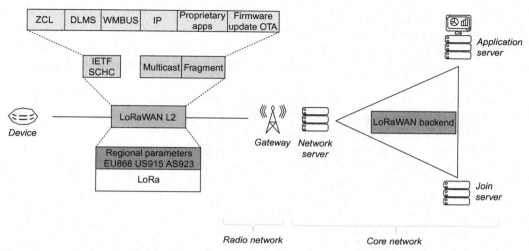

FIGURE 3–1 LoRaWAN architecture and protocols.

The LoRaWAN architecture has been expanded to include application stacks. Instead of developing yet another application stack, the LoRa Alliance chose to take popular stacks, such as the Zigbee Cluster Library, device language message specification (DLMS), Wireless M-Bus and ModBus, and adapt them by reducing nonessential options and using efficient encoding (i.e., compression) to match message flows and payloads within the constraints of ISM band use. Adaptation of DLMS also lent support to IETF's SCHC protocol, which clears the way to supporting Internet protocol (IP)-based applications.

Fig. 3–1 depicts the various network elements that make up the LoRaWAN network architecture as well as the protocols and specifications to support it. Starting from a single specification (the link layer), the LoRaWAN design has been growing on several fronts. The LoRa Alliance pays special attention to keeping complexity as low as possible, while being able to address evolving market needs. The constraints posed by low-power wide-area network (LPWAN) applications and ISM band use are unique and challenging. A ground-up design following a solid design philosophy is a must for the success of any LPWAN system architecture. Open standards backed by open-source implementations and low-cost hardware are also essential tools for this purpose. The LoRa Alliance has excelled in leveraging these elements to build a future-proof technology while nurturing an ever-growing ecosystem around it. IoT is an ecosystem play, and both LoRaWAN technology and the LoRa Alliance are open to new contributions to drive this effort forward.

3.2 LoRaWAN link layer

LoRaWAN networks are laid out in a star-of-stars topology in which gateways relay packets between devices and a central network server (NS) (see Fig. 3–2). The NS, in turn, routes packets received by network gateways to an associated application server and vice versa. Communication is generally bidirectional, although uplink communication from a device to the network and application servers is expected to be the predominant traffic. Moreover, uplinks can be received by multiple gateways, that is, there is no fixed association between devices and gateways.

Communication between devices and gateways uses single-hop LoRa or frequency-shift keying (FSK) radio transmissions, whereas LoRa transmissions are distributed over various frequency channels and data rates. Selecting the data rate, which ranges from 0.3 to 50 kbps, is a trade-off between communication range and packet transmission duration. Lower data rates have a longer range but consume more airtime. Communications with different data rates do not interfere with each other. To maximize both the battery life of devices and the overall network capacity, the LoRaWAN network infrastructure can manage the data rate and radio-frequency (RF) output power for each device individually by means of an adaptive data rate (ADR) scheme as illustrated in Fig. 3–3. Depending on gateway distribution, ADR may also allow the network to influence the number of gateways receiving uplinks from a device and thereby its connectivity degree (receive redundancy).

Devices may transmit over any available channel at any time using any available data rate, as long as the following rules are observed:

1. The device changes channels in a pseudo-random fashion for every radio transmission. The resulting frequency diversity makes the system more robust to interferences.
2. The device respects the maximum transmit duty cycle relative to the subband in which it is operating and compliant with local regulations.

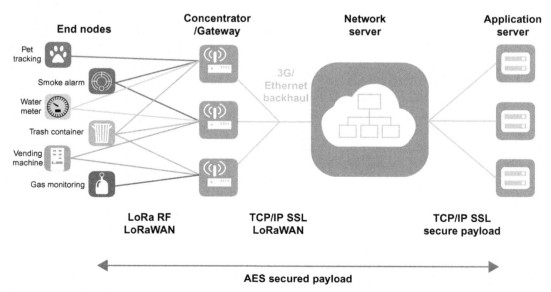

FIGURE 3–2 LoRaWAN network.

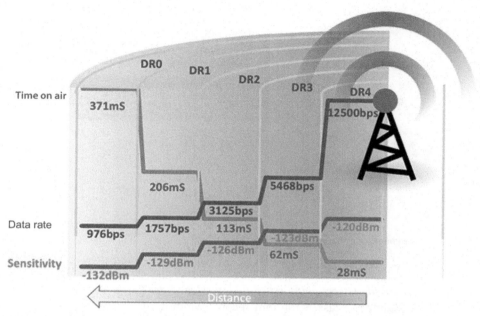

FIGURE 3–3 Data rates as a function of communication range and packet transmission duration.

3. The device respects the maximum transmit duration (or dwell time) relative to the subband in which it is operating and compliant with local regulations.

Maximum transmit duty cycle and dwell time per subband are region-specific and defined in the LoRaWAN Regional Parameters specification together with further regulatory requirements such as listen before talk (LBT) where applicable.

For secure radio transmissions, the LoRaWAN protocol relies on symmetric cryptography using session keys derived from device-specific root keys. In the backend, the device's root keys and the associated key derivation operations are stored by a join server (JS) during an over-the-air activation procedure. Alternatively, device-specific session keys may be manufactured directly into the device, known as activation by personalization.

LoRaWAN devices generally follow an ALOHA-type communication pattern, where devices may operate in one of the following three classes as shown in Fig. 3–4.

Class A: Bidirectional devices. Class A devices allow bidirectional communication, whereby each device's uplink transmission is followed by two short downlink receive windows. The transmission slot scheduled by the device is based on its own communication needs with a small variation on a random time basis (ALOHA-type protocol). Class A operation is the lowest power device system for applications that require only downlink communication from the NS shortly after the device has sent an uplink transmission. Downlink communications from the NS at any other time must wait until the next scheduled uplink.

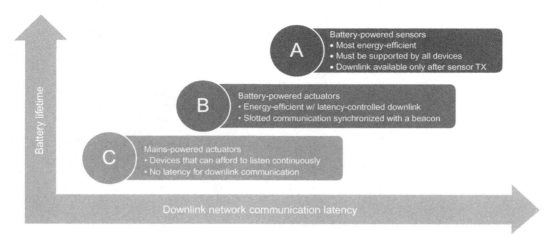

FIGURE 3–4 ALOHA communication classes.

Class B: Bidirectional devices with scheduled receive slots. Class B devices allow more receive slots. In addition to Class A random receive windows, Class B devices open further receive windows at scheduled times. In order for the device to open its receive window at the scheduled time, it receives a time-synchronized beacon from a gateway.

Class C: Bidirectional devices with maximum receive slots. Class C devices have nearly continuously open receive windows that close only when transmitting. They use more power than Class A or Class B devices, but feature the lowest latency for server-to-end-device communication.

Note that devices may switch from Class A to Class B operation mode and vice versa as needed.

3.3 Scaling LoRaWAN networks

IoT devices are projected to outnumber human-centric devices such as cell phones 10- to 100-fold by 2025. To accommodate this trend, the capacity and scaling of LoRaWAN deployments have been a topic of investigation both within academia and industry. Several recent publications have presented initial results regarding LoRaWAN capacity [6–8]. We have extended the methodology, in particular LoRaSim [9], to provide quantitative hindsight on the behavior of a LoRaWAN network as network usage scales and the network is densified.

3.3.1 LoRaWAN star topology with receive diversity is the key to scaling in an unlicensed spectrum

LoRaWAN deployments use a star topology with a frequency reuse factor of 1. This simplifies network deployment and ongoing densification because there is no need for frequency pattern planning or reshuffling as more gateways are added to the infrastructure. It also facilitates seamless collaboration between public and private networks.

Compared to mesh technologies, the single-hop-to-network infrastructure minimizes power consumption because nodes do not have to relay communication from other nodes. Another advantage is that initial network deployment in sparse mode with low node density is possible, compared with a mesh that requires minimum node density.

However, the most important design feature of LoRaWAN is its receive diversity. As usage of unlicensed spectra grows, background radio noise known as the "noise floor" is increasing. Some experts predict that unlicensed networks will inevitably face increasing packet loss and therefore cannot guarantee quality of service (QoS) in the long term. But this is not in fact inevitable. LoRaWAN networks can adapt to noise by leveraging multiple reception gateways operating simultaneously for each end device. LoRaWAN networks uplink messages that can be received by any gateway (RX macro-diversity). Such uplink macro-diversity significantly improves network capacity and QoS because it is very unlikely that destructive interference will occur at all antennas simultaneously. As a result, LoRaWAN networks are expected to cope with increasing noise much better than earlier mesh networks, where each node is managed by only one next-hop receiver at a time, and which, on the contrary, suffer catastrophic degradation due to the cumulative effect of increasing packet loss at each hop, as shown in Fig. 3—5.

3.3.2 Role of adaptive data rate

LoRaWAN also supports ADR, which allows NS to change dynamically the parameters of end devices such as transmit power, frequency list, spreading factor, and uplink repeat rate. Careful adjustment of transmission power is necessary not only to ensure that devices will use the lowest power necessary to communicate with the gateway, but also to minimize unnecessary noise for adjacent radio cells, and thus prevent the closest devices from having a shadowing effect on devices located at the cell edge. This is known as the far-near effect.

The spreading factor defines the relation between symbol rate and chip rate. A higher spreading factor increases sensitivity and range, but also prolongs the airtime of a packet and will likely raise the risk of a collision. LoRaWAN uses six different orthogonal spreading factors numbered 7—12.

To minimize device energy consumption and airtime, an NS will try to maximize the data rate for a given link budget between device and best antenna, while preserving a minimum level of macro-diversity, for instance, when the network also provides geolocation services. A given message transmission may be repeated while varying the channel (carrier frequency) at each transmission to offer frequency diversity. Such repetitions not only prolong airtime, but also make reception much more robust in the presence of impediments such as channel interference and collisions. Finding the best combination of repetition, data rate, and transmit power is a complex optimization process, which was our main motivation for building simulation models.

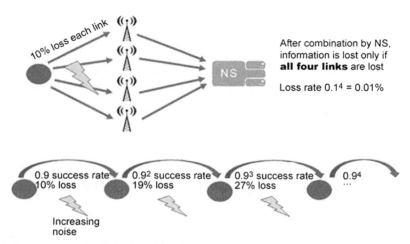

FIGURE 3–5 Mesh versus LoRaWAN behavior with rising noise level.

3.3.3 LoRaWAN network capacity and scaling

The Okumura–Hata model is the radio propagation model to simulate results as pictured in Fig. 3–6. It is valid for deployments for which the base station is installed at high outdoor locations compared to the height of surrounding buildings. Devices are assumed to have a 0 dBi antenna, which is typical for small devices. Our simulation uses 16 channels and targets a 10% packet error rate. Fig. 3–6 shows the random position of devices and color coding for their optimal spreading factor in a hexagonal network.

Fig. 3–7 shows the maximum capacity results for the network and a single cell. It confirms that capacity scales easily with the density of network gateways.

These results show that the future of LoRaWAN networks, particularly in urban environments where the noise floor is expected to rise due to increased traffic, is going toward microcellular networks, for example, with receivers integrated in triple-play modems. Macrodiversity provides not only higher capacity, see Fig. 3–8, but also greater resilience to interference and lower power consumption for end devices.

LoRaWAN provides a horizontal connectivity solution to address the wide-ranging needs of IoT applications for LPWAN deployments. However, these benefits are only possible with intelligent NS algorithms proprietary to network solution vendors.

3.4 LoRaWAN regional parameters

A LoRaWAN device or network uses unlicensed ISM bands. Therefore for a device to operate anywhere in the world, it must comply with the regulatory requirements for that region or territory. There were 195 countries in the world as of January 2019, plus several dozen other entities, 193 of which are members of the United Nations (UN).

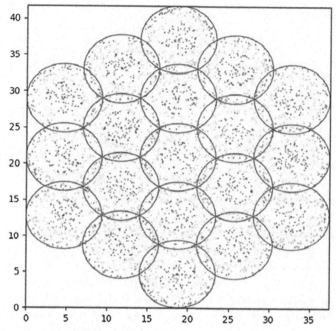

FIGURE 3–6 Simulated network layout with hexagonal tiling.

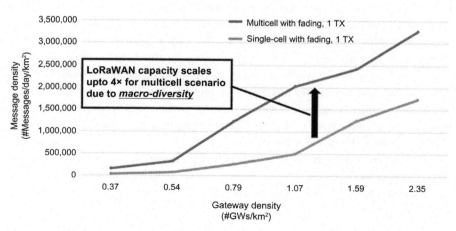

FIGURE 3–7 Impact of multicell deployment on capacity.

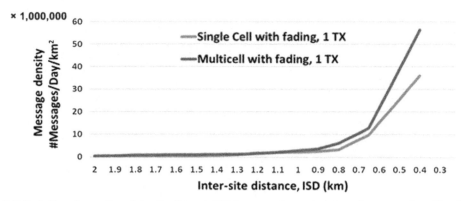

FIGURE 3–8 Evolution of capacity as intersite distance (ISD) becomes shorter (number of messages (in millions) vs ISD).

The International Telecommunication Union manages the common spectral use for the UN, but the final frequency assignment authority is granted by the governments of each country or region. However, many small or developing countries have no frequency assignment authority in place. This creates a complicated scenario for worldwide LoRaWAN deployments. For the global LoRaWAN network to be successful, an understanding is needed throughout target countries and regions to deploy LoRaWAN such that regulatory compliance can be obtained and LoRaWAN channel plans can be created.

To define LoRaWAN channel plans in every country or region, the LoRa Alliance has formed a working group that creates, manages, and maintains LoRaWAN regional parameter specifications. This document is a companion document to the various LoRaWAN Layer 2 (L2) protocol specifications and combines all regional parameter aspects defined in the LoRaWAN protocol. Separating the regional parameters from the protocol specification allows new regions to be added without impacting the LoRaWAN L2 document. The LoRaWAN regional parameters specification document is not intended to be an authoritative source of regional governmental requirements, so devices and networks must comply with any specific laws and regulations of the country or region in which they operate. Owing to such different regional regulatory entities, it is not possible to provide a single channel plan for the entire world. Nevertheless, the LoRaWAN regional parameters working group has combined several countries into various regional plans to reduce the number of LoRaWAN regions. AS923 is an example of this.

If a country or region is not yet covered by one of the 12 current LoRaWAN regions, it may be possible to include it in an existing region, depending on its regulatory use of ISM bands. This is shown in the cross-reference table in the LoRaWAN regional parameters specification. Countries currently using the ISO 3166-1 country name codes are shown in the LoRaWAN channel plan coverage map as shown in Fig. 3–9.

The LoRaWAN regional parameters specification document defines the following parameters:

- channels to be used when a device joins the network;
- default channels;

135 countries with
LoRaWAN channel plans

January 2019

>100 countries with LoRaWAN deployments

FIGURE 3–9 LoRaWAN channel plan coverage map.

- duty cycle or dual time restrictions;
- whether LBT is required;
- data rates;
- maximum transmit power and power configuration;
- channel plan and channel mask;
- maximum payload size versus data rate;
- receive window frequencies and receive window mapping; and
- Class B beacon and default downlink channel.

The worldwide regulatory environment is constantly changing, and this requires that the LoRaWAN regional parameters specification be updated regularly to stay abreast of these changes.

3.5 Activation and roaming

In the early days of the LoRa Alliance, there was only one interoperability specification called the link layer, and it dealt with only two end points: the end device and the network. As the technology, implementations, and deployments began to mature, the need to decompose the "network" became apparent, which gave way to defining three distinct network elements that form the core LoRaWAN network: join servers, network servers, and application servers.

Join servers store end-device credentials to perform mutual authentication before that end device is admitted to the network. The procedure involving a cryptographic handshake and end-device configuration with network parameters is called the activation or join procedure.

Network servers terminate the LoRaWAN link layer on the network side and act as bridges that transport data traffic to application servers and signaling traffic to JS.

Application servers are the end points for data traffic. The application payload of the uplink frames sent by an end device is consumed by application servers, and the application payload of downlink frames sent by NS is sourced from application servers.

Owing to the constraints of ISM band use, which specify that frames be very small in size and number, LoRaWAN uses symmetric cryptography. This requires that a symmetric device-specific root key is available in the network before the device attempts to activate itself. Unfortunately, it is rarely known at the time of device manufacture which network(s) the device will use throughout its lifetime. Therefore a solution was needed that can provision device credentials once on a centralized JS and allow the device to be activated on any network or NS anywhere in the world (see Fig. 3−10). The LoRaWAN backend interface specification made this possible by defining and implementing a standard interface between NS and JS.

The activation procedure uses a long-term, device-specific root key shared between the end device and a JS to generate ephemeral session keys that are sent from a JS to a NS for link-layer security and also to an application server for end-to-end encryption.

Secure elements can be used on end devices when it is necessary to protect high-value applications such as water, electric, or gas metering from tampering. Similarly, hardware-grade security can be achieved on the network side by using a hardware security module to protect the device keys on the JS side. Separating the JS from the network helps to reduce device manufacturing costs by eliminating per-network personalization, and it also yields a new business model for operating JS as a service.

Roaming is another important functionality enabled by the backend interface specification. LoRaWAN roaming allows networks to combine radio coverage for providing connectivity to

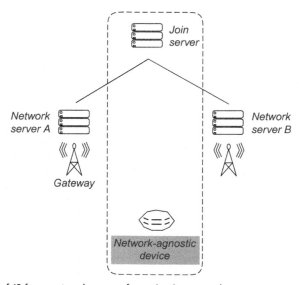

FIGURE 3−10 Separation of JS from network servers for activation procedure.

end devices, even to those outside their home network. In addition to this traditional use of roaming (network extension), LoRaWAN also provides a unique type of roaming that allows numerous overlapped networks to provide densified coverage, known as network densification.

The LoRaWAN architecture allows radio gateways and even NS to serve an end device without holding state information for that device. This allows uplink frames transmitted by an end device to be demodulated by several radio gateways and received by several NS at the same time, a feature called macro-diversity. Each NS receiving the uplink frame can forward the frame to the home network of the end device, which can be determined from its device address. This potentially translates into the device's home network and zero or more visited networks cooperatively receiving traffic from end devices. When radio gateways from numerous networks serve a device jointly, this has the same effect as network densification by a single operator. Consequently, the packet error rate is reduced, and an end device can transmit at a higher data rate while using less power, thanks to the ADR mechanism. Higher data rates mean greater capacity and less interference for the networks within the coverage of the device. Finally, more radio gateways receiving the same frame helps to improve LoRaWAN's geolocation capability. Therefore LoRaWAN roaming is beneficial to all parties involved: the end device, the home network, and the visited networks (see Fig. 3−11).

Roaming is completely transparent to the end device, which is why it is also called passive roaming. The aforementioned benefit of passive roaming fueled by macro-diversity is a unique feature of LoRaWAN. Competing LPWAN technologies have no counterpart for various reasons: Cellular IoT technologies such as narrow-band IoT require a state on the base

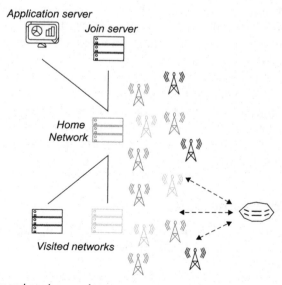

FIGURE 3–11 Macro-diversity and passive roaming.

station and the visited networks. Hence they lack the technical means to achieve this, and the SigFox business model limits the number of networks in a country to one, meaning that it is not open to other networks for cooperation.

LoRaWAN roaming accommodates not only mobile devices such as trackers but also fixed devices such as smog detectors. A fixed device may be permanently roaming within the coverage of a visited network. This is the typical case for international solution providers that want their own core network without having radio networks in most or any of the countries they serve.

Both roaming and activation features require core networks to interconnect, which is practical among a few networks. However, as the number of networks rises, the mesh of interconnecting these networks becomes more difficult to set up and maintain. This issue is overcome by the well-established "peering hub" concept, where the owner of a given NS or JS can interconnect its infrastructure with a peering hub and then be able to set up a logical connection easily with any of the other networks on that hub or another interconnected hub.

Roaming and activation have been designed to be low-cost, highly scalable, simple, and able to accommodate future extensibility design goals. They cater not only to large multinational operators, but also to small private networks (e.g., single gateway and NS combinations in homes). As this book goes to press, numerous networks are already using these facilities, most of which operate via the very first peering hub.

3.6 Network-based and multitechnology geolocation

3.6.1 Geolocation is a massive opportunity

If there is one killer IoT app, it is most likely geolocation. *Research and Markets* predicts that revenues from geolocation IoT will reach $49 billion by 2021.[2] The applications are endless: asset management, fleet management, antitheft software, vehicle rental, logistics and parcel tracking, worker safety, elderly and disabled care, pet and animal tracking, to name but a few. However, traditional geolocation technologies are still too expensive for a majority of applications, particularly considering the total cost of ownership (TCO), which also takes into account battery replacement costs. Elasticity of demand to price for geolocation is particularly high, and for geolocation IoT to become reality, there must be a dramatic reduction of TCO. As we will see in this section, LPWANs, and particularly LoRaWAN, are a key enabler technology for such a cost reduction.

3.6.2 Vast choice of technologies

As illustrated in Fig. 3–12, there are numerous potential technology combinations that will serve the need for asset geolocation. Horizontal bars indicate how much energy is necessary,

---

[2] *Research and Markets'* report entitled "Geo IoT Technologies, Services, and Applications Market Outlook: Positioning, Proximity, Location Data and Analytics 2016–2021."

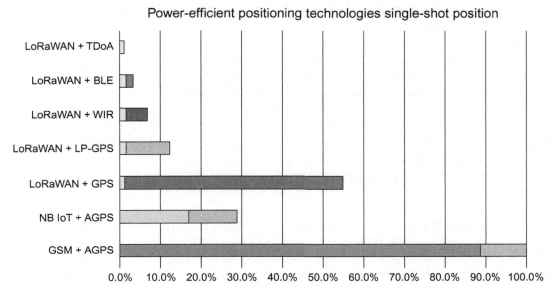

FIGURE 3–12 Energy use per WAN/geolocation technology combination.

translated into battery size or frequency of battery replacement, which are both key drivers of geolocation TCO.

It is immediately striking that using LoRaWAN as a wide-area backhaul technology for transferring location data to an application is a significant improvement over traditional cellular networks, and it remains significantly superior to more recent low-power cellular technologies such as narrow-band IoT. There are several geolocation technologies for a given backhaul communication technology.

GPS and other global navigation satellite systems (GNSS). In normal mode without server assistance, GNSS are a relatively power-hungry technology due to long synchronization latency. They also require good radio conditions during the synchronization period because the data stream of satellite position data uses a modulation that requires a high signal-to-noise ratio. When a device switches from indoors to outdoors, long gaps can occur before a fix can be obtained.

Assisted GPS (AGPS) and other low-power GPS (LP-GPS) optimizations. These optimizations rely on server-side processing and the ability to convey up-to-date satellite information to the GNSS module via a WAN. They eliminate the need for a strong signal during synchronization and significantly decrease synchronization time, which reduces energy consumption to 25% or even less compared to unoptimized GPS.

Combining LoRaWAN and AGPS is difficult because conventional AGPS cannot be used due to LoRaWAN's limited bandwidth and the fact that it cannot transmit the entire satellite information to the GNSS module. However, specific algorithms exist that can leverage the ability of LoRaWAN networks to provide an extremely accurate timestamp for uplink packets, making it possible to obtain a fix with only three GPS satellites, as opposed to four required by conventional AGPS. As a result, the combination of LoRaWAN + LP-GPS technology is extremely efficient.

WiFi and Bluetooth (BLE/BTLE) are attractive options for indoor geolocation.

Triangulation based on time difference of arrival (TDoA) is an attractive feature of LoRaWAN networks, thanks to the relatively broad bandwidth of LoRa channels (125 kHz or more), which allows packets to be timestamped with nanosecond accuracy. As no processing is required in the device, this is by far the most energy-efficient technology that features the lowest TCO of all options. However, owing to multipaths (i.e., RF signals may bounce, therefore the received signal does not necessarily travel in a straight line), the accuracy of geolocation is inferior to that of GPS, see Fig. 3−13.

3.6.3 LoRaWAN multitechnology geolocation is a game changer

Each geolocation technology has a sweet spot of use cases:

- GNSS is ideal for high-frequency, high-resolution use cases, where the GNSS module can remain synchronized.
- Assisted low-power GNSS is ideal if no other geolocation technology is available and random fixes are necessary from time to time.
- WiFi is ideal for urban environments and within an enterprise campus.
- Bluetooth is ideal for indoor geolocation.

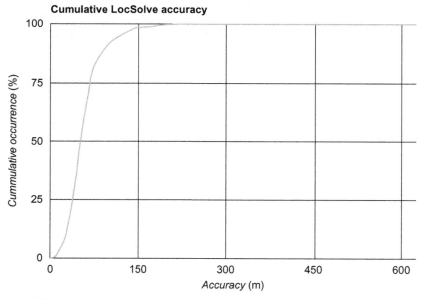

FIGURE 3–13 Cumulative distribution function of TDoA geolocation accuracy.

- TDoA is ideal for ultralow-power, high-frequency, and low-resolution applications such as geofencing or asset tracking, especially for countries with a nationwide LoRaWAN network enabled with TDoA geolocation, such as KPN in the Netherlands.

Most real-life applications are a combination of the above scenarios, and therefore an intelligent multiradio tracker, combined with a cloud application capable of selecting the most appropriate fix mode dynamically, is able to reduce the TCO of geolocation significantly. This also requires a low-power bidirectional communication network, for which LoRaWAN appears to be a perfect fit.

3.7 Using LoRaWAN for firmware upgrade over the air

3.7.1 Introduction

The ability to perform FUOTA is becoming a requirement for most IoT devices. The details of FUOTA procedures are tightly linked to the MCU architecture used by the device. However, FUOTA systems require an efficient and secure file delivery system in order to push the firmware patch file reliably to potentially many devices simultaneously. LoRaWAN provides a secure and reliable file distribution system using multicast, a service provided by a set of application-layer packages. This chapter presents two such packages on which the LoRaWAN file distribution service is based, namely the remote multicast setup and fragmented data block transport.

Performing a secure FUOTA is a complex task that usually involves the following steps.

1. Generate a binary firmware patch file specific to the device platform to be upgraded.
2. Sign this binary file with a private key to guarantee its integrity and authenticity.
3. Push that binary file to the group of devices to be upgraded.
4. Each device must then authenticate the file and its origin by verifying the signature.
5. Each device then installs the firmware patch.
6. Finally, each device reports the status of the upgrade operation (succeeded or failed), and that information is collected by a central device management platform to monitor the status of the device's fleet.

Steps 1, 2, 4, and 5 are device-specific and depend on many design choices, including but not limited to:

- device MCU architecture (ARM, X86, etc.);
- device memory size;
- whether the device uses an OS;
- whether the device uses a secure boot-loader;
- whether the device firmware was designed for partial upgrades; and
- whether the device features a cryptographic hardware accelerator.

These constraints make it impossible to standardize a complete FUOTA process for the diversity of devices currently being connected to LoRaWAN networks. However, a FUOTA system requires an efficient and reliable file-transport service to push the firmware patch file to a potentially very large group of devices through the LoRaWAN network.

An additional challenge is the intrinsically low bit rate of the LoRaWAN network. The implemented solution overcomes this challenge by leveraging the unique radio multicast capability of LoRaWAN. This means that a given radio packet transmitted by the network (containing a fragment of the firmware patch file) may be received by many devices. Therefore the patch file does not have to be transmitted for every single device but, ideally, only once for all devices.

3.7.2 LoRaWAN FUOTA principle

This section presents the operation of two application-layer packages: the multicast remote setup and the fragmented data block transport. These application-layer packages have been defined by the FUOTA working group within the LoRa Alliance, and the specifications are publicly available.

On the device side, these packages live on the application layer. They use the LoRaWAN MAC layer to transmit and receive packets to and from the network. There might be other packages coexisting on the device application layer, and there is usually some user application code as well. Each package on the device corresponds to a service implementation on the server side. For example, the clock synchronization package communicates with a clock synchronization service implemented in the backend as shown in Fig. 3—14. For the sake of simplicity, all backend services related to FUOTA are grouped into a single instance called a device management service. Each of these packages uses a different port to communicate over the air. In this way, the different data streams corresponding to each package can easily be separated in both uplink and down-link directions.

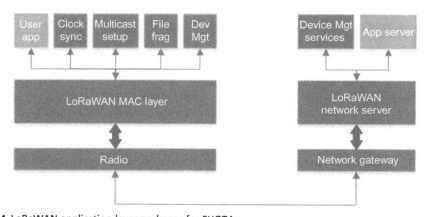

FIGURE 3–14 LoRaWAN application-layer packages for FUOTA.

3.7.3 LoRaWAN multicast groups

The first step of a LoRaWAN FUOTA process is to assemble the list of devices to be updated and then to group them into a LoRaWAN multicast group. This multicast group can be created dynamically and set up over the air, which is performed using the primitives provided by the remote multicast setup package. This section will cover the functionalities implemented in that package.

Every LoRaWAN device is equipped with a unicast identity consisting of a 32-bit device address and a unicast security context (a set of AES128 keys) to uniquely identify and authenticate a given device on the network. In addition, a device may be a member of up to four multicast groups.

A LoRaWAN multicast group is defined by a:

- Multicast group address
- Multicast security context
  - McAppSKey: A multicast application key to encrypt the payloads delivered to that group
  - McNwkSKey: A multicast network session key to compute the message integrity check field of the packets sent to the group
- Multicast frame counter

If a device is a member of more than one multicast group, the addresses of those groups must be different in order to differentiate them.

The multicast security context is identical for all devices belonging to a given group, but different multicast groups have different security contexts. Obviously, they have different addresses, and they must use different keys.

The remote multicast setup package provides commands to perform the following operations:

- Query the implementation version of the package.
- Query the list and status of the multicast groups currently defined in a device.
- Create or modify a multicast group in a device.
- Delete a multicast group definition in a device.
- Define a Class C or Class B broadcast session and associate it with a multicast group.

The McGroupSetup command creates the complete context of a multicast group in a device. This command embeds a mechanism to transport multicast group keys securely over the air, which is achieved by transporting the encrypted multicast keys using a device-specific encryption (or transport) key. It also provides the minimum and maximum valid frame counter values for this multicast context. Limiting the range of valid multicast frame counter values improves the security of the process by limiting the lifetime of a multicast group. However, merely creating the multicast group context in each device of the group is not sufficient to operate a multicast group.

LoRaWAN devices consume extremely little power and typically run on batteries. They usually wake up only when they have data to transmit and then immediately go back into sleep mode. Therefore to transmit a frame to an entire group of devices, the system must first ensure that the receivers of all the devices in the group are in active mode at the same time and for the same duration.

First, all the devices of a multicast group must share the same time. In LoRaWAN networks, the reference time is the GPS time, also called the GPS epoch. It was chosen over the more popular UTC time because GPS time can be represented simply by an unsigned 32-bit counter incremented by 1 and exactly 1 at every second, independently of leap seconds.[3]

The clocks of LoRaWAN devices can be synchronized to the network's time by several different means.

- The device may periodically query the network time using the corresponding MAC layer command (DeviceTimeReq command available in LoRaWAN1.0.3 and upward).
- The device may listen to the periodic Class B beacon, which contains the GPS time.
- The device may use a dedicated package called application-layer clock synchronization.

The decision to use one of these techniques is a trade-off of power consumption, application requirements, and timing precision requirements and is beyond the scope of this chapter.

Once the devices' clocks are synchronized, the device management service must program the devices to open their receivers simultaneously and use the same radio parameters (same channel and data rate). For this purpose, the remote multicast setup package allows the user to define either a Class C or a Class B multicast session.

A Class C multicast session is basically defined by a start time, duration, and a set of radio parameters. Once the device's clock reaches the start time, the device programs its radio using the provided radio parameters and turns on its receiver. The network can broadcast packets using the same radio parameters at any moment during the Class C session. This method is simple and does not require a very accurate timing synchronization because a few seconds of drift between devices is inconsequential. However, for the duration of the Class C session, the devices of the group must keep their receiver continuously active, meaning that this method is not power-efficient in regions where the network is obliged to respect a low transmission duty cycle. Europe, for example, has a network duty cycle of <10%, which means that 90% of the device's energy is wasted during a Class C multicast session.

This is why the package also supports Class B multicast sessions. In a Class B multicast session, all devices of the group must first synchronize onto the Class B beacon (broadcast by the network every 128 seconds) before the session starts. Once the session has started, the devices of the group open periodic reception slots simultaneously for the duration of the session.

---

[3] See https://en.wikipedia.org/wiki/Leap_second for an in-depth explanation of the leap-second insertion in UTC time.

This enables the devices to duty-cycle their receivers while spending most of their time in deep-sleep mode. Note that the network might not necessarily use every slot to transmit information. However, Class B mode is slightly more complicated to implement than Class C mode, and it requires the network to broadcast a Class B beacon. LoRaWAN is one of the very few radio protocols that enable true multicast transmissions over the air. For example, cellular networks emulate multicasting by setting up as many unicast sessions as devices in the group, but the data are actually sent as many times as there are devices in the group.

The multicast remote setup package makes no assumptions about the nature of the data frames being transmitted by the network during a Class C or B multicast session. These could be individual control frames, for example, turning an entire group of streetlights on or off with a single command. Therefore in order to transport a file efficiently during a multicast session, a way must be implemented to fragment a file efficiently and ensure complete delivery to every device of the group. This is the reason for the fragmented data block transport over LoRaWAN package presented in the next section.

3.7.4 Transporting a file to a multicast group

The use of radio multicast tremendously improves the efficiency of the network because a frame is transmitted only once and can be received by all devices of a multicast group, instead of having to transmit this frame once to each device of the group via unicast. However, it poses the challenge of how to cope with missed receptions.

Let us consider the following example. A network broadcasts 100 fragments of a file to a group of 1000 devices assembled in a single multicast group. The radio channel is interfered with and therefore each device may randomly lose 10% of the frames being broadcast by the network. If the network were communicating with only one device (unicast), this would be inconsequential, and a lost fragment might have to be repeated. This happens with a 10% probability anyway. Therefore on average it would take 110 transmissions to push 100 fragments to the device successfully. However, in the case of a multicast group, the situation is far worse.

On average, the first fragment broadcast by the network is received by 900 out of 1000 devices, meaning that 100 devices do not receive it. Therefore the network retransmits the fragment, and on average 10 out of the 100 remaining devices fail to receive it. The network transmits yet again, and one device still misses it. Therefore it takes approximately four transmissions of every fragment to ensure that every device of the group has received it. This situation worsens if the number of devices in the group or the packet error rate increases. Clearly, a different approach must be taken to guarantee that all devices of the group have received the entire file without having to repeat every fragment.

To overcome this problem, the fragmented data block transport package implements an erasure correction code. The principle is as follows. The file to be sent is first split into N fragments of equal length such that each fragment can fit into a single LoRaWAN broadcast payload. In our example, this step yields $N = 100$ fragments of equal length. From these N unencoded fragments, the erasure code generates $N + M$ encoded fragments, where

$(N + M)/N$ is the redundancy ratio, which must be greater—by a certain margin—than the packet loss rate we seek to compensate. In our example, assuming a typical 10% packet loss rate, we select a 20% redundancy ratio. Those encoded fragments are designed such that any subset of N encoded fragments out of the $N + M$ generated fragments can be used to rebuild the N original file fragments and therefore rebuild the complete file.

To continuing our example, the device management service would split the file into 100 fragments and generate 120 encoded fragments. These 120 fragments are then broadcast by the network during a Class B or Class C multicast session. Every listening device may lose a different subset of fragments during the multicast session. However, as soon as a device receives at least 100 of the 120 encoded fragments transmitted by the network, it can reconstruct the complete file. Whereas the repetition method would lead to 400 broadcast frames on average, our encoding scheme allows the network to achieve the same success rate by broadcasting only 120 frames.

In addition to the required encoding/decoding methods described above, the fragmented data block transport package provides the following functionalities. It can

- Query the implementation version of the fragmentation package.
- Create a fragmentation session.
- Request a device or group of devices to provide the status of a fragmentation session.
- Delete a fragmentation session.

Prior to sending a file to a device or group of devices, a fragmentation session must be created in order to provide the device(s) with the details of the file to be broadcast.

For this purpose, the FragSessionSetupRequest command is issued with the following parameters:

- A fragmentation session ID. A device may have up to four fragmentation sessions active simultaneously, so this ID is used to differentiate them.
- Fragment size expressed in bytes.
- Number of fragments required to reassemble the file being transported in this session (ranging from 1 to 65,535).
- Number of padding zeros contained in the last fragment if the file length is not an exact multiple of the fragment length.
- File descriptor, which is a freely allocated four-byte field. If the transported file is a firmware patch image, this field might, for example, encode the version of the transported firmware to allow compatibility verification on the end-device side. The encoding of this field is application-specific.

Once that command has been acknowledged by all the devices of the group, the encoded fragments of the file may be transmitted. Although multicast is more efficient when transmitting the same file to a group of devices, the fragmentation package also allows fragments to be sent using a unicast downlink. Once all the coded fragments have been broadcast, the server may use the FragSessionStatusRequest command to query the status of each device. This command may be sent as a unicast or a multicast frame.

FIGURE 3–15 Mutual authentication and end-to-end encryption provided by LoRaWAN security.

The devices respond with the following information:

- Status of the fragmentation session: File successfully reconstructed, or an error code issued if the file could not be reassembled.
- Total number of fragments received for this fragmentation session.
- Number of missing fragments before the file can be reassembled.

Additional coded fragments may be sent to devices that were unable to reassemble the file entirely. Those fragments may be sent as unicast downlinks, or a new multicast session may be created if too many devices are involved.

3.8 Security

Security is a fundamental requirement of all IoT applications and therefore was designed into LoRaWAN from the very beginning, in line with the general LoRaWAN design criteria: low power consumption, low implementation complexity, low cost, and high scalability. Nevertheless, the LoRaWAN security design adheres to state-of-the-art principles. That is, it uses standard, well-vetted algorithms, and end-to-end security. Furthermore, as devices are deployed in the field for long periods of time—sometimes for many years—security must be future-proof.

The fundamental properties supported by LoRaWAN security are mutual authentication, integrity protection, and confidentiality, see Fig. 3–15.

Mutual authentication is established between a LoRaWAN device and the LoRaWAN network as part of the network join procedure. This ensures that only genuine and authorized devices will join genuine and authentic networks. LoRaWAN MAC and application packets are origin-authenticated, integrity-protected, replay-protected, and encrypted. This protection, combined with mutual authentication, ensures that network traffic is not altered, originates from a legitimate device, is not comprehensible to eavesdroppers, and has not been captured and replayed by rogue actors.

LoRaWAN security furthermore implements end-to-end encryption for application payloads exchanged between devices and application servers. In fact, LoRaWAN is one of the very few IoT networks that implement end-to-end encryption. In some traditional cellular networks, traffic is encrypted over the air interface, but it is transported as plain text in the operator's core network. Consequently, end users are burdened by having to select, deploy, and manage an additional security layer, generally implemented by some type of VPN or application-layer encryption security such as TLS. Such an approach is not well suited in LPWANs, where over-the-top security layers add considerable power consumption, complexity, and cost.

The security mechanisms mentioned above rely on the well-tested and standardized AES cryptographic algorithms [10,11]. These algorithms have been analyzed by the cryptographic community for many years, are NIST-approved, and widely adopted as security best practices for constrained nodes and networks. In particular, LoRaWAN security uses the AES cryptographic primitive combined with several modes of operation: cipher-based message authentication code (CMAC) for integrity protection and counter-mode encryption (CTR). In other words, all LoRaWAN traffic is protected using two session keys as illustrated in Fig. 3−16. Each payload is encrypted by AES-CTR and carries a frame counter to avoid packet replay and a message integrity code computed with AES-CMAC to avoid packet tampering.

Each LoRaWAN device is personalized with a unique 128-bit AES key called an AppKey and a globally unique identifier (EUI-64-based DevEUI), both of which are used during the device authentication process. Allocation of EUI-64 identifiers requires the assignor to have an organizationally unique identifier issued by the IEEE Registration Authority. Similarly, LoRaWAN networks are identified by a 24-bit globally unique identifier assigned by the LoRa Alliance.

3.9 LoRaWAN certification

The purpose of the LoRaWAN certification is to ensure device interoperability and guarantee that all LoRaWAN end devices work under any network conditions and in any public or private network.

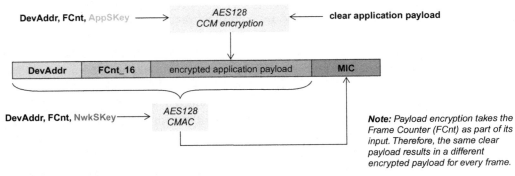

FIGURE 3−16 LoRaWAN security mechanism based on standardized Advanced Encryption Standard cryptographic algorithms.

As the LoRaWAN technology is aimed at providing very low-cost devices, the LoRaWAN certification program has been designed to test all protocols defined in the LoRaWAN (L2) specification in a cost-effective manner. It achieves this by running tests within a fully automatic test harness that requires no expensive test equipment. Instead, it uses a standard LoRaWAN gateway and a PC running an NS and the test control layer (TCL). To keep the cost of certification competitive, testing is completed within 1 day, not weeks or months as with other technologies. The DUT and gateway (GW) are housed in a shielded box to avoid other radio transmissions in the ISM band that might affect the test results. Moreover, as end devices are likely to be small radio devices, certification has been designed to use only the LoRaWAN radio connection on the device to be certified (DUT) and no additional test connections (Fig. 3−17).

The LoRa Alliance Certification Committee has created a comprehensive test procedure for DUT that will ensure the correct LoRaWAN behavior of end devices and include testing all MAC commands and functional testing of the device. The certification test sends and receives more than 1000 messages between the NS and the end device.

The LoRa Alliance certification program does not test the regulatory requirements of the ISM band for the country or region in which it is operating but, as these requirements can change the behavior of the device, there is a slightly different certification test requirement specification for the different ISM bands.

The LoRa Alliance controls and operates the certification program and has appointed several authorized test houses (ATH) throughout the world to perform certification. These ATH

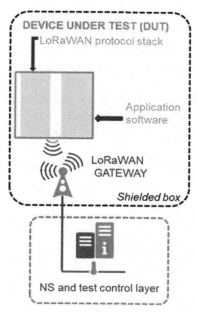

FIGURE 3–17 LoRaWAN certification.

can provide testing at locations near the device developer, even if the device is to be certified for a different region.

To allow the device manufacturer to pretest devices before sending them to an ATH, the LoRa Alliance has commissioned and made available to LoRa Alliance members a LoRaWAN certification test tool, which can be used at their own premises with a PC running a NS, a TCL, and a local gateway.

These protocols not only verify adherence to the LoRaWAN L2 specification, but also test the RF performance of the device, which is critical to its usability and reliability. The LoRa Alliance Certification Committee has produced a comprehensive set of guidelines on how to measure the RF performance of a device. These guidelines can be used by device manufacturers to test their devices before sending them to an ATH for official certification.

When a device has been certified by an ATH, the results are sent to the LoRa Alliance, which issues a certificate to the manufacturer and adds the device to the master list of certified products available on the LoRa Alliance website (LoRa-Alliance.org).

Author Contribution

3.1 Technical overview of LoRaWAN specifications: Alper Yegin, Actility, Lannion, France.

3.2 LoRaWAN link layer: Dr. Thorsten Kramp, Formerly with Semtech (International) AG, Rapperswil-Jona, Switzerland.

3.3 Scaling LoRaWAN networks: Pierre Dufour, Rohit Gupta, Ramez Soss, Olivier Hersent, Actility, Lannion, France.

3.4 LoRaWAN regional parameters: Derek Hunt, LoRa Alliance Certification Committee Chairman.

3.5 Activation and roaming: Alper Yegin, Actility, Lannion, France.

3.6 Network-based and multitechnology geolocation: Olivier Hersent, Actility, Lannion, France.

3.7 Using LoRaWAN for firmware upgrade over the air: Nicolas Sornin, CTO, Semtech (International) AG, Rapperswil-Jona, Switzerland.

3.8 Security: Dr. Thorsten Kramp, Formerly with Semtech (International) AG, Rapperswil-Jona, Switzerland.

3.9 LoRaWAN certification: Derek Hunt, LoRa Alliance Certification Committee Chairman.

References

[1] LoRa Alliance®, LoRaWAN 1.0.3 Specification, 2018.

[2] LoRa Alliance®, LoRaWAN 1.0.3 Regional Parameters Rev A, 2018.

[3] LoRa Alliance®, LoRaWAN Backend Interfaces 1.0 Specification, 2017.

[4] LoRa Alliance®, LoRaWAN Fragmented Data Block Transport Specification v1.0. 0, 2018.

[5] LoRa Alliance®, LoRaWAN Remote Multicast Setup Specification v1.0. 0, 2018.

[6]   M. Bor, U. Roedig, T. Voigt, J. Alonso, Do LoRa Low-Power Wide-Area Networks Scale? in: Proceedings of 19th ACM International Conference on Modeling, Analysis and Simulation of Wireless and Mobile Systems, MSWiM, 2016.

[7]   T. Voigt, M. Bor, U. Roedig, J. Alonso, Mitigating Inter-Network Interference in LoRa Networks, in: Proceedings of International Conference on Embedded Wireless Systems and Networks, EWSN, 2017.

[8]   F. Van den Abeele, J. Haxhibeqiri, I. Moerman, J. Hoebeke, Scalability analysis of large-scale LoRaWAN® networks in ns-3, University of Ghent. <https://arxiv.org/abs/1705.05899>.

[9]   LoRaSim. LoRa-Alliance.org.

[10]   J.H. Song, R. Poovendran, J. Lee, T. Iwata, The AES-CMAC algorithm, RFC 4493, RFC Editor, June 2006.

[11]   NIST, FIPS 197, Advanced Encryption Standard (AES), November 2001.

4

Radio channel access challenges in LoRa low-power wide-area networks

Congduc Pham[1], Ahcène Bounceur[2], Laurent Clavier[3], Umber Noreen[2], Muhammad Ehsan[1]

[1]UNIVERSITY OF PAU, LIUPPA, PAU, FRANCE [2]UNIVERSITY OF BREST, LABSTICC, BREST, FRANCE [3]IMT LILLE, LILLE, FRANCE

4.1 Purpose of this chapter

Recently, low-power wide-area networks (LPWANs) play a key role in the Internet-of-things (IoT) maturation process. Under the LPWAN broad term are a variety of technologies enabling power-efficient wireless communication over very long distances. For instance, technologies based on ultranarrowband modulation (UNB)—for example Sigfox—or chirp spread spectrum (CSS) modulation—for example LoRa—have become de facto standards in the IoT ecosystem. Most of LPWAN technologies can achieve more than 20 km in line of sight (LOS) condition. In a typical long-range one-hop connectivity scenario, the gateway is the single interface to Internet servers through cellular/asymetric digital subscriber line/ Ethernet/Wi-Fi technologies depending on what is available locally. Devices typically communicate directly to one or more gateways, which removes the need for constructing and maintaining a complex multihop network. Recent deployment tests with LoRa gateways located on top of high building show more than 6 km range in urban scenarios for smart city applications [1]. A large city can easily be covered with less than 10 gateways. Indoor smart building applications are also enabled by the easy coverage of buildings several stories high. Communication to high-altitude balloons has also been realized successfully [2,3] and tests with low-orbit satellites are on the way [4]. These very versatile technologies definitely provide a better connectivity answer for battery-operated IoT devices by avoiding complex synchronization and costly relay nodes to be deployed and maintained.

Given the incredible worldwide uptake of LPWANs for a large variety of innovative IoT applications, including multimedia sensors, it is important to understand the challenges behind large-scale and dense LPWAN deployment, especially because both Sigfox and LoRa networks are currently deployed in unlicensed bands. This situation is most likely not going

LPWAN Technologies for IoT and M2M Applications. DOI: https://doi.org/10.1016/B978-0-12-818880-4.00004-1

to change, at least in the next few years, as working in the unlicensed band allows for a much quicker uptake of the technology.

This chapter has a particular focus on LoRa technology as it can be deployed in a private and ad-hoc manner, making experimental deployments much easier. Existing studies on LoRa scalability and radio channel access mechanisms for LoRa LPWAN will be reviewed and promising approaches will be presented in more detail. The chapter will also provide to the readers useful information on the LoRa physical layer, as well as on promising interference mitigation techniques that can be applied, such as capture effect (CE) and successive interference cancellation.

The chapter will also give a large part on experimental results. The authors have conducted a large number of LoRa LPWAN experimental studies, as well as real-world deployments of both IoT test-beds and IoT production networks in the context of three R&D projects (two EU H2020 projects—WAZIUP and WAZIHUB—and one national ANR project—PERSEPTEUR) in addition to numerous academic and industrial collaborations. Models and large-scale simulations are also considered and the chapter will present the open-source CupCarbon simulation tool, which especially addresses the modeling and performance evaluation of radio physical layers. The original feature of CupCarbon over other existing open-source simulation environments is its particular focus on radio propagation modeling in urban environments by taking into account three-dimensional (3D) maps of the simulated areas.

We hope that the chapter will provide a proper balance between research and experimental results. It therefore definitely targets students, academicians, researchers, and industry professionals who need a broad understanding of the technologies, their challenges, the existing solutions along with their limitations, and feedbacks from real-world deployments, as well as insights on how larger-scale scenarios can be evaluated.

4.2 Review of LoRa physical layer

LoRa utilizes bandwidth of usually 125 kHz to broadcast a signal. Using bands that are not too narrow allows LoRa to exhibit some robustness against some characteristics of the channel such as frequency selectivity and Doppler effect, to name a few. The transmitter generates chirp signals by varying their frequency over time and keeping phase between adjacent symbols constant. Receiver can decode even a severely attenuated signal 19.5 dB below the noise level [5]. The main characteristics of LoRa modulation depend on spreading factor (*SF*), coding rate (*CR*), and bandwidth (*BW*). Spreading factor $SF = log_2(R_c/R_s)$ is the ratio between symbol rate (R_s) and chip rate (R_c). LoRa employs six orthogonal spreading factors, 7−12. *SF* provides a tradeoff between data rate and range through higher receiver's sensibility. Along with the spreading factors, forward error correction (FEC) techniques are used in LoRa to increase the receiver's sensibility further. Code rate *CR* defines the amount of FEC in LoRa frame. LoRa offers *CR* = 0, 1, 2, 3, and 4, where *CR* = 0 means no encoding. The choice of higher *SF* and *CR* values dramatically increase the time on air (ToA), T_{air}. Using a higher value for *BW* will reduce T_{air} but lowers the receiver's sensibility. LoRa provides several bandwidth values from 7.8 to 500 kHz with 125, 250, and 500 kHz being the most used. Taking these parameters into account, the useful bit rate R_b equals:

$$R_b = \frac{4 \times SF \times BW}{(4 + CR) \times 2^{SF}} \quad (\text{bits}/s)$$

(4.1)

4.2.1 LoRa physical layer (PHY) structure

LoRa is a Semtech proprietary technology and is not fully open. This section gives the analysis on the working of PHY in LoRa, according to our understanding. Fig. 4−1 shows the block diagram of LoRa transceiver, which is briefly explained in the following paragraphs.

4.2.1.1 Encoding
First, the binary source input bits pass through an encoder. The output of encoder depends on the choice of *CR* value. Encoding reduces the packet error rate in the presence of short bursts of interference. LoRa uses Hamming codes for FEC. These are linear block codes and are easy to implement. LoRa uses coding rates CR of 4/5, 2/3, 4/7, and 1/2, which means if the code rate is denoted as k/n, where k represents the number of useful information bits, and encoder generates n output bits, then $(n - k)$ are the redundant bits. If we assume *CR* values between 1, 2, 3, and 4 for coding rates 4/5, 4/6, 4/7, and 4/8, respectively, then the error detection and correction capabilities are as shown in Table 4−1.

4.2.1.2 Whitening
The output of the encoder passes through the whitening block. Whitening is an optional step in LoRa, which can be implemented by Manchester encoding to induce randomness. Here, the purpose of whitening is to make sure that there are no long chains of 0s or 1s in the payload.

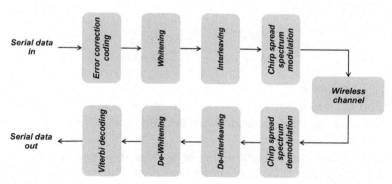

FIGURE 4–1 LoRa PHY layer architecture.

Table 4–1 Error detection and correction capabilities of LoRa.

Coding rates	Error detection (bits)	Error correction (bits)
4/5	0	0
4/6	1	0
4/7	2	1
4/8	3	1

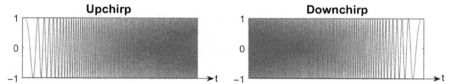

FIGURE 4–2 Upchirp and downchirp signal.

4.2.1.3 Interleaving

The output of whitening block is passed to the interleaving block. The interleaver uses diagonal placing method to scramble each $4 + CR$ codeword and sends it to the spreading block.

4.2.1.4 Chirp spread spectrum modulation

This block here spreads each symbol over an upchirp, according to the SF value used. For example, for $SF = 7$ and $SF = 12$, 128 and 4096 chips/symbol will be used, respectively. The relationship between the symbol rate $R_s = BW/2^{SF}$ and chip rate R_c is:

$$R_c = 2^{SF} \times R_s \quad \text{and} \quad R_c = \frac{2^{SF} \times BW}{2^{SF}}, \text{ so}, \quad R_c = BW \text{ chips}/s$$

It takes much larger BW for transmission than required for the considered data rate. Chirp signal is a sinusoidal signal with either linearly increasing or decreasing frequency. A linear chirp waveform can be expressed as:

$$c(t) = \begin{cases} \exp(2\pi j(at + b)t), & \dfrac{-T_s}{2} \leq t \leq \dfrac{T_s}{2} \\ 0, & \text{otherwise} \end{cases} \tag{4.2}$$

with $at + b = f_{\min} + \dfrac{f_{\max} - f_{\min}}{T_s} t$

where $f_{\max}$ and $f_{\min}$ are the maximum (125 kHz in our case) and minimum frequency, respectively. T_s is the symbol duration. An upchirp and a downchirp are shown in Fig. 4–2.

Each symbol of SF bits can be represented by shifting the frequency ramp based on the symbol value. So each coded chirp is obtained by a cyclic shift in an upchirp, as illustrated in Fig. 4–3. Circular shift in raw upchirp is expressed in Eq. (4.3).

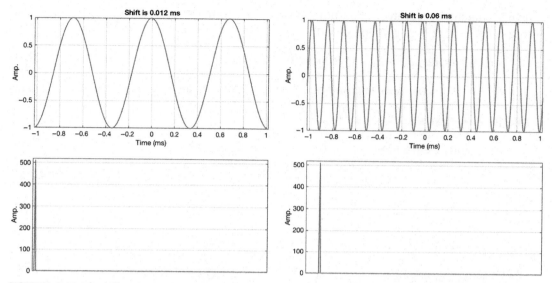

FIGURE 4–3 Matched filter output.

$$c(t) = \begin{cases} \exp(2\pi j(a(T_s + t - \Delta t) + b)(Ts + t - \Delta t)), & \dfrac{-T_s}{2} \le t \le \dfrac{-T_s}{2} + \Delta t \\[2ex] \exp(2\pi j(a(t - \Delta t) + b)(t - \Delta t)), & \dfrac{-T_s}{2} + \Delta t \le t \le \dfrac{T_s}{2} \\[2ex] 0, & \text{otherwise} \end{cases} \quad (4.3)$$

where Δt is the shift in time that depends on the symbol value.

4.2.1.5 Chirp spread spectrum demodulation

The matched receiver for a linear chirp is performed by multiplication with the downchirp, as shown in Fig. 4–3.

This process can be represented mathematically for $\Delta t = 0$. Note that, in case of symbol value "0," the transmitter will send an upchirp:

$$y(t) = \exp(2\pi j(at + b)t) \times \exp(-2\pi j(a(T_s - t) + b)(T_s - t)) \quad (4.4)$$

$$y(t) = \exp(2\pi j(at^2 + bt - aT_s^2 + 2aT_s - at^2 - bT_s + bt)) \quad (4.5)$$

$$y(t) = \exp(2\pi j(2atT_s + 2bt - aT_s^2 - bT_s)) \quad (4.6)$$

$$y(t) = \exp(2\pi j(2t(aT_s + b) - aT_s^2 - bT_s)) \quad (4.7)$$

The matched receiver output for any other symbol would be for $-T_s/2 \le t \le -T_s/2 + \Delta t$ (when $\Delta t \ne 0$):

$$y_i(t) = \exp(2\pi j(a(t + T_s - \Delta t) + b)(t + T_s - \Delta t)) \times \exp(-2\pi j(a(T_s - t) + b)(T_s - t)) \tag{4.8}$$

$$y_i(t) = \exp(2\pi j(2t(b - a\Delta t - 2aT_s) + a\Delta t^2 - 2a\Delta t T_s - b\Delta t)) \tag{4.9}$$

And for $\frac{-T_s}{2} + \Delta t \le t \le \frac{T_s}{2}$:

$$y_i(t) = \exp(2\pi j(a(t - \Delta t) + b)(t - \Delta t)) \times \exp(-2\pi j(a(T_s - t) + b)(T_s - t)) \tag{4.10}$$

$$y_i(t) = \exp(2\pi j(2t(b - a\Delta t + aT_s) + a\Delta t^2 - b\Delta t - aT_s^2 - bT_s)) \tag{4.11}$$

Hence,

$$y_i(t) = \begin{cases} \exp(2\pi j(2t(b - a\Delta t - 2aT_s) + a\Delta t^2 - 2a\Delta t T_s - b\Delta t)), & \dfrac{-T_s}{2} \le t \le \dfrac{-T_s}{2} + \Delta t \\[2ex] \exp(2\pi j(2t(b - a\Delta t + aT_s) + a\Delta t^2 - aT_s^2 - bT_s - b\Delta t)), & \dfrac{-T_s}{2} + \Delta t \le t \le \dfrac{T_s}{2} \\[2ex] 0, & \text{otherwise} \end{cases} \tag{4.12}$$

Then, the output signal is analyzed to identify the presence of the sharp narrow peak in frequency domain, which is at "3" and "15" in Fig. 4−3. The sharp narrow peak occurs at the time index corresponding to the constant value of the coded chirp.

4.2.1.6 Packet structure and time on air

Fig. 4−4 shows the packet structure used by LoRa. LoRa offers a maximum packet size of 256 bytes. More details on the LoRa packet structure can be found in [5]. For the purpose of this chapter, the main part of interest is the preamble that is a sequence of constant upchirps, two downchirps, and a quarter of upchirp.

The receiver uses the preamble to start synchronizing with the transmitter. The LoRa packet ToA can be defined as:

$$T_{\text{air}} = T_{\text{preamble}} + T_{\text{payload}} \tag{4.13}$$

where T_{preamble} is the preamble duration and T_{payload} is the payload duration that also includes optional header and cyclic redundancy check (CRC) fields. Without going into the

PHY Frame	Preamble	Header	Header CRC	Payload	Payload CRC
		Coding rate = 4/8		Coding rate = 4/(4 + CR), where CR = 0,1,2,3, or 4	
Size	Min. 4.25 symbols	2 bytes	2 bytes	Max. 255 bytes	2 bytes

FIGURE 4−4 LoRa PHY frame format.

details of exact ToA computation, which can be found in [5], one can say that *SF* and *BW* have direct influence on the ToA of the LoRa packet, as these parameters typically define the symbol rate: higher *SF* increases ToA, while higher *BW* decreases ToA at the cost of lower receiver's sensibility.

4.2.2 PHY performance

Fig. 4−5 shows the bit error rate (BER) performance of LoRa-based PHY, plotted over the signal-to-noise ratio (SNR). Vertical axis shows different bit error rate values and horizontal axis shows SNR values. Fig. 4−5 shows the impact of *SF*: increasing *SF* decreases noticeably the BER at the cost of a reduced data rate and increased ToA. These results have been produced with $CR = 0$ and $BW = 125$ kHz.

Table 4−2 shows the data rates associated with the spreading factor *SF* and the coding rate *CR*. It can be easily noticed that *CR* also affects the data rate. But increasing *SF* and *CR* helps combat the harsh wireless conditions.

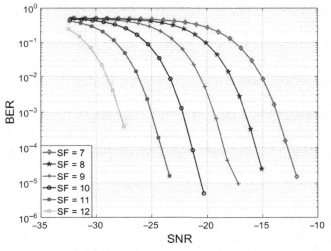

FIGURE 4–5 BER with different *SF* values and $CR = 0$.

Table 4–2   Data rates offered in LoRa.

SF	Data rate ($CR = 0$)	CR	Data rate ($SF = 7$)
7	6.8 kbps	0	6.8 kbps
8	3.9 kbps	1	5.4 kbps
9	2.1 kbps	2	4.5 kbps
10	1 kbps	3	3.9 bps
1	671 bps	4	3.4 kbps
12	366 bps	−	−

4.2.3 Interference in LoRa

A LoRa network does not define any particular channel access control, and it is therefore similar to a pure so-called ALOHA system. One of the limitations of ALOHA-based systems is its blind transmission strategy that allows the transmitters to transmit whenever there is a frame to send without carrier sensing. The vulnerable time in pure ALOHA-based network is twice the frame time, for example $2 \times T_{air}$ in LoRa. That means, a packet will be destroyed by any overlapping transmission starting in the time window that starts one packet time before the transmission of the packet and closes at the end of the transmission of the packet. Hence, the throughput of such network is: [6]

$$\eta = Ge^{-2G} \tag{4.14}$$

"2" in the superscript of exponential is because the vulnerable time is twice the frame time T_{air}. G represents average number of transmission attempts during frame time.

If we consider a large ALOHA LoRa network, transmission attempts occur according to the Poisson process with average rate G attempts per slot. If we assume equal-length packets and only nodes with same SF will collide, then G can be defined as the average number of attempts per time frame:

$$G = N \times p_i \times \lambda_i \times T_{air} \tag{4.15}$$

where T_{air_i} is the ToA of one LoRa frame (stated in Eq. 4.13), λ_i is the packet generation rate of all N end-devices using SF_i in a network, and p_i the probability that an end-device uses SF_i. The maximum throughput of pure ALOHA system can be defined by taking the derivative of Eq. (4.14) with respect to average traffic G and setting it equal to zero, $d/dG(Ge^{-2G}) = 0$, gives $G = 1/2$. Substituting this value into Eq. (4.14) gives:

$$\eta = \frac{1}{2e^{-1}} = 0.1839 \tag{4.16}$$

So, pure ALOHA-based network can give 18.39% of maximum efficiency. For LoRa modulation, this efficiency can be increased by improving the receiver technique. Fig. 4−6 shows a collision occurring between two consecutive LoRa symbols.

I_1 and I_2 are the symbols from the interference signal. R_s is the symbol of the ongoing transmission. The expression for $-T_s/2 \leq t \leq -T_s/2 + \Delta T$ can be written as:

$$S_{i_1}(t) = S(t) + \begin{cases} \exp(2\pi j(a(\Delta T - t + \Delta t_1) + b)(\Delta T - t + \Delta t_1)), & (1) \\ \exp(2\pi j(a(\Delta T - t + \Delta t_1 - T_s) + b)(\Delta T - t + \Delta t_1 - T_s)), & (2) \\ 0, & \text{otherwise} \end{cases} \tag{4.17}$$

Where conditions (1) and (2) are, respectively, $-T_s/2 \leq t \leq -T_s/2 + \Delta t_1$ and $-T_s/2 + \Delta t_1 \leq t \leq T_s/2$. Then, the expression for $\frac{-T_s}{2} + \Delta T \leq t \leq T_s/2$ can be written as:

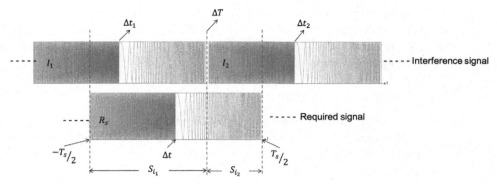

FIGURE 4–6 Interfering LoRa symbols at receiver.

$$S_{i_2}(t) = S(t) + \begin{cases} \exp(2\pi j(a(t + T_s - \Delta T + \Delta t_2) + b)(t + T_s - \Delta T + \Delta t_2)), & (3) \\ \exp(2\pi j(a(t - \Delta T + \Delta t_2) + b)(t - \Delta T + \Delta t_2)), & (4) \\ 0, & \text{otherwise} \end{cases} \quad (4.18)$$

Here, conditions (3) and (4) are, respectively, $-T_s/2 \le t \le -T_s/2 + \Delta t_2$ and $-T_s/2 + \Delta t_2 \le t \le T_s/2$. At this point, multiplication with downchirp $D(t)$ is performed, and the expression would be:

$$y_{i_1} = S_{i_1}(t) \times D(t)$$

$$y_{i_2} = S_{i_2}(t) \times D(t)$$

$$y_{i_1}(t) = y_i(t) + \begin{cases} \exp(2\pi j(2at(T_s - \Delta T - \Delta t_1) + a(\Delta T^2 + \Delta t_1^2 - T_s^2 + 2\Delta T\Delta t_1) \\ \quad + b(\Delta T + \Delta t_1 + T_s))), & (1) \\ \exp(2\pi j(2at(2T_s - \Delta T - \Delta t_1) + a(2\Delta T\Delta t_1 - 2T_s\Delta T - 2s\Delta t_1 \\ \quad + \Delta T^2 + \Delta t_1^2) + b(\Delta T + \Delta t_1 - 2T_s))), & (2) \\ 0, \quad \text{otherwise} \end{cases} \quad (4.19)$$

$$y_{i_2}(t) = y_i(t) + \begin{cases} \exp(2\pi j(2t(2aT_s - a\Delta T + a\Delta t_2 + b) + a(2T_s\Delta t_2 - 2T_s\Delta T \\ \quad - 2\Delta T\Delta t_2 + \Delta T^2 + \Delta t_2^2) + b(\Delta t_2 - \Delta T))), & (3) \\ \exp(2\pi j(2t(aT_s - a\Delta T + a\Delta t_2 + b) + a(\Delta T^2 + \Delta t_2^2 - T_s^2 \\ \quad - 2\Delta T\Delta t_2) + b(\Delta t_2 - \Delta T - T_s))), & (4) \\ 0, \quad \text{otherwise} \end{cases} \quad (4.20)$$

The frequency response of Eqs. (4.19) and (4.20) will give sharp narrow peaks at time Δt, Δt_1 and Δt_2 that correspond to the constant value of the coded chirps, as shown in Fig. 4−7. The receiver can make the decision by selecting the peak with the highest amplitude value.

The receiver can successfully differentiate between the interferers based on their received power values. The problem occurs when one or more interferers carry either the same value

($\Delta t = \Delta t_1 = \Delta t_2$) so that their power adds up or have an equivalent or more power than the useful one. In Fig. 4–8, first interferer has the same received power as the useful user and in Fig. 4–9, all interferers contain the same value ($\Delta t = \Delta t_1 = \Delta t_2$), which causes an increase in peak value in the frequency domain.

4.2.4 Orthogonality of LoRa transmissions

The spreading factor SF in LoRa is the ratio between symbol rate R_s and chip rate R_c: $SF = log_2(R_c/R_s)$. LoRa employs six spreading factors from 7 to 12 that provide some sort of orthogonality in data transmission (collision-free) occurring on the same frequency. This orthogonality property can be explained as follows: different SF will give a different chirp rate, which is the change in the frequency with respect to time. As CSS uses frequency chirps with linear variation of frequency over time, when plotting frequency against time the chirp rate will be the slope of the line. Therefore, theoretically, different SF will give different slopes that provide the orthogonality property.

However, when using a LoRa combination of BW and SF, not all combinations are orthogonal because some of them define the same chirp rate (same slope) [7]. This can be further explained as follows: the slope (chirp rate) can be defined as slope $= (f_{max} - f_{min})/T_s$, where T_s is the symbol duration. As the symbol rate is $R_s = BW/2^{SF}$ then $T_s = 1/R_s = 2^{SF}/BW$. With $f_{max} - f_{min} = BW$ we have slope $= BW * (BW/2^{SF}) = BW^2/2^{SF}$.

FIGURE 4–7 Coded chirps at 30, 100, and 128 (different TX power). $SF = 8$.

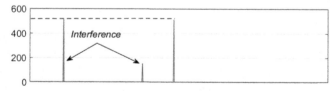

FIGURE 4–8 Coded chirps at 30, 100, and 128 (same TX power). $SF = 8$.

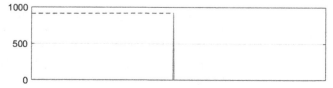

FIGURE 4–9 FFT of three coded chirps at 128. $SF = 8$.

SF		7	8	9	10	11	12	7	8	9	10	11	12	7	8	9	10	11	12
	BW	125	125	125	125	125	125	250	250	250	250	250	250	500	500	500	500	500	500
7	125	x								x								x	
8	125		x								x								x
9	125			x								x							
10	125				x								x						
11	125					x													
12	125						x												
7	250							x								x			
8	250								x								x		
9	250	x								x								x	
10	250		x								x								x
11	250			x								x							
12	250				x								x						
7	500													x					
8	500														x				
9	500							x								x			
10	500								x								x		
11	500	x								x								x	
12	500		x								x								x

FIGURE 4–10 Orthogonality of *SF* and *BW* combination. x = not orthogonal.

For instance, if we take SF7BW125, SF9BW250, and SF11BW500, we have the same chirp rate that is 122070312.5. Fig. 4−10 shows all BW and SF combinations indicating those that are not orthogonal by an "x."

It is also worth mentioning that [8] reported some issues with different *SF* values when there is high power difference such as in near-far conditions.

4.3 Dealing with interferences in LoRa

When dealing with dense deployment scenarios, one of the most immediate questions is "How does LoRa scale?". This question has been addressed by a number of articles [9−14], to name a few. Before providing some answers, it is necessary to distinguish scalability provided by the LoRa lower layers from the scalability provided by applying higher-level mechanisms or regulations and also scalability provided by dedicated hardware such as in LoRaWAN-compliant gateways based on the SX1301 radio concentrator capable of simultaneous listening on eight frequency channels, each channel also capable to handle different *SF* values.

When considering only the LoRa lower layers that mainly consist in the physical layer without any Listen Before Talk (LBT, or Carrier Sense) mechanism, a simple scalability study can use the simple ALOHA model shown in Section 4.2.3. If we add *SF* orthogonality as explained previously, we can consider that each orthogonal *BW* and *SF* combination provides an independent ALOHA channel. Here, the issue is how to distribute or assign a given (*BW*, *SF*) combination to devices. We will discuss this issue and present the adaptive data rate (ADR) mechanism proposed by the higher-level LoRaWAN specification in Section 4.3.1.

At the highest level, limitation of device's duty cycle imposed by regulation agencies, or by a user community with a so-called "fair share" approach such as the one proposed by TheThingsNetwork, can be a solution to reduce uplink traffic and thus the collision probability. This will be explained in more detail in Section 4.3.2.

4.3.1 Impact of adaptive data rate

The LoRaWAN specifications [15] (v1.1 at the time of writing) propose an ADR mechanism for optimizing data rates, airtime, and energy consumption in the network. By optimizing, that is mainly reducing, airtime, contention on the radio medium is also reduced and thus the collision probability. Basically, LoRa gateways can use information, such as RSSI and/or SNR, from uplink messages from a given device to determine the "margin" that is still available to correctly demodulate messages. This margin is used to determine how much the device can increase its data rate, therefore reducing the airtime. Practically, airtime can be reduced by using higher value for *BW* and/or smaller value for *SF*. After determining the new parameters, a downlink command message will be sent by a gateway to a device to set the new data rate.

The agility of the ADR mechanism has been studied in Ref. [16]. The authors reported that the convergence time can be very high because many packets need to be received by gateways. As network size increases the convergence time increases, as well due to high contention on the radio medium: "from around 200 minutes for a 100-node network to more than 3000 minutes for a 4000-node network." The convergence time also increases significantly when the link quality degrades. The explanation by the authors is as follows: devices need lots of time to move from lower to higher value of *SF* to regain connectivity because the process unfortunately requires devices to lose sufficient number of sent packets before moving to higher *SF*.

For static devices, as it is most likely the case in a large variety of IoT deployment, the ADR mechanism can indeed help to reduce the ToA of uplink transmissions. However, as many devices can converge to use the same LoRa parameters, collision probability may increase, leading to lower link quality, thus impacting convergence time as observed in [16]. We believe there is a risk that the ADR mechanism will make gateways and devices oscillating from one parameter combination to another without taking into account that more variety in (*BW*, *SF*) combinations provides more orthogonality, making the system closer to *N* independent ALOHA channels. These issues certainly need more studies in the future to see how smarter parameter assignment strategies can be proposed as slightly addressed in [12].

There are also some interesting works on scheduling LoRa transmissions when the network size is not too large [17−19].

4.3.2 Impact of duty-cycle limitation

The flexibility of long-range transmission in the unlicensed bands comes at the cost of stricter legal regulations such as limited duty-cycling, for example maximum transmission time per hour. For instance, in Europe, electromagnetic transmissions in the unlicensed EU 863−870 MHz industrial-scientific-medical (ISM) band used by Semtech's LoRa technology falls into the ETSI's short-range devices (SRD) category. The ETSI EN300-220-1 document [20] specifies for Europe various requirements for SRD, especially those on radio activity. Basically, a transmitter is constrained to 1% duty cycle (i.e., 36 s/h) in the general case. This duty-cycle limit applies to the total transmission time, even if the transmitter can change to another channel. Note that this duty-cycle limitation approach is also adopted in China in the 779−787 MHz ISM band. US regulations in the 902−928 MHz ISM band do not specify duty cycle but rather a maximum transmission time per packet with frequency hopping constraints. LBT along with adaptive frequency agility can be used to go beyond the 1% duty-cycle limit but then additional restrictions are introduced: the ToA for a single transmission cannot exceed 1 s. If this 1 s limit is respected, then the transmitter is allowed to use a given channel for a maximum T_x on time of 100 s over a period of 1 hour for any 200 kHz bandwidth.

Therefore, due to regional restrictions on operation in licensed free ISM bands, frame generation rate λ_i usually depends on the duty cycle D and ToA. If we assume that deployed network locates in Europe, then according to ETSI 1% duty cycle applied on the usage of each subband [21].

$$\lambda_i = \frac{D}{T_{\text{air}_i}} \quad \text{or} \quad \lambda_i = \frac{1}{T_{\text{of } f_i} + T_{\text{air}_i}} \tag{4.21}$$

Higher specifications such as LoRaWAN can define $(T_{\text{of } f_i} = (T_{\text{air}_i}/D) - T_{\text{air}_i})$ to be the minimum period during which a device cannot access the medium due to the duty-cycle restriction. In this case Eq. (4.14) becomes:

$$\eta_i = p_i \lambda_i N T_{\text{air}_i} e^{(-2 p_i \lambda_i N T_{\text{air}_i})}, \quad i \in SF \tag{4.22}$$

The term e^{-2G} is called probability of successful transmission P of a frame and $0 \le P \le 1$. For a LoRa network, it can be defined as:

$$P_i = e^{(-2 p_i \lambda_i N T_{\text{air}_i})}, \quad i \in SF \tag{4.23}$$

The outage probability will be:

$$P_{\text{out}_i} = 1 - P_i, \quad i \in SF \tag{4.24}$$

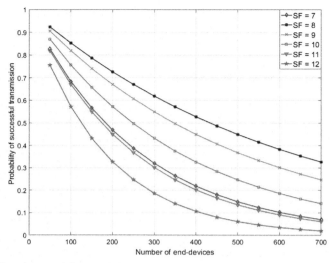

FIGURE 4–11 Probability of successful transmission.

For simplicity, we are going to consider that different spreading factors do not interfere, and we are interested in the limit in successful transmissions in a LoRaWAN system. If the probabilities p_i for an end-device to use SF_i are the same, then the definition of $T_{\text{of } f_i}$ will make the load on each SF similar. In Fig. 4–11, we illustrate the probability of successful transmission depending on the number of users and SF when $p_{7->12} = [0.19, 0.08, 0.10, 0.14, 0.20, 0.28]$. With the increase in the network size, the probability of successful reception drops due to an increase in collisions in the network but is higher than a traditional ALOHA system thanks to the duty-cycle limitations. Our analytic results shown in Fig. 4–11 are consistent with simulation results presented in Ref. [13].

4.3.3 Interference mitigation: capture effect

LoRa PHY is a kind of frequency modulation that manifests capture effect (CE). In the past, many theoretical studies on CE have been performed to increase the packet reception rate (PRR) of a network, in the presence of a collision. But in the context of LoRa, not much research has been done. Practical studies in Refs. [9,13,14,22] have shown CE for LoRa-based system. In Ref. [23], authors presented capture study on equal power collisions in the pure ALOHA-based 802.15.4 system. In Ref. [24], authors state that their collision detection approach can differentiate between a packet collision and packet loss for 802.15.4-based system.

In general, the receiver keeps monitoring for the new potential preambles and if its signal-to-interference noise ratio (SINR) is above a given ratio, receiver stops ongoing reception and resynchronizes with the new packet and demodulates the signal. We are going to characterize CE in the next section (the probability for a packet to be decoded despite the presence of one interferer). Note that for the sake of simplicity, we only consider two-packet collision scenario. This analysis can be extended to three or more packet collisions. The capture characteristics of any radio transceiver depend on the modulation, decoding schemes, and its hardware design and implementation. In a radio-frequency interference environment, a particular signal X can be successfully decoded if:

$$\text{SINR}_X = \frac{P_X}{\sum P_I + \sigma^2} > Th \qquad (4.25)$$

where P_X is the source signal strength, $\sum P_I$ is the aggregate interference strength from the other active users in the network, σ is the channel noise coefficient, and Th is the minimum SINR threshold required to successfully decode signal X. When two or more packets collide, with CE, it is still possible to receive one of them. CE enables the receiver to decode a packet that satisfies Eq. (4.25), even if it arrives during the reception of an ongoing packet.

For a LoRa modulation, as shown in Figs. 4−8 and 4−9, only the strength of the strongest interferer will matter as long as the simultaneous number of interferers is not too high and the probability to have interferers with a shift that falls at the same time remains low. So Eq. (4.25) will become:

$$\text{SIR}_X = \frac{P_X}{P_I} > Th \qquad (4.26)$$

In the literature [14,22−25], usually, two capture scenarios are taken into account. Both capture scenarios are shown in Fig. 4−12.

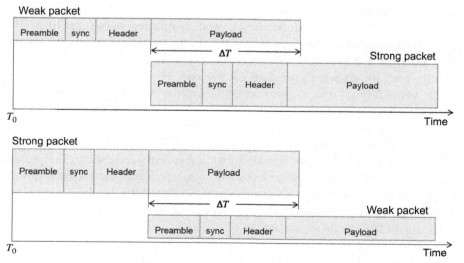

FIGURE 4−12 Capture Scenarios: (top): Stronger First and (bottom): Stronger Last.

- Decoding the First: During the reception of a packet, a second packet arrives and creates collision. In this case, receiver synchronizes with the first packet and tries to perform successful reception.
- Decoding the Last: Another scenario would be to decode the packet that arrives later. This necessitates to be able to detect the preamble of the second packet and then to correctly decode the packet.

It is worth mentioning that a recent work proposes a mechanism able to decode both transmissions either when there is a small time difference between the two signals or when they are well synchronized [26].

4.3.3.1 Capture effect simulations

We first conducted simulations and generated random collisions at the receiver by generating two LoRa packets with time difference ($T_0 \leq \Delta T \leq T_{air}$). The first signal arrives at T_0 and the second signal arrives after a random duration, within the T_{air} of the first packet. The transmission of an interfering packet can start at any time, and overlapping length ΔT of both packets varies randomly. The goal here is to identify under which power settings the collision detection and successful reception will work. In both cases, PRR is measured at the receiver, in simple steps as follows:

- Preamble detection: if preamble detection is valid, then it passes for sync word detection.
- Sync word detection: after the receiver detects the preamble it searches for the sync word and finds the starting of the header.
- Validation of header and payload: if header and payload data are not corrupted, then it is considered as successful frame reception.

The packet structure used in these simulations is shown in Fig. 4−4. The preamble consists of four upchirps, two downchirps, and a quarter of an upchirp. However, increase in preamble duration can improve the detection probability. An explicit header is used with a 2-byte CRC. The header is encoded with $CR = 4$. The payload is 20 bytes long, with no channel encoding $CR = 0$ and $SF = 8$. Channel coding is used to improve the reliability of the communication system by adding redundancy in the transmit data.

Fig. 4−13 presents the capture results. The probability of successful reception is calculated with 1000 packets transmissions for each power setting on random overlapping lengths ΔT. The x-axis shows signal-to-interference (SIR) power and y-axis shows the probability of successful reception with capture. Note that we do not expect that the received power ordering is known a priori. From Fig. 4−13, we can assume that if the received power difference between two interferers is around 1 dB, the receiver can successfully decode the strong packet. Thus, Eq. (4.26) can be expressed as:

$$\text{SIR}_X = \frac{P_X}{P_I} \geq 1 \ \text{dB} \tag{4.27}$$

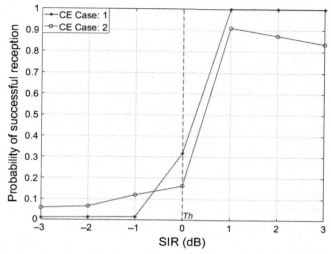

FIGURE 4–13 Capture results with $SF = 8$, $CR = 0$, and $BW = 125$ kHz.

Successfully decoding one of the colliding packet can significantly increase the system throughput of any network. In the presence of CE, the total throughput of the system of N nodes can be expressed as:

$$\eta_{CE} = G \times [P(\text{no collison}) + P(\text{collison}) \sum_{i=1}^{G} P(\text{SIR}_i > Th)] \tag{4.28}$$

$P(\text{no collison})$ and $P(\text{collison})$ are probability of no collision and probability of collision at the receiver, respectively. Th is a threshold value set on signal to interferer i power ratio SIR_i of received signal.

In a pure ALOHA network, a node can successfully transmit a frame if no other node has a frame to transmit during two consecutive frame times (vulnerable time $2T_{\text{air}}$). The probability of a node having no frame to send is $(1 - p)$. The probability that none among the rest of $N - 1$ nodes have a frame to send will be $(1-p)^{N-1}$. The probability that none of the $N - 1$ nodes have a frame to send during the vulnerable time is $(1-p)^{2(N-1)}$. Then the probability of being alone of a particular node will be $P = p(1-p)^{2(N-1)}$.

4.3.3.2 Capture effect experimentations

In Refs. [13,14], the authors experimentally proved the possibility of successful reception of concurrent transmissions using LoRa's modulation. They concluded that there are two important things to keep an eye on. First, the start time of the collision and second is the interfering signal strength. The authors concluded that when the received signal strength indication (RSSI) from the interfering signal is the same or lower than the signal being interfered, and that if the interfering transmission starts after the preamble of the transmission being interfered, then the interfered transmission will be received correctly. They found that to synchronize with a transmitting node, the receiver only needs six symbols of the preamble to be received without collision.

More detailed experimentations with very accurate timing have been performed in Ref. [13]. The authors also tested the case when the RSSI of the interfering transmission is higher at the receiver than the interfered transmission. They found that if the interfering transmission or interfering signal starts after the end of the preamble and header time, the transmission being interfered will be received with wrong payload CRC. However, in case the last six symbols of the transmitter's preamble can be received correctly, the receiver can synchronize with the transmitter and the reception can be successful. It is also important to note that the authors only used the 125-kHz channel bandwidth, and they think that additional experiments are required for other bandwidth channels.

In order to get accurate timing, authors in Ref. [13] experimented with devices placed close together and all connected to a timing unit. We performed additional experiments to get results in a real setting with more LoRa transmission parameters.

4.3.3.2.1 Capture effect setting

Our experimentation setting consists of two transmitters (one master and one slave node) and two receivers (gateways) as depicted in Fig. 4–14: the master node is at around 25 m from the gateway and the slave node was placed at a distance of around 150 m from the same gateway. Tests are performed outdoor in LOS conditions.

To synchronize the transmitter nodes, the master node continuously sends a message to the slave node. On receiving a message from the master, the slave acknowledges the reception of the message by sending an ACK. The slave node synchronizes with the master's clock by taking the message reception time and subtracting the ToA of the message from it. The master node on receiving the ACK performs the same action: it takes the ToA of the ACK and removes it from the time of reception of the ACK, hence synchronizing with the slave clock. Once the nodes are synchronized, they start broadcasting a message every 25,000 ms. We switch on an LED at the beginning of a transmission to visually check that the nodes have successfully synchronized.

Once the nodes are synchronized, to analyze the CE, the nodes start transmitting at the same time. Then, every 10 messages, we add a predefined delay at the slave node (the delay is approximately 1/8th of the ToA of the transmitted message): the first 10 messages (round 0) are sent at the same time by both nodes, the next 10 messages (round 1) are sent by the slave with a delay and so on. If t_{master} is the transmission time at the master, the slave will send its message at $t_{slave} = t_{master} + r * delay$, where r is the round number. The delay is introduced after every 10 messages until there is no transmission overlap.

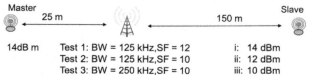

FIGURE 4–14 Experimentation setting.

We tested three different LoRa settings by varying bandwidth *BW* and the spreading factor *SF*. We also performed the tests with different maximum power settings for each transmitting node. We also performed the tests with different maximum power settings for each transmitting node: first with both the transmitters having the same maximum output power of 14 dBm, then we reduce the maximum output power of the slave node to 12 dBm for the second test and 10dBm for the third test. Finally, we did a final test with both the transmitters having maximum output power of 10 dBm. Code rate *CR* is kept same for all the experiments. Both transmitters have the same payload, which is 240 bytes, which will remain constant throughout the experiment. The master and slave nodes are Arduino Nano boards each equipped with a LoRa inAir9 radio module. All the communication took place at 868 MHz frequency band. Two gateways are used: one is an Arduino Nano with the same inAir9 radio module and one is a LoRaWAN RAK831 gateway with an SX1301 radio concentrator running a simple util_pkt_logger program.

4.3.3.2.2 Results Test 1

For the first group of tests, we have *BW* set to 125 kHz and *SF* to 12. For a payload of 240 bytes, the ToA is 8870 ms. The slave uses a delay increment of 1000 ms every 10 messages. Fig. 4—15 shows the results when we put the slave node at 14 dBm (same as master), then at

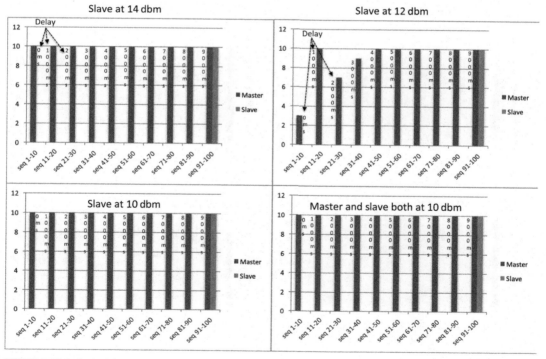

FIGURE 4–15 LoRa test 1.

12 and 10 dBm. In these results, we clearly see that we were able to receive most of the messages from the master, hence proving the CE. When the slave reaches a cumulative delay of 9000 ms, there is no overlap anymore and gateways receive from both transmitters.

4.3.3.2.3 Results Test 2

For the second group of tests, *SF* is now 10, while *BW* remains at 125 kHz. ToA is now 2206 ms and the slave uses a delay increment of 300 ms. Results are summarized in Fig. 4−16 and again confirm the CE as 89% of the messages were received.

4.3.3.2.4 Results Test 3

For the third group of tests, *BW* is now set to 250 kHz and *SF* remains at 10. Toa is 1100 ms and the slave uses a delay increment of 300 ms. The results were found very similar to Test 1, with 94.5% of the messages received with no error.

4.3.3.2.5 Test in indoor conditions

We also performed the three previously described scenarios in indoor conditions. While the main results remain the same confirming the CE for the strongest and first transmitted messages, there is more instability in the results and no packets can be received in many cases.

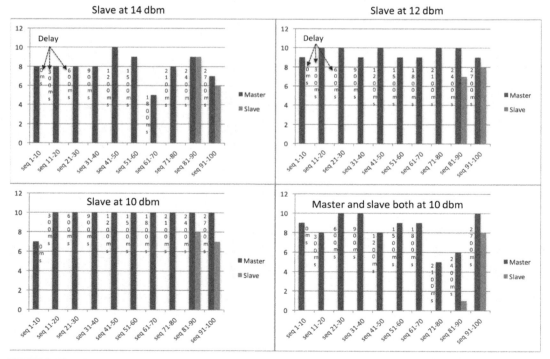

FIGURE 4−16 LoRa test 2.

4.3.4 Interference mitigation: interference cancellation

As we have described earlier, in LoRa, collision occurs due to the simultaneous arrival of two or more packets with the same *SF* at the receiver. According to CE, the strongest packet can be received successfully and the other packet will be considered as interference. However, successive interference cancellation facilitates the recovery of weaker packet too, as well. First, stronger signal is decoded normally; decoded signal is then subtracted from the combined signal. Then, weaker signal is extracted from the residue [27]. In case of multiple interferences, this can lead to an iterative process. The strongest signal is detected first from the received signal and then the next strongest and so on. After each signal's decoding, the received signal for that user can be reconstructed by recreating the transmit signal and applying an estimate of the channel to it. This can be subtracted from the composite received signal, which then allows subsequent users to experience a cleaner signal. A block diagram of CE and successive interference cancellation (SIC) is shown in Fig. 4−17.

There is very little work on SIC for LoRa to the best of our knowledge. In Ref. [28], the authors use simulation to optimize gateway placement when interference cancellation can be realized. In Ref. [29], the authors investigated SIC for UNB networks due to its specific interference behavior. Here, we try to provide some simple analytic results to show SIC potential benefits when considering SIC as pure receiver technique, which means it does not require any type of modification on the transmitter.

Let us assume a LoRa network consisting of a gateway node (receiver) and N transmitters scattered around the gateway node in Poisson field $\Phi = \{(L_i, H_i)\} \subset \mathbb{R}^d \times \mathbb{R}^+$. Where L_i represents the location of each transmitting node and H_i is the channel attenuation coefficient. The SIR-

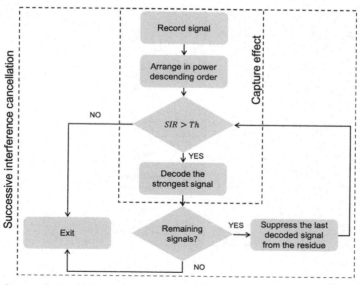

FIGURE 4–17 Block diagram for CE and SIC.

based successful decoding of single user is defined in Eq. (4.26). Which says that a particular signal X at $L \in \Phi$ can be decoded successfully if its SIR is greater than some threshold. Any signal x_r can be decoded successfully from the residue of received signal $Y(t) = \sum_{\text{active users}} X_i(t)H_i(t)$, if its signal to residual interference plus noise power ratio SI_rR_x is:

$$SI_rR_x = \prod_{}^{n} \frac{P_i}{P_{i+1}} \geq Th \qquad (4.29)$$

Figs. 4−18 and 4−19 show the probability of successful decoding of the packets and throughput of LoRa-based system with collisions, consecutively. We have varied the network size N. The assumption is that all the nodes are using the same SF, and their packet generation rate is given in Eq. (4.21). It can be observed that CE and SIC can significantly improve

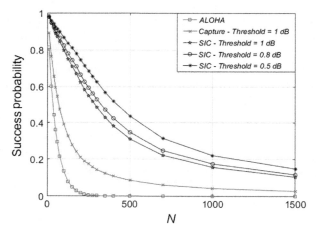

FIGURE 4–18 Probability of successful transmission.

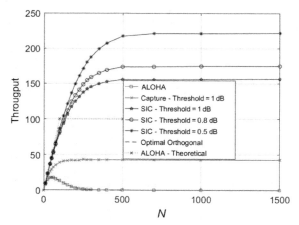

FIGURE 4–19 Throughput.

network performance. This case can be considered a worst-case scenario because no channel coding is used which would significantly improve the performance.

4.4 Channel access: sharing the bandwidth

Sharing the wireless channel is a joined task of the PHY and media access control (MAC) layer. While sharing at PHY layer is usually done with advanced signal coding/modulation schemes, sharing at MAC layer usually assumes that simultaneous transmissions at PHY layer will result in interferences leading to so-called packet collisions. Therefore, MAC layer sharing mechanisms usually implement either a time-division multiple access (TDMA) method or a competition-based approach where nodes compete to get the channel. We focus here in the latter category as TDMA methods need a high synchronization level (note that Zorbas et al. [17] did propose a TDMA-like scheduling mechanism for LoRa). In the competition-based category, random access protocols such as the early ALOHA and various variants of carrier-sense multiple access (CSMA) are widely used in wireless networks because of their relative simplicity and their distributed operation mode that does not require coordination nor overwhelming signaling overheads. There has been a notable amount of research done on the performance of ALOHA and CSMA in wireless networks. It is beyond the scope of this paper to go through all these contributions but interested readers can refer to Refs. [30−32] as a starting point.

4.4.1 Review of media access control mechanisms

4.4.1.1 IEEE 802.11

Among many CSMA variants, the one implemented in the IEEE 802.11 (Wi-Fi) is certainly one of the most used in wireless networks thanks to the worldwide success of Wi-Fi technology and is therefore quite representative of the approach taken by most of random access protocols with so-called backoff procedure. Fig. 4−20 illustrates the IEEE 802.11 CSMA

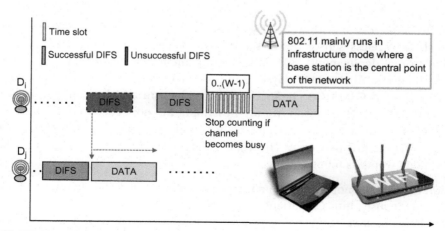

FIGURE 4–20 IEEE 802.11 DCF CSMA/CA.

mechanism used in the basic distributed coordinated function (DCF) mode, which is the common operation mode of Wi-Fi networks with a base station. In this basic mode, the optional received signal strength indication mode is not used. The basic DCF IEEE 802.11 CSMA/CA (collision avoidance) works as follows:

- Collision detection is not used since a node is unable to detect the channel and transmit data simultaneously, thus CA variant.
- A node senses the channel to determine whether another node is transmitting before initiating a transmission.
- If the medium is sensed to be free for a DCF interframe space (DIFS) time interval, the transmission will proceed (green DIFS).
- If the medium is busy (red DIFS), the node defers its transmission until the end of the current transmission and then it will wait for an additional DIFS interval before generating a random number of backoff slot time chosen in the range $[0, W - 1]$. W is called the backoff window or contention window.
- The backoff timer is decreased as long as the medium is sensed to be idle, and frozen when a transmission is detected on the medium, and resumed when the channel is detected as idle again for more than DIFS.
- When the backoff reaches 0, the node transmits its packet.
- The initial W is set to 1. W is doubled for each retry (exponential backoff) until it reaches a maximum value.
- If the maximum number of retries is reached, report error to higher layers.

The random backoff timer is applied after a busy channel because it is exactly in that case that the probability of a collision is at its highest value. This is because several users could have been waiting for the medium to be available again.

4.4.1.2 IEEE 802.15.4

Closer to the domain of IoT, IEEE 802.15.4 was for many years the standard for low-power devices such as wireless sensor networks (WSNs). Being a short-range technology (about 100−200 m in real deployment), 802.15.4 is complemented at higher layers with multihop routing mechanisms. At the MAC layer, IEEE 802.15.4 proposes both nonbeacon-enabled mode with unslotted CSMA/CA channel access mechanism and beacon-enabled networks with slotted CSMA/CA. Here, again, we are describing the nonbeacon-enabled mode as the beacon-enabled needs a coordinator and higher level of synchronization that is definitely not suited for LoRa IoT networks. The IEEE 802.15.4 nonbeacon-enabled with unslotted CSMA/CA mode works as follows:

- Collision detection is not used since a node is unable to detect the channel and transmit data simultaneously, thus CA variant.
- Before a transmission, a node waits for a random number of backoff periods chosen in the range $[0.2^{BE} - 1]$. backoff exponent (*BE)* is set to 3 initially.

- If at the end of the waiting time the medium is sensed to be free (clear channel assessment, CCA) the transmission will proceed.
- If the medium is busy, the node defers its transmission, increases *BE* until it reaches a maximum value, and waits for additional $[0.2^{BE} - 1]$ backoff periods.
- If the maximum number of retries is reached, report error to higher layers.

Compared to IEEE 802.11, IEEE 802.15.4 always implements a backoff timer prior to any transmission and simply increases the backoff timer interval each time the channel is found busy for the same packet, without constantly checking the channel to know when it is going back to idle. There are several reasons for these differences. One reason is that simply increasing the backoff timer interval is less energy-consuming than determining the end of the current transmission, especially if the transmission of a packet can take a long time (802.15.4 usually runs at 250 kbps while 802.11 runs at 11 Mbps and above). Another reason is that the node and traffic density of IEEE 802.15.4 networks is expected to be much smaller than those of Wi-Fi networks. There is an additional reason 802.15.4's CSMA is different from 802.11's CSMA: 802.15.4 for WSN mainly runs under mesh topology (i.e., P2P and without central coordinator) with a shorter radio range (i.e., low transmit power); therefore the spatial reuse is higher, contributing again to decrease the traffic density at any given point in the network.

Again, there has been a huge amount of research in improving the basic 802.15.4 MAC protocol to better support multihop and duty-cycled low-power WSN. For instance, and to name a few, sensor mac [33] which introduces synchronization features to have common active periods and berkeley mac [34] and X-MAC [35], both with low-power listening capabilities. Readers can refer to Ref. [36] for a survey of MAC protocols for WSN (Fig. 4−21).

4.4.2 Clear channel assessment in LoRa

Before investigating what CSMA approach can be adapted for LoRa, it is necessary to know how a LoRa channel can be defined busy or idle to implement a carrier-sense mechanism.

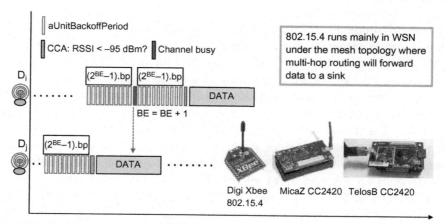

FIGURE 4–21 IEEE 802.15.4 nonbeacon unslotted CSMA.

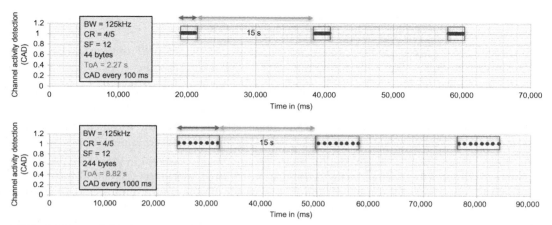

FIGURE 4–22 Test of the LoRa CAD mechanism.

As LoRa reception can be done below the noise floor, the use of the RSSI is not reliable enough. For CCA, there is a special channel activity detection (CAD) procedure that can be realized by a LoRa chip. To assess the CAD reliability, we experimentally tested how the CAD procedure can detect real transmissions on the wireless medium. We use a dedicated device to constantly perform CAD procedure and another device to send periodic messages. Fig. 4–22 shows two cases: (1) 44-byte message (40 bytes payload + 4-byte header) every 15 s with a CAD procedure every 100 ms and (2) 244-byte message (240 + 4) every 15 s with a CAD procedure every 1000 ms. As can be seen in Fig. 4–22 the LoRa CAD procedure can correctly detect all the LoRa transmission, and not only the preamble.

4.4.3 Adaptation from 802.11

As a first attempt toward a CSMA protocol for LoRa, we start by adapting the previously shown 802.11 CSMA protocol and not the 802.15.4 one, although 802.15.4 is widely used in WSN and early IoT implementation, for two reasons. The first reason is that LoRa network architecture is mainly a single-hop star topology from devices to gateway, which is very similar to the Wi-Fi topology with a base station. Therefore the concept and the management of the 802.11's random backoff timer after a busy channel looks efficient for such an environment. The second reason for not starting from 802.15.4 comes from its initial random waiting without channel sensing method that is more suitable for low-density networks than for high-density networks that will definitely be the case for LoRa networks.

To adapt the 802.11 CSMA protocol, we first need to define how the DIFS operation can be implemented. Usually, IFS should be related somehow to the symbol period T_{sym}. For LoRa, T_{sym} depends on BW and SF as follows: $T_{\text{sym}} = 2^{SF}/BW$. For instance, LoRa mode 1 use $BW = 125$ kHz and $SF = 12$; therefore $T_{\text{sym}}^{\text{mode}-1} = 2^{12}/125,000 = 0.032768$. In Ref. [37], it is reported that the CAD duration is between $1.75T_{\text{sym}}$ and $2.25T_{\text{sym}}$ depending on the spreading factor, see Fig. 4–23. We performed some experimental tests to verify the real

LoRa mode	BW/SF	Tsym (ms)	CAD duration (Tsym)45%	CAD duration (ms)	Experimental measures	
					min value	max value
1	BW125SF12	32.768	1.86	60.948	60	62
2	BW250SF12	16.384	1.86	30.474	29	31
3	BW125SF10	8.192	1.77	14.500	14	16
4	BW500SF12	8.192	1.86	15.237	15	16
5	BW250SF10	4.096	1.77	7.250	7	8
6	BW500SF11	4.096	1.81	7.414	7	9
7	BW250SF9	2.048	1.75	3.584	3	5
8	BW500SF9	1.024	1.75	1.792	1	3
9	BW500SF8	0.512	1.79	0.916	1	1
10	BW500SF7	0.256	1.92	0.492	0	1

FIGURE 4–23 Theoretical CAD duration and experimental measures.

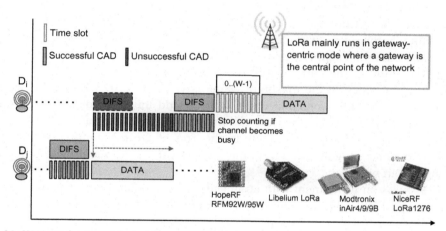

FIGURE 4–24 CSMA mechanism. *Adapted from IEEE 802.11.*

CAD duration against what is given in Ref. [37]: Fig. 4−23 also shows the minimum and the maximum values measured with a 1-ms accuracy clock (the Arduino millis() function). We can see that the measured CAD durations are quite consistent.

In our current implementation DIFS does not depend directly on T_{sym} but on the duration of the CAD mechanism; therefore we assign an integer number of CAD to DIFS. Our communication library provides a low-level doCAD(counter) function that takes an integer number of CAD, that is counter, performs sequentially the requested number of CAD, and returns to 0 if all CAD have been successful (no channel activity). If one CAD detects activity the function exits with a value greater than 0. The DIFS procedure shown in Fig. 4−24 works that way and once a failed CAD has been observed the node exits the DIFS procedure and continuously checks for a free channel.

In Fig. 4−24, DIFS is assigned nine CAD, which gives a duration of about 9×61 ms $= 549$ ms for LoRa mode 1. At this point of the study, the duration of DIFS is not really important as we only need to be able to assert a free channel for a given duration. The value of nine CAD provides enough time to detect channel activity and also provides the possibility to define a much shorter timer (using three CAD for instance), such as the 802.11's DIFS, to implement priority schemes is needed, and still be able to detect channel activity. Then, the random backoff timer is also defined as a number of CAD because the channel should be checked in order to freeze or continue the decrease of the backoff timer. The upper bound, W, of the random backoff timer can be set in relation to the number of CAD defined for DIFS. For instance, if DIFS = 9 CAD then W can be defined as $n \times$ DIFS. For instance, if $n = 2$ then $W = 2 \times 9 = 18$ CAD.

It is also possible to double W for each retry (exponential backoff) until it reaches a maximum value. However, while 802.11 initiates a retry when no ACK is received after a given time, the usage of acknowledgment is not common in LoRa as it is very costly for the gateway (the gateway is considered as a normal node and therefore its radio duty cycle can be limited by regulations). Therefore there is no such retry concept with unacknowledged transmissions. Nevertheless, when 802.11 doubles W for each retry the underlying assumption for the transmission errors is a denser channel. Here, we can follow the same guideline and double W each time the channel cannot be found free for an entire DIFS, starting from the second DIFS attempt. In the current implementation, we set $W = 18$ CAD initially, and we can double it three times so the maximum value is $W = 144$ CAD which will give a maximum wait timer of 8784 ms for LoRa mode 1. If we add the value of the successful DIFS which is 9 CAD, that is 549 ms, then the maximum total wait time after a busy channel is about 9333 ms. which correspond roughly to the ToA of the maximum LoRa packet size. This property remains roughly true for all the defined LoRa modes and therefore can avoid waiting longer than necessary.

Fig. 4−25 shows an experiment with an image sensor sending four image packets (about 240 bytes per packet), while another device (interactive device) is sending medium-size

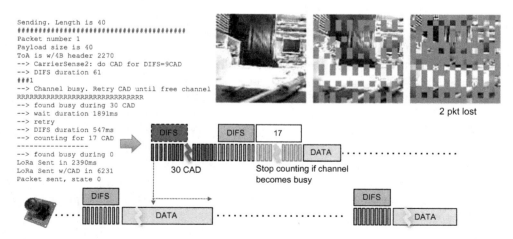

FIGURE 4–25 Experimental test of the proposed CSMA adaptation.

messages of 40 bytes. The output is from the interactive device and it can be seen that the adapted CSMA protocol can nicely avoid the collision by deferring the transmission of the interactive message. In the illustrated experiment, transmission is deferred only once before transmission succeeds as the time between two image packets is greater than a DIFS plus the random backoff timer of 17 CAD.

Fig. 4−25 also shows the received image without any packet loss and two examples of received images when there is no channel access mechanism (pure ALOHA). In all our tests, the proposed CSMA protocol adapted from 802.11 and further referred to as $CSMA_{802.11}^{LoRa}$, totally avoids packet losses for both the image sensor and the interactive device.

4.4.4 Channel activity detection reliability issues

By testing further the CSMA mechanism in various long-range deployment, we observed a fast decrease of the CAD's reliability when distance increases: although a transmission can be successful at several kilometers, CAD starts to not reliably detect the whole transmission when the distance to the sender is about 1 km (with dense vegetation, CAD reliability can start to decrease even at 400 m). Fig. 4−26 shows CAD reliability with the same traffic pattern previously shown in Fig. 4−22 but with the sender and the device performing CAD separated by 400 m with some trees between them. As can be seen, the CAD procedure fails to detect channel activity many times during an ongoing transmission.

This CAD unreliability issue in real-world deployment scenario has a huge negative impact on the CS mechanism. For instance, in the previously proposed CSMA adaptation from 802.11, it is not possible anymore to rely on CAD to detect when the channel will become really free after a busy state nor to rely on a successful DIFS as a free channel indication to start transmission. However, what can be observed in Fig. 4−22 and verified by the tests that we performed is that during a long transmission the probability that all CAD attempts fail is quite low. In all our tests, and up to 1 km in non line of sight conditions, there have always been some successful CAD during any transmission.

4.4.5 A solution to protect long messages

The CAD reliability issue raised previously calls for a different approach to prevent collisions. First, the previous DIFS is extended to the ToA of the longest LoRa packet in a given LoRa setting, for example 9150 ms for 255 bytes when $BW = 125$ kHz and $SF = 12$. During this

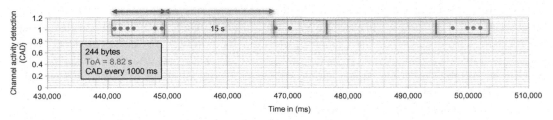

FIGURE 4–26 CAD fails to detect the activity of ongoing transmissions.

extended DIFS(ToA$_{max}$), CAD procedure is performed periodically (for instance every 1000 ms as in Fig. 4−22—bottom). The purpose of DIFS(ToA$_{max}$) is to maximize the probability to detect an ongoing transmission, which can possibly be a long message with many unsuccessful CADs, thus appearing by mistake as a short message.

Then, when a CAD fails during a DIFS(ToA$_{max}$), instead of continuously waiting for a free channel followed by a DIFS plus random backoff timer where CAD is checked constantly; here, there is a simple constant waiting period (pure delay) of ToA$_{max}$. Again, the purpose of the constant delay of ToA$_{max}$ is to avoid performing CAD and transmission retries during the transmission of a possible long message, as a successful CAD does not guarantee a free channel. After the delay, the transmitter will try again to see a free channel for at least a DIFS(ToA$_{max}$) and the process continues until a maximum number of retries have been performed.

In all our experiments with the new proposed CSMA protocol, we can totally avoid packet losses for both the image sensor and the interactive device even when the nodes are 100s of meters away from each other. However, clearly, this is achieved at the cost of a much higher latency [38].

4.5 Studying large-scale LoRa deployments

The rapid growth in the field of WSNs and IoT entails the need of creating new simulators that have more specific capabilities to tackle interference and multipath propagation effects that are present in the wireless environment. Finding a suitable simulation environment that allows researchers to verify new ideas and compare proposed solutions in a virtual environment is a difficult task. Most of the existing open-source simulators are mainly used to study routing protocols and they offer limited real-time interference and radio propagation modeling features for smart cities and IoT applications.

The CupCarbon open-source simulator is the main simulation kernel of the French ANR project PERSEPTEUR that aims to develop models and algorithms for accurate simulations of signal propagation and interference in a 3D urban environment [39]. This simulator runs under the Java environment and can be downloaded from the Internet (http://www.cupcarbon.com). The main idea behind proposing CupCarbon is to keep simulation time short while taking into account a realistic evaluation of the wireless interference in a 3D environment with an accurately simulated radio channel. It supports wireless communication interference models such as Gaussian and α-stable models [40−42] and includes a 3D ray-tracing channel model. Focusing on physical layer, CupCarbon provides visualization of the impact of wireless interference and signal propagation. In addition to LoRa physical layer, CupCarbon also simulates the PHY layers of ZigBee (IEEE 802.15.4) and Wi-Fi (IEEE 802.11), making it suitable to study smart cities scenarios using all these wireless technologies (Fig. 4−27).

A WSN/IoT network can be prototyped with the CupCarbon's intuitive graphical interface that embeds OpenStreetMap framework to allow sensor nodes to be directly placed on the

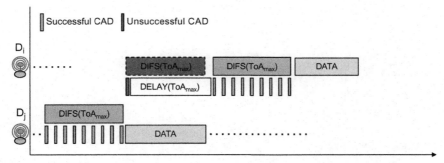

FIGURE 4–27 New CSMA proposition.

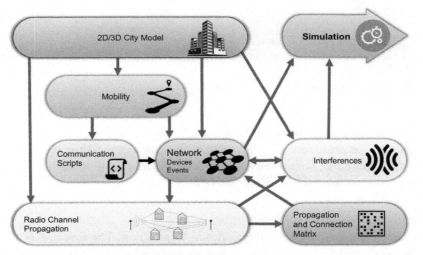

FIGURE 4–28 Main components of the CupCarbon architecture.

map. In addition, each sensor node can be individually configured by its command line with the script called SenScript. From SenScript, it is also possible to generate codes for hardware platforms such as Arduino.

4.5.1 The CupCarbon architecture

Fig. 4–28 shows the main modules of the CupCarbon simulator and we are going to present the main modules relevant for LoRa networks.

4.5.1.1 Two-dimensional/three-dimensional city model module

2D/3D visualization of the urban environment and the deployed network is an important part of a WSN/IoT simulation. The 3D environment helps to obtain an accurate deployment

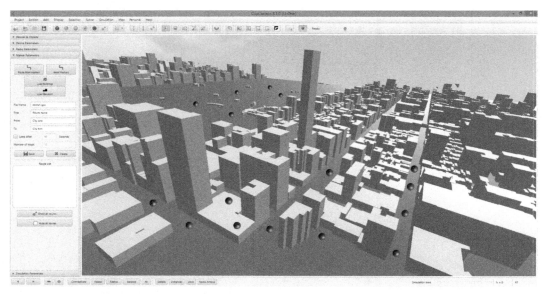

FIGURE 4–29 Example of a 3D display of a city in CupCarbon.

in which the elevation can be taken into account. Fig. 4–29 shows an example of a city displayed in the 3D environment of the CupCarbon simulator. The 3D environment of CupCarbon is composed of ground elevation, buildings, and various objects such as sensor nodes. More details can be found in Refs. [43,44].

4.5.1.2 Radio channel propagation module

Two radio propagation models are integrated into CupCarbon. The first one is a 2.5D based on a point to zone acceleration structure called visibility tree [45,46]. It provides the estimation of channel attenuation and the channel impulse response according to a large number of the receivers. The second model is a full 3D ray-tracing associated with a Monte Carlo algorithm. Resulting data help to determine the quality of the wireless channel between groups of the nodes and are used to decide whether a communication link can be established or not, see Fig. 4–30.

4.5.1.3 Interference module

This block is the core contribution of CupCarbon. It can be further divided into two categories:

1. PHY layer: a significant originality of CupCarbon is to propose very realistic models of many WSN/IoT physical layers. The evaluation of a link quality can now be based on accurate transmission conditions that consider the radio channel and the data encoding.

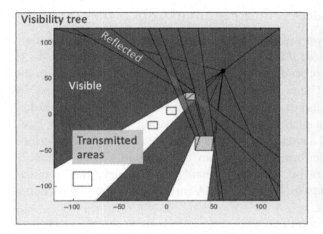

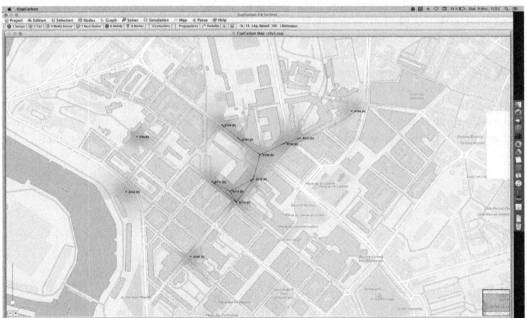

FIGURE 4–30 Channel propagation module.

2. Interference models: interference is a significant limiting factor in dense networks and CupCarbon proposes Gaussian or α-stable models. Radio in congested areas on the map can choose the impulsive interference model (α-stable model) while less congested areas can use the Gaussian model instead.

4.5.2 LoRa PHY/media control access integration in CupCarbon

Data from the MAC layer to the PHY layer are provided by the user from SenScript program. Then, a radio technology can be assigned: IEEE 802.11 or IEEE 802.15.4 or LoRa. Fig. 4−31 shows the block diagram of the whole procedure. With LoRa, the user must also assign the spreading factor SF between 7 and 12. The receiver technique should also be mentioned. By default, the receiver technique is none and the receiver will work normally without considering CE or successive interference cancellation. Default LoRa setting also uses the basic ALOHA approach and a node transmits a packet as soon as it has data to send. As described previously, ALOHA limits the system performance as it can only give 18.39% of maximum efficiency. However, CE and SIC can be used to enhance the system performance. More elaborated channel access methods such as the CSMA variants described in Section 4 can also be selected.

Fig. 4−32 shows the CupCarbon environment for the assignment of the parameters in case of selecting LoRa communication protocol. It shows interconnected sensor nodes placed arbitrarily. All the nodes are assigned the LoRa protocol. Nodes should be in the radio range of each other to communicate. The LoRa gateway can be defined as one of the nodes.

For the transmission, first the transmitter generates an upchirp signal using the input SF. By default the CR, SF, and BW values are considered to be 0, 8, and 125 kHz, respectively. Then, there are interleaving and modulation operations. After applying the transmission functions, the system calculates the parameters of the selected interference model, the Gaussian model, or the α-stable model, and applies the interference model to the transmitted signal. The receiver generates a downchirp signal using the appropriate SF and performs the demodulation.

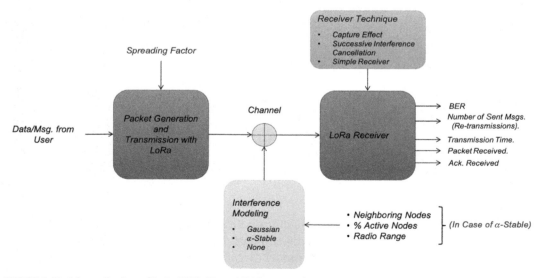

FIGURE 4–31 Schematic view of LoRa PHY, CE, and SIC in CupCarbon.

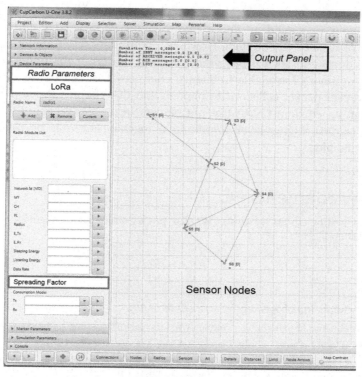

FIGURE 4–32 Radio parameter settings for LoRa in CupCarbon.

4.6 Conclusions

This chapter focused on LoRa LPWAN technology and presented the main issues and challenges of radio channel access in dense LoRa networks. Results from analytic analysis, simulations, and experimentations have been presented to illustrate the behavior of LoRa networks and try to assess its scalability in a dense environment. When considering only the LoRa physical layer without any LBT nor CSMA mechanism a deployed network is similar to a simple ALOHA system, which is known to have very small efficiency. However, this chapter also presented some positive effects such as CE and successive interference cancellation techniques. In addition, although not perfect, LoRa's *SF* quasiorthogonality can practically increase the scalability of deployed gateways, especially when they are based on radio concentrators capable of listening on at least eight frequencies, each of them accepting multiple *SF* values simultaneously. At the time of writing, 16-channel gateways are already available. At a higher level, legal regulations in many countries drastically limit the duty cycle or the ToA of a single transmission in current ISM license-free bands in order to prevent saturation of the radio channel. Some regulations also impose very tight frequency hopping constraints.

At the time of writing, given the relatively small number of deployed LoRa devices in the world, PRR is usually very good. In dense environments such as large cities, it is expected that the number of devices can be high. However, community-based initiatives also showed that the gateway density in such urban areas can also be very high, therefore increasing to some extent the number of available frequency channels. Last but not least, none of these measures can prevent packet collisions to happen and this chapter also reported how diffi-cult it is to design an efficient CSMA-like mechanism because a reliable CCA is difficult to achieve. In this context, higher-level coordination mechanisms may be needed to further improve LoRa scalability.

4.7 Acknowledgments

This work was supported by the EU H2020 RIA WAZIUP project under grant agreement no. 687607 and by the ANR PERSEPTEUR project.

References

[1] TTN mapper. <https://ttnmapper.org/> (accessed 06.04.19).

[2] D. Akerman, High altitude ballooning. <https://www.daveakerman.com/> (accessed 06.04.19).

[3] Ground breaking world record. <https://bit.ly/2wOtad4> (accessed 06.04.19).

[4] T. Telkamp, LoRa transmission from low orbit satellite. <https://www.thethingsnetwork.org/article/lora-transmission-from-low-orbit-satellite> (accessed 06.04.19).

[5] Semtech. <http://www.semtech.com/images/datasheet/an1200.22.pdf> (accessed 15.11.17).

[6] S. Andrew, D.J.W. Tanenbaum, Computer Networks, Prentice-Hall, 2010, ISBN: 9780133485936.

[7] B. Reynders, S. Pollin, Chirp spread spectrum as a modulation technique for long range communication, in: 2016 Symposium on Communications and Vehicular Technologies (SCVT), 2016, pp. 1−5. Available from: https://doi.org/10.1109/SCVT.2016.7797659.

[8] D. Croce, M. Gucciardo, S. Mangione, G. Santaromita, I. Tinnirello, Impact of LoRa imperfect orthogo-nality: analysis of link-level performance, IEEE Commun. Lett. 22 (4) (2018) 796−799.

[9] A. Rahmadhani, F. Kuipers, When LoRaWAN frames collide, in: ACM WiNTECH, 2018.

[10] F. Adelantado, X. Vilajosana, P. Tuset-Peiró, B. Martínez, J. Melià-Seguí, T. Watteyne, Understanding the limits of LoRaWAN, IEEE Commun. Mag. 55 (9) (2017) 34−40. Available from: https://doi.org/10.1109/MCOM.2017.1600613.

[11] O. Georgiou, U. Raza, Low power wide area network analysis: can LoRa scale? IEEE Wirel. Commun. Lett. 6 (2) (2017) 162−165. Available from: https://doi.org/10.1109/LWC.2016.2647247.

[12] F.V.D. Abeele, J. Haxhibeqiri, I. Moerman, J. Hoebeke, Scalability analysis of large-scale LoRaWAN net-works in ns-3, IEEE Internet Things J. 4 (6) (2017) 2186−2198. Available from: https://doi.org/10.1109/JIOT.2017.2768498.

[13] J. Haxhibeqiri, F.V.D. Abeele, I. Moerman, J. Hoebeke, LoRa scalability: a simulation model based on interference measurements, Sensors 17 (6) (2017) 1193. Available from: https://doi.org/10.3390/s17061193.

[14] M.C. Bor, U. Roedig, T. Voigt, J.M. Alonso, Do LoRa low-power wide-area networks scale? Proceedings of MSWiM 2016, ACM, 2016, pp. 59−67.

[15] LoRaWAN 1.1 specification. <https://lora-alliance.org/resource-hub/lorawantm-specification-v11> (accessed 06.04.19).

[16] S. Li, U. Raza, A. Khan, How agile is the adaptive data rate mechanism of LoRaWAN? in: IEEE Global Communications Conference, GLOBECOM 2018, 9−13 December 2018, Abu Dhabi, United Arab Emirates, 2018, pp. 206−212. Available from: https://doi.org/10.1109/GLOCOM.2018.8647469.

[17] D. Zorbas, K.Q. Abdefadeel, V. Cionca, D. Pesch, B. O'Flynn, Offline scheduling algorithms for time-slotted LoRa-based bulk data transmission, in: IEEE 5th World Forum on Internet of Things (WF-IoT), 2019.

[18] B. Reynders, Q. Wang, P. Tuset-Peiró, X. Vilajosana, S. Pollin, Improving reliability and scalability of LoRaWANs through lightweight scheduling, IEEE Internet Things J. 5 (3) (2018) 1830−1842.

[19] J. Haxhibeqiri, I. Moerman, J. Hoebeke, Low overhead scheduling of LoRa transmissions for improved scalability, IEEE Internet Things J. 6 (2018) 3097−3109.

[20] ETSI, Electromagnetic compatibility and radio spectrum matters (ERM); short range devices (SRD); radio equipment to be used in the 25 MHz to 1000 MHz frequency range with power levels ranging up to 500 mW; part 1: technical characteristics and test methods, 2012.

[21] V241 EE. Electromagnetic compatibility and radio spectrum matters (ERM) short range devices (SRD) radio equipment to be used in the 25 MHz to 1000 MHz frequency range with power levels ranging up to 500 mW, technical report, European Telecommunications Standards Institute, Sophia Antipolis Cedex, France, 2012.

[22] M. Bor, J. Vidler, U. Roedig, LoRa for the Internet of things, in: Proceedings of the 2016 International Conference on Embedded Wireless Systems and Networks, 2016, pp. 361−366.

[23] S. Kosunalp, P. Mitchell, D. Grace, T. Clarke, Experimental study of the capture effect for medium access control with ALOHA, ETRI J. 37 (2) (2015) 359−368. Available from: https://doi.org/10.4218/etrij.15.0114.1369.

[24] K. Whitehouse, A. Woo, F. Jiang, J. Polastre, D. Culler, Exploiting the capture effect for collision detection and recovery, The Second IEEE Workshop on Embedded Networked Sensors, 2005, EmNetS-II, 2005, pp. 45−52. Available from: https://doi.org/10.1109/EMNETS.2005.1469098.

[25] T. Voigt, M. Bor, U. Roedig, J. Alonso, Mitigating inter-network interference in LoRa networks, in: Proceedings of the 2017 International Conference on Embedded Wireless Systems and Networks (EWSN'17), 2017.

[26] N.E. Rachkidy, A. Guitton, M. Kaneko, Decoding superposed LoRa signals, in: IEEE LCN, 2018.

[27] J.G. Andrews, Interference cancellation for cellular systems: a contemporary overview, IEEE Wirel. Commun. 12 (2) (2005) 19−29. Available from: https://doi.org/10.1109/MWC.2005.1421925.

[28] H. Tian, M.A. Weitnauer, G. Nyengele, Optimized gateway placement for interference cancellation in transmit-only LPWA networks, Sensors 18 (11) (2018) 3884. Available from: https://doi.org/10.3390/s18113884.

[29] Y. Mo, C. Goursaud, J. Gorce, On the benefits of successive interference cancellation for ultra narrow band networks: theory and application to IoT, in: IEEE International Conference on Communications, ICC 2017, 21−25 May 2017, Paris, France, 2017, pp. 1−6. Available from: https://doi.org/10.1109/ICC.2017.7996900.

[30] M. Kaynia, N. Jindal, Performance of ALOHA and CSMA in spatially distributed wireless networks, in: Proceedings of IEEE International Conference on Communications (ICC), 2008.

[31] Y. Yang, T.S.P. Yum, Delay distributions of slotted ALOHA and CSMA, IEEE Trans. Commun. 51 (2003).

[32] F.A. Tobagi, Distribution of packet delay and interdeparture time in slotted aloha and carrier sense multiple access, J. Assoc. Comput. Mach. 29 (1982).

[33] W. Ye, J. Heidemann, D. Estrin, Medium access control with coordinated adaptive sleeping for wireless sensor networks, IEEE/ACM Trans. Netw. 12 (3) (2004) 493–506.

[34] J. Polastre, J. Hill, D. Culler, Versatile low power media access for wireless sensor networks, in: Proceedings of the 2nd International Conference on Embedded Networked Sensor Systems, SenSys '04, 2004, pp. 95–107. ISBN 1-58113-879-2.

[35] M. Buettner, G.V. Yee, E. Anderson, R. Han, X-MAC: a short preamble MAC protocol for duty-cycled wireless sensor networks, in: Proceedings of the 4th International Conference on Embedded Networked Sensor Systems, SenSys '06, 2006, pp. 307–320.

[36] A. Bachir, M. Dohler, T. Watteyne, K.K. Leung, MAC essentials for wireless sensor networks, IEEE Commun. Surv. Tutor. 12 (2) (2010) 222–248.

[37] Semtech, SX1272/73—860 MHz to 1020 MHz low power long range transceiver, rev.2-07/2014, 2014.

[38] C. Pham, Investigating and experimenting CSMA channel access mechanisms for long-range LoRa IoT networks, in: Proceedings of the IEEE WCNC, 2018.

[39] M. Saoudi, F. Lalem, A. Bounceur, R. Euler, M.T. Kechadi, A. Laouid, et al., D-LPCN: a distributed least polar-angle connected node algorithm for finding the boundary of a wireless sensor network, Ad Hoc Netw. J. 56 (1) (2017). Available from: https://doi.org/10.1016/j.adhoc.2016.11.010.

[40] A. Bounceur, O. Marc, M. Lounis, J. Soler, L. Clavier, P. Combeau, et al., CupCarbon-Lab: An IoT emulator, in: 2018 15th IEEE Annual Consumer Communications Networking Conference (CCNC), 2018, pp. 1–2. Available from: https://doi.org/10.1109/CCNC.2018.8319313.

[41] M. Kamal, M. Lounis, A. Bounceur, M.T. Kechadi, CupCarbon: a multi-agent and discrete event wireless sensor network design and simulation tool, in: SimuTools, 2014.

[42] CupCarbon, ANR project PERSEPTEUR, CupCarbon simulator. <http://www.cupcarbon.com>. (accessed 15.11.15).

[43] CupCarbon: ray-tracing technique with 3D visualization. <https://www.youtube.com/watch?v = 8HBhjVNlZgI>. (accessed 30.04.19).

[44] CupCarbon: integration of 3D maps. <https://www.youtube.com/watch?v = 5hyEXrVOPQU>. (accessed 30.04.19).

[45] T. Alwajeeh, P. Combeau, A. Bounceur, R. Vauzelle, Efficient method for associating radio propagation models with spatial partitioning for smart city applications, 2016, pp. 1–7. Available from: https://doi.org/10.1145/2896387.2901918.

[46] T. Alwajeeh, P. Combeau, R. Vauzelle, A. Bounceur, A high-speed 2.5D ray-tracing propagation model for microcellular systems, application: smart cities, in: 2017 11th European Conference on Antennas and Propagation (EUCAP), 2017, pp. 3515–3519. Available from: https://doi.org/10.23919/EuCAP.2017.7928760.

5

An introduction to Sigfox radio system

Christophe Fourtet[1], Benoît Ponsard[2]

[1]CO-FOUNDER AND CHIEF SCIENCE OFFICER, SIGFOX, LABÈGE, FRANCE [2]DIRECTOR OF STANDARDIZATION, SIGFOX, LABÈGE, FRANCE

5.1 Internet of things, a new usage for the radiocommunication industry

The radiocommunication era started in 1895 with the first Marconi's experiments. Hundred years later, radiocommunication evolution is largely driven by cellular systems and their billions of subscribers. Cellular systems started with analog, and then went to digital. In the late 1990s, digital 2G systems have been used for machine-to-machine (M2M) communication. M2M takes advantage of existing data bearers, primarily designed to address human-to-human (H2H) or human-to-machine (H2M) usage. They are not optimized for the emerging usage of Internet of things (IoT). Many definitions exist for IoT [1] with a quite strong focus on network. Here we give an object-centric definition for the IoT, which is "the ability, for objects, to connect the Internet without being considered, near or far, as computer."

This definition implies that communication is only a side feature of a connected object. Implementation of IoT communication must be easy and cost-effective, particularly if there is no electrical power natively in the object to be connected. Compared to H2H, H2M, and M2M, IoT is a new usage, with new constraints for radiocommunication technologies:

- Low cost: In an object, communication function must have only a marginal cost, compared to the total cost of the object, as it is an additional function only. At time of printing, volume price for Sigfox modules is as low as 2$. Target price is estimated to be as low as 0.2$, with mass production.
- Low volume of data: Connected objects transmit information such as sensed data, status, index, and alarms and receive commands or parameters. Compared to machines that may be complex and may require M2M communication, connected objects have a single function, which communicate infrequent small application packets [2].
- Massive number of connected objects: Thanks to low-cost and easy-to-use IoT communication function, it becomes possible to connect many types of objects, resulting in much higher density of connections per square kilometer that is usually seen in cellular systems. In urban area, density of connected objects may be over 50k per square kilometer [2].

LPWAN Technologies for IoT and M2M Applications. DOI: https://doi.org/10.1016/B978-0-12-818880-4.00005-3

5.2 Low-power wide-area network: a new paradigm in radio network engineering

Low-power wide-area network (LPWAN) acronym was forged by Machina Research in June 2013 [3] to refer to emerging communication systems. Whereas IoT is about usage, LPWAN is about radio technology. LPWAN novelty resides in mixing two contradictory notions: low power and wide-area network. LPWANs have a series of key characteristics, as follows:

- Massively asymmetrical: LPWAN base stations are in small numbers to reduce infrastructure cost. Therefore, each base station may serve tens of thousands of connected objects.
- Connected objects transmitting at low power: Use of radio transmitter, with tens of milliwatt of transmit power, is essential to get low complexity in communication modules and to reduce the cost of the connection function. Low transmit power means also small batteries and, sometimes, even no battery thanks to energy harvesting techniques.
- High sensitivity in base stations: As a consequence of low power in connected objects, high sensitivity in base stations is a must to get wide-area connectivity with one-hop communication. Multihop connection is not the preferred option to get wide-area coverage, because of energy spent in signaling.
- Complexity moved from objects to base stations: Low power in connected objects means that protocol complexity must be kept minimal in objects also. In LPWAN, complexity balance between objects and network is opposite to the trend of the past decade in telco system, where telephone sets become more and more complex, going from electromechanical objects to analog electronics, then to digital, and now with sophisticated signal processing.
- No strong latency requirement: LPWANs address IoT usage where application is neither time-critical, nor real-time; several seconds for uplink or downlink latency are affordable in most of IoT use cases.
- No high-speed mobility requirement: LPWAN technologies address application that are stationary or at walking/cycling speed (e.g., up to 30 km/h).

These system characteristics are new technical constraints for radio design engineers. Therefore, they require new techniques to be solved. Ultra-narrow band (UNB) is a disruptive answer, based on revisited technology developed before World War II.

5.3 Ultra-narrowband: a disruptive way to use radio spectrum

5.3.1 Tuning: an old answer to capacity challenge

Since the beginning of radiocommunication, engineers have put a lot of effort in receiver tuning. Tuning is about adjusting a receiver so that its center frequency is aligned with one of the transmitters you want to listen to. Its objective is to have several radio transmissions occurring at the same time in the same place and without interfering each other.

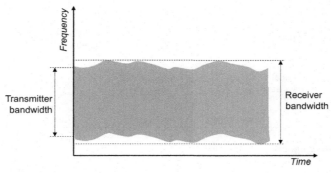

FIGURE 5–1 Conventional narrowband system.

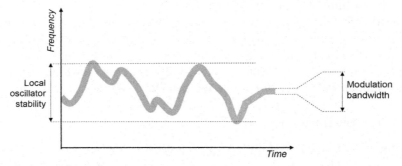

FIGURE 5–2 UNB modulation bandwidth.

In the past, tuning was about adjusting an inductance or a capacitor to match frequencies of local oscillators in transmitter and receiver. In practice, this process has to be permanent, because local oscillator of transmitter and receiver is not perfect and may drift quite significantly over transmission duration. When relative drift of transmitter (Tx) and receiver (Rx) oscillators is lower than the transmitter bandwidth (see Fig. 5−1), most part of signal energy resides within receiver bandwidth. Tuning is manageable with known techniques such as low drift oscillators or adaptive frequency control in receivers.

5.3.2 The 1-ppm limit

Paradoxically, receiver tuning becomes much more difficult with progress in microelectronics. Availability of integrated fractional-N phase-locked loops (PLLs), which may have a frequency synthesis step as low as 1 Hz, makes possible modulation rates of a few tens of symbols per second even with carrier center frequency up to a few gigahertz. In such case, local oscillator fluctuation of the PLL is clearly larger than the modulation bandwidth (see Fig. 5−2). Definition of UNB captures this paradox: UNB is a radiocommunication system that exhibits a modulation bandwidth lower than oscillator stability.

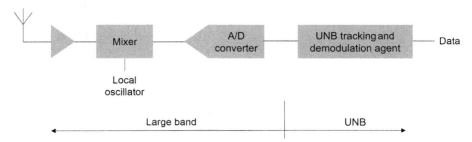

FIGURE 5–3 SDR for UNB reception.

When applied to subgigahertz technologies, UNB may be defined as a modulation bandwidth smaller than 1 ppm of carrier center frequency.

In UNB communication systems, receiver tuning issue reaches a new level of complexity. Its answer cannot be in conventional receiver architecture: the answer is a software-defined radio (SDR) receiver. SDR listens for a large spectrum range with a high dynamic range and a software agent searches for the signal of interest and continuously tracks its carrier center frequency, as a human radio operator would do (see Fig. 5–3).

5.3.3 Ultra-narrowband benefits for low-power wide-area networks

As such, having an SDR receiver is nothing new, but using SDR to overcome local oscillator unstability of transmitter is, from our best knowledge, a new approach in radiocommunication systems. When used along with UNB modulation scheme, it brings new opportunities and/or benefits as detailed in the following subsections.

5.3.3.1 Frequency channel allocation revisited

UNB modulation in transmitter and SDR receiver enlarge drastically the frequency space for sharing medium access. In conventional narrowband systems, operating frequency band is split into a couple of communication channels, each of them having an occupied bandwidth from tens to hundreds of kilohertz. In subgigahertz spectrum, UNB brings thousands of pseudo-channels. As an example, the 25 mW unlicensed frequency band in Europe [4], ranging from 868.0 to 868.6 MHz, brings 6000 pseudo-channels of 100 Hz width. This large amount is a real game changer: instead of implementing mechanism to share a scarce frequency resource from central point, objects may select randomly carrier center frequency, without excessive collision rate. Even if hundreds of objects transmit at the same time, reception in base station is made possible with corresponding number of demodulation agents (see Fig. 5–4).

5.3.3.2 Capacity given by base station processing power

UNB modulation and SDR receiver in base station give new opportunity for capacity increase, especially when base stations have a large coverage. Ref. [5] shows theoretical

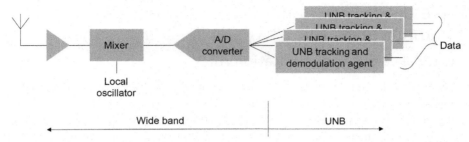

FIGURE 5–4 SDR implementation for UNB capacity.

time/frequency duality of narrowband communication when used in random unslotted time-frequency ALOHA networks. But, when using unlicensed spectrum, where regulation limits maximum power in transmission, UNB benefits from unlimited time to expand transmission. Furthermore, base station capacity leverages high dynamic range in SDR receivers because it is possible to receive messages from objects close to a base station along with messages far from base station. When looking for increased capacity, Ref. [6] shows the benefit of successive interference cancellation.

5.3.3.3 Complexity pushed back into core network

As explained in Sections 5.3.3.1 and 5.3.3.2, use of UNB removes a lot of complexity in objects. They do not need to care about channel selection and fine frequency tuning, because everything is managed by base stations and in the core network, where almost unlimited processing power is available. This balance of complexity is quite new in the telecommunication where more processing power in terminals supports the quest for more throughput and more multimedia capabilities in cell phones.

5.3.3.4 Ultra-narrowband robustness in unlicensed spectrum

In essence, LPWAN may be deployed in licensed or unlicensed spectrum. In practice, LPWANs go for the later because they have no license fees. Unlicensed spectrum, known as short range device (SRD) in Europe and instrumental, scientific and medical (ISM) in the United States, may be shared by a number of different systems, as long as they comply with technical constraints for sharing the spectrum. As these constraints are always minimalist, many different systems and technologies may coexist in SRD/ISM bands. As a consequence, spectrum occupancy in unlicensed bands is unpredictable: it is up to each system/technology to cope with interferences.

UNB brings robustness to LPWANs in unlicensed spectrum where many systems may coexist. Its high injection factor focuses Tx energy in a small spectrum interval allowing good signal-to-noise ratio. High dynamic SDR receivers leverage this in base stations.

5.4 Triple diversity ultra-narrowband, the Sigfox communication rules

5.4.1 Protocol versus communication rules

Triple diversity UNB (3D-UNB) is the name of communication rules for radio interface of the Sigfox solution (see Fig. 5–5). 3D stands for triple diversity, that is, diversity in time, frequency, and space. Table 5–1 gives 3D-UNB key numerical figures. A fully featured description of Sigfox communication rules is publicly available on the Internet [7]. The following subsections address key characteristics of 3D-UNB and give rationale for their design choices.

Thanks to UNB, Sigfox designed a very simple radio interface with low level of complexity in objects. The clue is randomness in managing time, frequency, and space. Randomness allows absence of coordination between transmitter and receiver. This is the reason why Sigfox radio interface is more a set of communication rules than a full communication protocol.

5.4.2 Uplink communication rules

5.4.2.1 Six steps to build an uplink radio burst

As any other communication systems, construction of a 3D-UNB uplink radio burst requires several steps from applicative level to physical level (see Fig. 5–6).

5.4.2.2 Small payload size for Internet of things usage

In communication protocols, payload size is a hint on the need of fragmentation for carrying large application messages. Multimedia applications benefit from variable payload sizes (ranging up to 1500 bytes, and even more with jumbo frames) that are available in IP-based

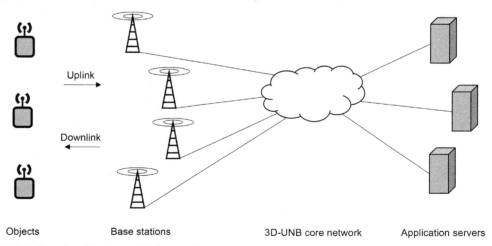

FIGURE 5–5 Overall architecture of Sigfox solution.

Table 5–1 3D-UNB characteristics in uplink and downlink.

Characteristics	Uplink (UL)	Downlink (DL)
Payload size	0 bit to 12 bytes	8 bytes (fixed length)
Total MAC/LINK overhead	8–10 bytes	2 bytes
Total PHY overhead	6 bytes	18 bytes
Modulation rate	100 baud or 600 baud	600 baud
Modulation scheme	Differential binary phase shift keying	Gaussian frequency shift keying
Transmit power	25 mW in Europe	
	150 mW in United States[a]	500 mW[b]
Frequency band	Subgigahertz unlicensed bands	

[a]Other values may exist in other countries, since maximum transmit power is defined by national regulations.
[b]Total transmit power may be balanced between multiple downlink messages, transmitted at the same time.

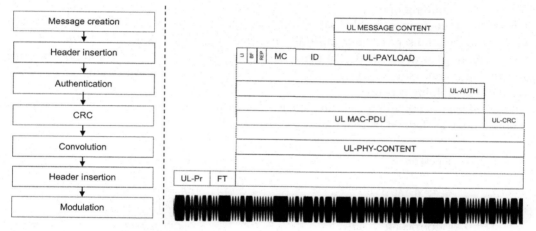

FIGURE 5–6 Uplink communication stack.

networks. IoT usage has different constraints for LPWAN systems. Instead of being able to carry largely variable size messages, IoT constraints is more about small application packets with limited overhead for the sake of energy saving.

Payload of 3D-UNB application messages goes from 0 to 12 bytes. This is enough to carry environmental data as well as a GPS position. Twelve bytes of application data are transmitted in a radio burst of only 2.08 seconds at 100 baud (see Table 5–3), with 1.8 mJ of energy per useful application bit, when using most recent radio modules rated at 25 mW/3.3 V in 25 mW transmission mode.

5.4.2.3 Replay attack protection with rolling counter

MAC/LINK header contains a message counter (MC) that is incremented after each message transmission. It is used in authentication algorithm as a rolling counter for replay attack protection. 3D-UNB core network rejects all messages received outside a sliding acceptance window. Replaying an old message is always possible but lag to do it must be greater than roll-over time, which is given by formula:

$$Roll\_over\_time = \frac{2^{MC\_length}}{Message\_rate}$$

With an MC length of 12 bits and a message rate of 140 messages per day in Europe, because of regulations, roll-over time of MC counter is 29 days. This value is large enough to get a reasonable replay attack protection for data collected by IoT sensors.

5.4.2.4 Convolution code for local or remote combining

For each uplink message, an object may transmit one or three uplink frame(s) (see Fig. 5–7). Transmitting one frame results in the lowest object power consumption, whereas transmitting three frames for the same message provides increased resiliency. Moreover, multiple transmissions give an opportunity to have combining algorithms in base station (local combining) or in 3D-UNB core network (remote combining). Local combining benefits from frequency diversity and convolution codes, whereas remote combining adds spatial diversity to combining processing.

To get efficient combining algorithms, each frame in a message is encoded with a different polynomial, as defined in Table 5–2.

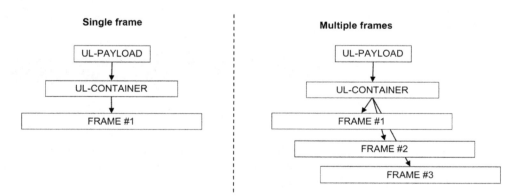

FIGURE 5–7 Principle of single or multiple transmission.

Table 5–2 Polynomials for convolution coding of uplink frames.

Frame emission rank	Polynomial
First	$R = 1$ (identify)
Second	$R + 1 + X + X^2$
Third	$R = 1 + X^2$

Table 5–3   Uplink frame type values.

Message type	UL Payload	LI value (binary)	UL-AUTH field size (bytes)	MAC-PDU size (bytes)	Frame length (bytes)	Frame rank	FT value (binary)	FT value (hexa)	Radio burst duration (@100baud)	Radio burst duration (@600baud)
Application message	empty	00	2	8	14	First	0b0000001101011	0x006B	1.12 s	187 ms
	1 bit (=0)	10	2			Second	0b0011011100000	0x06E0		
	1 bit (=1)	11	2			Third	0b0000000110100	0x0034		
	1 byte	00	2	9	15	First	0b0000010001101	0x008D	1.20 s	200 ms
	2 bytes	10	4			Second	0b0000011010010	0x00D2		
	3 bytes	01	3	12	18	Third	0b0001100000010	0x0302		
	4 bytes	00	2			First	0b0001101011111	0x035F	1.44 s	240 ms
	5 bytes	11	5			Second	0b0010110011000	0x0598		
	6 bytes	10	4	16	22	Third	0b0010110100011	0x05A3		
	7 bytes	01	3			First	0b0011000010001	0x0611	1.76 s	293 ms
	8 bytes	00	2			Second	0b0011010111111	0x06BF		
	9 bytes	11	5			Third	0b0011100101100	0x072C		
	10 bytes	10	4	20	26	First	0b0100101001100	0x094C	2.08 s	347 ms
	11 bytes	01	3			Second	0b0100101110001	0x0971		
	12 bytes	00	2			Third	0b0100110010111	0x0997		
Control message	5 bytes	11	5			First	0b0111101100111	0x0F67	1.76 s	293 ms
	6 bytes	10	4	16	22	Second	0b0111111001001	0x0FC9		
	7 bytes	01	3			Third	0b1000110111110	0x11BE		
	8 bytes	00	2							

5.4.2.5 Frame type, a multipurpose field

Frame type is an important field of MAC/LINK header. It is 13 bit long and carries two pieces of meta-information (see Table 5–3).

The first meta-information is the length of uplink MAC-PDU (see Fig. 5–6). Uplink MAC-PDU size is five discrete values only, depending on payload size (see Table 5–3).

The second meta-information is the frame rank, when multiple frames transmitted for an uplink message (see Section 5.4.2.4).

As frame type field is critical for decoding remaining parts of a frame, it is coded in 13 bits, and with a Hamming distance of at least five for all frame type values (see Table 5–4).

5.4.3 Downlink communication rules

5.4.3.1 Six steps to build a downlink radio burst

Construction of a 3D-UNB uplink radio burst requires six steps from applicative level to physical level (see Fig. 5–8). Full description is available in [7] and key characteristics of downlink 3D-UNB transmission are given in Table 5–1.

Fig. 5–8 depicts construction steps for downlink communication from applicative level to PHY level (i.e., from base station perspective). The key characteristics of 3D-UNB downlink communication rules are presented in the following subsections from object perspective, that is, from PHY level up to applicative level.

Table 5–4 Hamming distance of frame type values.

	0b1000110111110	0b0111111001001	0b0111101100111	0b0100110010111	0b0100101110001	0b0100101001100	0b0011100101100	0b0011010111111	0b0011000010001	0b0010110100011	0b0010110011000	0b0001101011111	0b0001100000010	0b0000011010010	0b0000010001101	0b0000000110100	0b0011011100000	0b0000001101011
0b0000001101011	7	7	6	8	5	6	7	6	7	5	8	5	6	5	5	6	6	**0**
0b0011011100000	9	5	6	10	7	8	5	6	5	5	6	9	6	5	7	6	**0**	6
0b0000000110100	5	11	8	6	5	6	5	6	5	7	6	7	6	5	5	**0**	6	6
0b0000010001101	6	6	9	5	8	5	6	5	6	6	5	6	7	6	**0**	5	7	5
0b0000011010010	6	8	9	5	6	7	10	7	6	6	5	6	5	**0**	6	5	5	5
0b0001100000010	7	7	6	6	7	6	5	8	5	5	6	5	**0**	5	7	6	6	6
0b0001101011111	6	6	5	5	6	5	6	5	6	8	7	**0**	5	6	6	7	9	5
0b0010110011000	5	5	10	6	7	6	5	6	5	5	**0**	7	6	5	5	6	6	8
0b0010110100011	6	6	5	5	6	9	6	5	6	**0**	5	8	5	6	6	7	5	5
0b0011000010001	10	6	7	7	6	9	6	5	**0**	6	5	6	5	6	6	5	5	7
0b0011010111111	5	7	6	6	9	10	5	**0**	5	5	6	5	8	7	5	6	6	6
0b0011100101100	6	6	5	9	8	5	**0**	5	6	6	5	6	5	10	6	5	5	7
0b0100101001100	7	5	6	6	5	**0**	5	10	9	9	6	5	6	7	5	6	8	6
0b0100101110001	8	6	5	5	**0**	5	8	9	6	6	7	6	7	6	8	5	7	5
0b0100110010111	5	7	6	**0**	5	6	9	6	7	5	6	5	6	5	5	6	10	8
0b0111101100111	9	5	**0**	6	5	6	5	6	7	5	10	5	6	9	9	8	6	6
0b0111111001001	10	**0**	5	7	6	5	6	7	6	6	5	6	7	8	6	11	5	7
0b1000110111110	**0**	10	9	5	8	7	6	5	10	6	5	6	7	6	6	5	9	7

FIGURE 5–8 Construction steps of downlink radio bursts.

5.4.3.2 Object-triggered downlink communication

Ability to have downlink communication at random is a request coming from H2H services, (i.e., incoming calls). In the case of fixed landline telephone, this feature requires almost no energy because the line current is only a few micro amps, when headset is on hook. In the case of cellular phones, this feature drains milliamps, because of permanent cell reselection and location update mechanisms. Having such a drain current is not acceptable in IoT objects for two reasons:

- It would require power source of significant size, which is not compatible with IoT concept where radio connection is just a side feature in the object
- In many cases, IoT usage requires always uplink communication, but downlink is either of seldom use or with no strong latency constraint.

For 3D-UNB systems, every downlink message is triggered by a previous uplink message, sent by object. Hence, it is up to an object to decide, on a per-message basis, whether to pull for a downlink message or not. This feature implements a bidirectional flag (BF) in uplink MAC/LINK header (see Fig. 5−6) for triggering downlink communication (see Fig. 5−9).

5.4.3.3 Relationship between uplink and downlink carrier center frequency

Section 5.3.3.1 introduces concepts and benefits of random selection of uplink carrier frequency for UNB communication and how inaccuracy of local oscillators is managed by SDR processing in base station. An equivalent issue exists for receiver tuning in downlink. Once again, work-around comes from base stations that use SDR also for downlink communication.

This is done by defining a fixed frequency gap between carrier center frequency of an uplink message and carrier center frequency of the onward downlink message

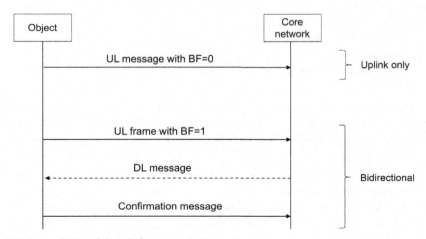

FIGURE 5–9 Uplink or uplink and downlink communications.

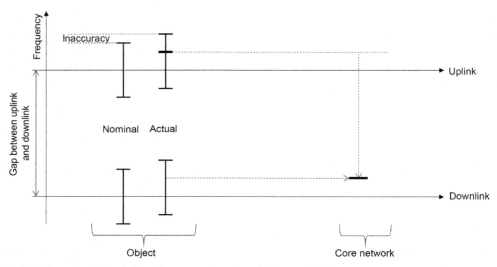

FIGURE 5–10 Compensation of object frequency inaccuracy in core network.

(see Fig. 5–10). Local oscillator inaccuracy is transparently compensated by core network, as long as it remains constant between uplink transmission interval and downlink reception interval. This short-term stability is easy to achieve even in low-cost/low-complexity radio hardware of IoT objects.

5.4.3.4 Downlink authentication reusing uplink context

Authentication of downlink incoming messages uses corresponding uplink context in three different ways.

The first authentication check is implicit: an object opens a time and frequency window where it expects a downlink message. Other downlink messages, which base station may transmit out of this window, will be ignored by object.

The second authentication check uses whitening function that core network applies to have evenly distributed symbol transitions in downlink radio bursts. Whitening function uses pseudorandom generator whose initialization value includes object identifier (ID) and MC value coming from corresponding uplink message. If de-whitening output contains too much errors, it is rejected.

The third authentication is more classic and based on AES128 cryptographic function. Downlink messages contain an authentication field (DL-AUTH in Fig. 5–8) that is evaluated with parameters derived from uplink context (see Fig. 5–11). This is possible because each downlink message is triggered by uplink message (see Section 5.4.3.2). Object ID and MC values are derived from corresponding uplink message; authentication key is known by core network and object.

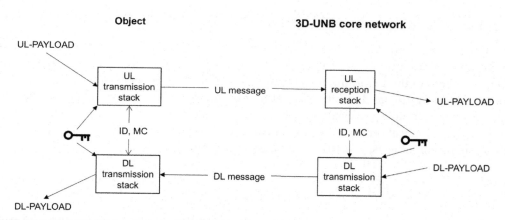

FIGURE 5–11 DL-AUTH field evaluated with uplink context.

5.5 Seven questions on Sigfox radio interface

5.5.1 Why Sigfox radio access network is not a cellular network?

Since their very early beginnings, cellular systems have been based on the concept of frequency reuse. Frequency reuse is very efficient to get capacity over very large coverage with a small set of frequencies, but it implies complex attachment procedures to detect and connect the network. Power consumption in cellular phones is significantly impacted by these procedures, particularly when moving.

On the contrary, in order to keep 3D-UNB as simple as possible, all 3D-UNB base stations listen for the same frequency band. There is no procedure in objects to retrieve the proper frequency to be used in a given area and no extra power consumption when objects are mobile. 3D-UNB objects just have to know the international telecommunication union (ITU) zone, where they are deployed and use the frequency intervals selected by 3D-UNB network. A 3D-UNB system is like built with a single worldwide cell.

5.5.2 Why is there no attachment procedure in Sigfox radio access network?

Attachment procedures were designed to be able to receive an incoming call at once, wherever mobile phones are in the network. In many IoT use cases, there is no need for downlink messages triggered only by core network. That is why there is no attachment procedure in 3D-UNB systems. Only implicit attachment is available: each time an object transmits an uplink message, the network can detect where the object is. If a downlink message is requested, core network is able to select adequate base station for downlink transmission. Benefit of implicit attachment is less power consumption and less complexity in objects.

5.5.3 What is cooperative reception?

Cooperative reception is a side benefit of noncellular nature of 3D-UNB system. As all base stations are on the same frequency, large overlap of base stations coverage exists. Whereas overlaps are kept to the minimum in cellular systems, they are beneficial to 3D-UNB systems because they bring spatial diversity. Several base stations may receive an uplink radio burst simultaneously. Multiple received packets are then de-duplicated in core network, before transmission to application server. Cooperative reception improves quality of service without extra complexity in objects; everything is done in core network.

5.5.4 Why is there no destination address in Sigfox radio bursts?

In local area networks, it is common to have source and destination address fields in each packet. This is because several nodes may share a common medium of communication. In 3D-UNB systems, there is no object-to-object communication: all messages go from objects to application servers or from application servers down to objects (see Fig. 5–12).

In uplink, destination of radio bursts is unambiguous: it is all base stations that are in the vicinity of a transmitting object. So, no destination address is needed, resulting in smaller radio bursts. In core network, messages are routed to application servers, thanks to a database that links object identifiers to application servers.

In downlink, radio frames do not have any address fields at all, because downlink transmission is triggered by an uplink message (see Fig. 5–10). Carrier center frequency and time interval for downlink transmission are defined by corresponding uplink message (see

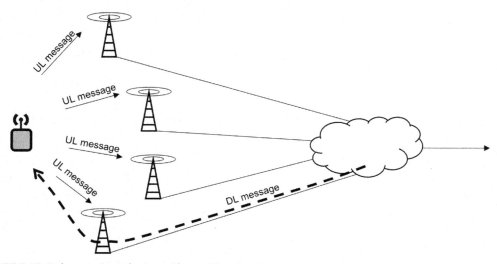

FIGURE 5–12 DL base station selection with cognitive algorithm.

Fig. 5−11), creating a kind of implicit address. Possible collision in downlink frequency/time slot is resolved by whitening function and authentication that use ID and MC values (see Section 5.4.3.4). So, no explicit address field is needed in downlink frames.

5.5.5 Why does Sigfox radio access technology use ALOHA for accessing the spectrum?

ALOHA [8] implements a minimalist medium-sharing technique: pure random access. The benefit is in its lightweight implementation. The drawback is its poor capability to withstand medium or high offered load. Success rate is good only when offered load is kept low or very low.

3D-UNB communication rules use ALOHA to access the spectrum for the following two reasons:

- It is easy to implement in low-cost/low-complexity objects and fully compliant with spectrum-sharing techniques of SRD/ISM unlicensed frequency bands.
- It is efficient thanks to UNB that gives extra capacity to the unlicensed frequency bands that are limited. In other words, it is as if ALOHA over UNB becomes a two-dimensional ALOHA: uplink messages are evenly distributed over time and frequency with a de facto reduction of offered load per frequency pseudo-channel.

5.5.6 Why is Sigfox radio access technology cognitive?

Cognition in telecommunications means that a system changes its parameters according to its environment or usage it sees. 3D-UNB systems are cognitive in two ways.

Base stations implement cognitive process in their SDR signal processing by allocating demodulating and decoding agents for each radio signal of interest in received spectrum. The more signals are received in parallel, the more demodulating and decoding agents run in parallel in a base station.

Core network implement cognitive algorithm to select base station for downlink transmission. Because of large coverage overlaps, several base stations are in communication range of an object. When a downlink communication is triggered by an uplink message, all surrounding base stations are a candidate for downlink transmission. Core network run cognitive algorithm to select the best base station for each downlink message (see Fig. 5−12). Various parameters may be included when selecting base station for downlink transmission such as load, interference level, frequency balancing, and available transmit power.

5.5.7 Why do Sigfox objects control the network and not the other way around?

Coordination of terminal nodes by a central base station is common in licensed radio systems, such as cellular or private mobile radio networks, because it helps to get most of the frequency bands, paid by operators. The drawback of this approach is additional signaling messages and/ or reception time in terminal nodes to acquire and keep synchronization with base stations.

IoT objects are drastically constrained in power consumption, so the less energy is consumed in network coordination, the better it is for objects. The simplest implementation is to have transmit frequency and time selected by objects, that is, without any prior synchronization with surrounding base stations. Avoiding such synchronization procedures is beneficial for power consumption in objects, as long as collision rate is kept reasonably low. This is possible thanks to UNB, which gives increased capacity (see Section 5.3.3), and 3D-UNB communication rules, which implement pure random access (see Section 5.5.5).

3D-UNB base station behavior depends on what they receive from surrounding objects. Base stations have to be ready to receive uplink radio packets, at any time and on any frequency selected by objects. This is why the network is controlled by objects.

5.6 Conclusion

This chapter presents Sigfox radio system, which is optimized for small infrequent messages of IoT connectivity. Its radio interface implements UNB over the air, SDR processing in base stations, and cognition in core network. UNB and SDR processing are beneficial for capacity and simplicity even when operated in unpredictable spectrum, because they bring robustness to interferers, selectivity, and sensitivity. Cognition brings spatial diversity in uplink and load optimization in downlink. In the future, it is expected that SDR will allow improved uplink reception capabilities (e.g., with successive interference cancellation) and improved downlink capacity thanks to optimization algorithms for base station selection. These improvements will add no complexity to objects, that will remain simple and low energy, which is critical for the IoT.

References

[1] S. Madakam, R. Ramaswamy, S. Tripathi, Internet of Things (IoT): a literature review, J. Comp. Commun. 3 (2015) 164–173. Available from: https://doi.org/10.4236/jcc.2015.35021.

[2] Short range devices; Low throughput networks (LTN) architecture; LTN architecture. ETSI TS 103 358 V1.1.1 2018-06.

[3] <https://machinaresearch.com/news/webinar-revolution-evolution-or-distraction-machina-researchs-view-on-emerging-low-power-wide-area-wireless-technologies-in-m2m/>. Consulted on March 2019.

[4] CEPT Electronic Communication Committee, ERC recommendation 70-03 relating to the use of short range devices (SRD). Tromsø 1997 and Subsequent amendments, Published October 5, 2018.

[5] C. Goursaud, Y. Mo, Random unslotted time-frequency ALOHA: theory and application to IoT UNB networks, in: 23rd International Conference on Telecommunications (ICT), May 2016, Thessaloniki, Greece, 2016, pp. 1–5. https://doi.org/10.1109/ICT.2016.7500489. <hal-01389362>.

[6] Y. Mo, C. Goursaud, J.-M. Gorce, Bénéfice de l'annulation successive d'interférence pour des reseaux ultra narrow band: theorie et application a l'IoT, in: ALGOTEL 2018—20emes Rencontres Francophones sur les Aspects Algorithmiques des Telecommunications, May 2018, Roscoff, France, 2018, pp. 1-4. <hal-01783923>.

[7] <https://build.sigfox.com/sigfox-device-radio-specifications>. Consulted on March 2019.

[8] N.M. Abramson, The aloha system: another alternative for computer communications, in: AFIPS Fall Joint Computing Conference, 1970.

6

NB-IoT: concepts, applications, and deployment challenges

Mona Bakri Hassan[1], Elmustafa Sayed Ali[2], Rania A. Mokhtar[1,3], Rashid A. Saeed[1,3], Bharat S. Chaudhari[4]

[1]DEPARTMENT OF ELECTRONICS ENGINEERING, SUDAN UNIVERSITY OF SCIENCE AND TECHNOLOGY (SUST), KHARTOUM, SUDAN [2]DEPARTMENT OF ELECTRICAL AND ELECTRONICS ENGINEERING, RED SEA UNIVERSITY, PORT SUDAN, SUDAN [3]DEPARTMENT OF COMPUTER ENGINEERING, TAIF UNIVERSITY, ALHAWIYA, TAIF, SOUTH AFRICA [4]SCHOOL OF ELECTRONICS AND COMMUNICATION ENGINEERING, MIT WORLD PEACE UNIVERSITY, PUNE, INDIA

6.1 Narrowband-Internet of Things overview

For the IoT future growth and development in the mobile industry, the Third Generation Partnership Project (3GPP) has standardized new class technologies for low-power wide-area network (LPWAN) applications. These standards are referred to as mobile IoT. They are designed for licensed spectrum and to support devices with requirements of low power consumption, long range, low cost, and security [1]. A variety of LPWA IoT applications have emerged and their requirements differ from each other. One LPWAN technology cannot address the requirements of all low-power IoT applications and for this reason, two complementary licensed 3GPP standards Narrowband-Internet of Things (NB-IoT) and Long-Term Evolution for Machines (LTE-M) have been proposed and are built on the LTE [2].

NB-IoT is a new standard known as massive LPWAN to support long range and low data rate for IoT applications. It has several characteristics such as ultralow power consumption, wide coverage, and massive connection. Also, NB-IoT has several novel characteristics for LPWAN deployment to overcome shortcomings such as poor security, poor reliability, and high operational and maintenance costs. NB-IoT is able to be loaded by major mobile equipment and module manufacturers, and indeed it will be existing to be adaptable with 2G, 3G, 4G, and 5G cellular networks [3]. It enables to operate traditional IoT businesses and opens up new opportunities for industry applications and other aspects. Additionally, it promises a strong market trend pointing at the growing demand for different smart applications.

6.1.1 History and standards

The rapid and strong growth of NB-IoT technology in the markets shows a large number of deployment plans with billions of connections and pushing the operator's revenue in trillions of dollars by 2020 be [4]. Currently, NB-IoT market is driven by many operators from Asia, Europe, and the United States [5]. According to market research on future narrowband IoT, the global NB-IoT market is expected to grow at compound annual growth rate of 50% between years 2016 and 2022 due to rapid development in IoT industry and rising demand of new cellular communication technologies, which are dedicated to IoT LPWA applications [6].

For any new technology standardization, an important issue is to ensure interoperability products and applications which will help to develop and regulate the operation of products by the technology and ensure security and privacy of the data for users. Different standards shown in Fig. 6−1 have been released by the national and international developing levels with the help of communication manufactures that have played a major role in setting the latest standard for NB-IoT [7]. The standards of narrowband M2M and narrowband LTE were released between years 2014 and 2015 where both technologies contributed to the emergence of what is known as narrowband cellular IoT (NB-CIoT). NB-CIoT with the standards of 3GPP has led to a clear vision of the use of NB-IoT technology.

3GPP has developed a standard for IoT including eMTC and EC-GSM-IoT. They also produced new specifications for cellular communication by the NB-IoT, and in 2016 they completed its first set of specifications on NB-IoT as radio standard developed for LPWAN to support IoT technologies [8]. With the completion of the NB-IoT, 3GPP has concluded a major effort in what is known as Release 13 to address the IoT market. The existence of new technologies including NB-IoT allows 3GPP operators to address their different market requirements of IoT.

6.1.2 Narrowband-Internet of Things concepts

NB-IoT is based on the characteristics of LTE, enabling easy deployment and integration with LTE networks. It is used for IoT applications with extended coverage and low power

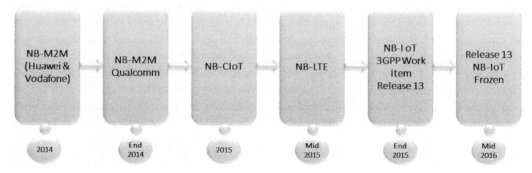

FIGURE 6–1 NB-IoT standard evolution.

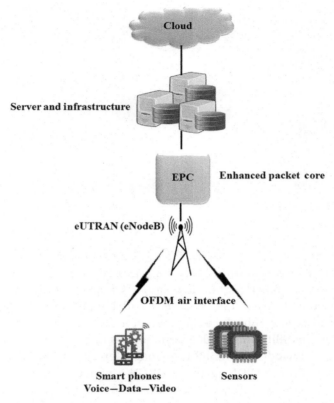

FIGURE 6–2 Basic LTE/LTE-M architecture.

consumption. LTE is the fourth generation cellular access network as defined by the 3GPP standard. It is a comparatively flat Internet Protocol (IP)-based standard to significantly enhance 3G performance via the use of OFDMA based radio access network (RNC) and packet-only core [9]. The basic architecture for LTE/LTE-M is shown in Fig. 6–2.

In LTE network, eUTRAN/eNodeB constitutes the radio network controller (RNC) and the packet core designated as enhanced packet core (EPC). The major entities in LTE access networks are defined by the 3GPP standards named 3GPP Release 16. As shown in Fig. 6–3, integrated base station/controller eUTRAN complex (eNodeB) provides radio access and management functions and the radio bearer (RB) to the user equipment (UE). Its functions include radio admission control (RAC), radio resource management (RRM), encryption, compression, air interface mobility, hand over (HO), RB control, connection mobility control, and dynamic allocation of resources to UE in uplink and downlink, scheduling, in addition to security. The operation is based on OFDMA in downlink and SC-FDMA in uplink with 15 kHz tone spacing [10].

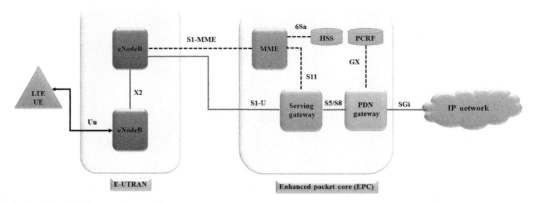

FIGURE 6–3 LTE/LTE-M major entities.

Data channels are defined to provide effective and efficient data transport over the LTE radio interface. These are formulated using resource block (RBs) which is standardized with a frequency of 180 kHz. The RAN supports a range of frequency bands, in particular, 1.4, 5, 10, and 20 MHz. EPC manages and handles various types of user and device traffic, supports and introduces new equipment and applications, and provides support for seamless mobility and service portability across wireless IP networks. The key elements in the EPC include the mobility management entity (MME), serving gateway (S-GW), packet data network (PDN) gateway (P-GW), policy and charging rules function, and home subscriber system (HSS) with a set of associated subsystems [11]. MME provides the control plane functions related to subscriber and session management, equipment management and tracking, in addition to location management. The S-GW provides packet data routing and is the user traffic mobility anchor. The P-GW is the default gateway to the PDN. It is responsible for packet filtering and QoS enforcement. The HSS stores and updates a database with user and device information, international mobile subscriber identity, and approved QoS profiles [11]. Transmission time interval (TTI) in LTE is the smallest unit of time in which eNB is capable of scheduling any user for uplink or downlink transmission and is 1 ms in LTE/LTE-M [12]. If a user is receiving downlink data, then during each 1 ms, eNB will assign resources and inform user where to look for its downlink data through Physical Downlink Control Channel (PDCCH) channel (see Fig. 6–4).

Hybrid automatic repeat request (HARQ) as shown in Fig. 6–5 is a process where receiver combines the new transmission every time with previous erroneous data. LTE/LTE-M implements the incremental redundancy HARQ version [13].

In HARQ when the receiver detects erroneous data, it doesn't discard it. Upon receiving a NACK, the sender sends the same data with a different set of coded bits. The receiver combines the previously received erroneous data with newly attempted data by the sender and the process is repeated until the receiver successfully decodes the correct data or the

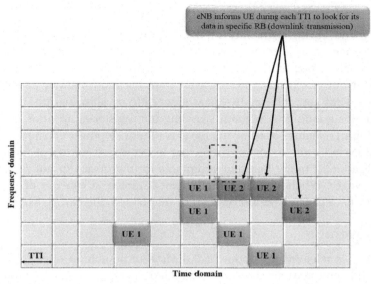

FIGURE 6–4 Transmission time interval (TTI) in LTE.

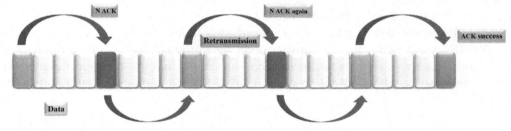

FIGURE 6–5 Normal HARQ in LTE.

number of transmissions exceeds a threshold. This operation can result in delays and too much control overhead in the case of poor radio conditions if the sender has to attempt many transmissions.

6.2 Narrowband-Internet of Things general features

The deployment of NB-IoT technology which made the concept of smart solutions more realistic has main features that make it more reliable in applications [14]. The most important features of NB-IoT based on which it works are connectivity, power efficiency, massive engagement, and security.

6.2.1 Low power consumption

Power efficiency and saving capability in the NB-IoT devices are one of the significant features in long-range massive connectivity. NB-IoT should be able to reduce the amount of energy used as much as possible to have long battery life in sensors. NB-IoT uses two types of mechanisms for power efficiency in massive communication, namely power saving mode (PSM) and eDRX. PSM can achieve 10 years battery life by allowing the devices to go into deep sleep mode for a long time without signaling but registering in prior. eDRX is considered as an existing LTE feature used by the devices to reduce power consumption by extending the terminal cycle in idle mode and reducing unnecessary startup of receiving cell [15]. For an IoT application, it might be quite acceptable for the device to not be reachable for a few seconds or longer. Whenever the device is not listening, it can use eDRX to switch off the radio receiver for a defined period of time, so that battery life can be extended.

6.2.2 Wide coverage

NB-IoT technology is released to support IoT devices to be able to operate in deep indoor, wide outdoor, and remote areas that need extended coverage. This brings opportunity for the new application classes including data acquisition and control of equipment located in manholes, pipelines, and other environments where existing communication infrastructure is otherwise unreachable. Though the penetration of the signal is improved, the devices are expected to work on the lower bounds of signal reception, so support for reliable data transport should be provided as a part of the connectivity solution. NB-IoT coverage power can reach 164 dB in independent deployment mode by introducing a set of techniques for enhancing the coverage, such as retransmission (200 times) and low-frequency modulation adoption, and also by taking advantage of the relaxed IoT requirements regarding data rate and latency [16].

6.2.3 High connection density

The scalability of NB-IoT network is necessary for many applications that support a massive number of devices and smart sensors that can reach approximately 10s of billions. 3GPP achieves massive scalability, a minimum of 1 M devices/km^2, to support the small devices with ultra-energy efficiency. NB-IoT as an existing technology for 3GPP is optimized for low power, deeper coverage, and higher device density of up to 50K devices per cell, while seamlessly coexisting with LTE services. Together, NB-IoT and LTE support a wide range of low-power IoT applications.

6.2.4 Privacy and reliability

NB-IoT technology operates in a licensed band and hence, it improves the reliability of transmission. The process of standardizing the technology, along with the input of several key worldwide carriers, mean that NB-IoT possesses all the concepts of reliability; in addition to the same security measures that are currently present in LTE networks. This make This makes the NB-IoT more secure and reliable. For data security, features such as secure authentication, signaling protection, and end-to-end data encryption are available. NB-IoT uses the user datagram (UDP) protocol since it consumes low power. The cloud server is responsible for authenticating and decoding of the data. NB-IoT can also use other security levels as defined in LTE [17].

6.3 Narrowband-Internet of Things fundamental theories and characteristics

The deployment of NB-IoT is based on many technical and theoretical characteristics that are considered as key features that can affect the performance of NB-IoT setup and ability of devices to operate in dynamic environments. The different characteristics of NB-IoT technology will be reviewed in the following sections to find out what the specifications are that will impact the performance of NB-IoT [18].

6.3.1 Narrowband-Internet of Things key technologies

NB-IoT performance in practical applications depends on measures of data rate, number of devices, and latency which are of the interest of operators, in addition to other key features that impact the performance of NB-IoT, such as channel overhead, radio environment, and interference. Also, allocating resources to massive IoT device is considered as one of the sensitive indicators that will limit the performance of NB-IoT since the spectrum resource availability is limited to low-frequency spectrum. For all these reasons, it is necessary to evaluate the theoretical and main technological characteristics of the NB-IoT according to the following considerations: signaling and data performance, connection analysis, latency analysis, and coverage enhancement to achieve the reliability and performance of NB-IoT deployment in different applications [18].

6.3.1.1 Signaling and data
The relationship between signaling and data is a key factor that determines the effect of simultaneous access of the massive IoT terminals in NB-IoT application and it depends on the terminal behaviors and network model where they can sometimes cause signaling overhead [19]. For this reason, the analysis of new physical and MAC layers must be considered to establish dynamic signaling overhead model, which can describe the relevance between signaling and data service, so that a theoretical guidance for joint pressure reduction of signaling and data and congestion prevention is provided.

Massive IoT terminals in NB-IoT applications will specify the range of data rate of the typical application since the data rate is the fraction of the channel capacity governed by signaling for data transmission. In addition, the data rate is affected by many factors such as efficiency of collision avoidance, control signaling overhead, channel utilization, and latency. By expanding the transmission time with low power use, the data rate in NB-IoT can be controlled to help in improving the power consumption of the terminals and devices but

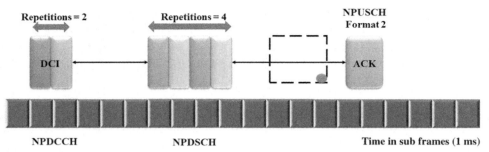

FIGURE 6–6 Channel repetitions process in NB-IoT.

deduction from coverage area. The coverage area enhancement can be extended with the help of repetitions process (see Fig. 6–6) [20].

Repetitions in NB-IoT is a technique involving repeating of the same transmission several times, which can achieve an extra coverage of up to 20 dB. Each repetition is self-decodable, the scramble code changes for each transmission and the repetitions are ACK-ed only once for all NB-IoT channels. In NB-IoT, random-access procedure is a process that represents the number of repetitions access between the NB-IoT devices and NB-IoT eNodeB in higher layer protocol interaction. The preamble repetitions can be repeated up to 128 times.

6.3.1.2 Connection analysis

In NB-IoT, the number of connections that can be reached depends on the burstiness of NB-IoT service in the application. Studies have focused on how the services are evenly distributed within a day for connections in NB-IoT-related technologies and its overload effect in accessing network when a large number of devices enter the network simultaneously. In LTE-M, the load pressure of LTE random-access channel (RACH) and overload control mechanisms such as classified controlled access, exclusive RACH resources, dynamic RACH resources allocation, exclusive back-off mechanism, time division access, and active paging mechanism are all taken into consideration to enhance the connection ability in massive accessing [21]. In order to improve capacity of NB-IoT, the researchers study with the maximal number of connections supported by NB-IoT RACH and the optimum resources allocation proportion for arbitrary random-access strength and total constrained bandwidth.

6.3.1.3 Latency analysis

The 3GPP showed that apart from the number of connections, a theoretical computing model for uplink access latency is also required. Many kinds of latencies related to deterministic processing and others are required for signal detection. For low-latency and high-reliability communication, software-defined networking-enabled network architecture is proposed by many researchers, which integrates different types of access technologies. Other studies are undertaken to investigate the statistical properties of NB-IoT random-access latency including mean value, variance, and PDF for arbitrary random-access strength excitation to improve the latency analysis theory for NB-IoT. NB-IoT aims at a peak latency of 10 seconds. That is 100 times longer than a normal WAN connection, but it should suffice for most applications using LPWAN. The latency could probably be lowered, but it might interfere with other design goals such as device density.

6.3.1.4 Coverage enhancement

Coverage enhancement can be achieved by improving the receiver sensitivity and spectral efficiency in NB-IoT. Adaptive modulation and enhanced encoding techniques are used in improving the performance analysis and optimal design of enhancement mechanisms. It is based on coverage class aiming to develop a kind of coverage class discrimination/ improvement mechanism and coverage enhancement technology based on dynamic statistical multiplexing. This methodology works on extracting an optimal threshold of coverage class by referring to received signal strength indicator (RSSI) and SINR determined by building penetration loss, dynamic adjustment of coverage class according to hybrid automatic repeat request (HARQ), and coverage enhancement [22]. For link enhancement, receivers can test packet error using error detection codes such as cyclic redundancy check (CRC), ACK/NACK messages and ARQ technique combined with channel coding under HARQ scheme. In HARQ process, the number of packets included into HARQ can be expressed as follows:

$$Nb = \alpha\beta Rd \tag{6.1}$$

where α represents the modulation index, β is the effective code rate, and Rd is a resource size for multicast packets. In NB-IoT, coverage enhancement can also be achieved by using signal repetitions, control channels, and uplink/downlink bandwidth reduction. In order to enhance the coverage, it is important to measure the signal-to-noise ratio (SNR) of the channel and transmission coding rate (CR) by the following equations.

$$SNR = \frac{Px}{L \cdot F \cdot No \cdot BW} \tag{6.2}$$

$$CR = \frac{b + CRC}{RU \cdot \frac{Symbols}{RU} \cdot \frac{Bits}{Symbol}} \tag{6.3}$$

In Eq. (6.2), Px denotes transmission power, L represents path loss, F is the receiver noise figure, and No is thermal noise density. In Eq. (6.3), CRC denotes size in bits of the cyclic redundancy check code, and RU is the number of resource units allocated to the UE. The metrics, signal-to-noise ratio, bandwidth utilization, and energy for repetitions and bandwidth reduction can be obtained by applying the equations that will be reviewed in Section 6.2.1.

For deep indoor coverage, NB-IoT requires a maximum coupling loss (MCL) of 164 dB which is 20 dB higher than LTE with the assumption that the number of devices in a cell is

55,000 in the long range between 10 and 15 km [23]. NB-IoT allows up to three coverage levels to be defined by a serving cell. Each coverage level is associated with a configuration that defines the number of repetitions to be used on each physical uplink/downlink channel and the UEs choose one among the three coverage levels based on the received downlink signal power.

6.3.2 Narrowband-Internet of Things technical properties

Spectrum bandwidth and modulation are the two basic factors that affect the performance of NB-IoT depending on operation modes and frame structure in uplink and downlink transmission modes. The following section will give a brief review of technical features related to the NB-IoT.

6.3.2.1 Spectrum bandwidth and modulation

In NB-IoT, UE supports only a narrow carrier bandwidth of 180 kHz, which ultimately leads to a reduction in the device complexity and cost compared with that of wide-band devices. The downlink transmission in NB-IoT is based on OFDMA as in LTE. The uplink design is slightly different from that of LTE. For the uplink, SC-FDMA multiple access scheme is used to support low complexity for UEs and a high number of simultaneous access, with normal 15 kHz for subcarrier spacing [24]. Fig. 6−7 shows the frequency-domain structure for NB-IoT.

Radio network technology for NB-IoT is based on ungradable software, which will offer three types of spectrum access bands: 2100 MHz for 3G and 4G, 1800 MHz for 2G and 4G, 900 MHz for 3G and 4G. To support these bands, there are a number of chipsets and modules designed by different manufacturers such as QUALCOMM, Intel, Ublox, Neul, and Quectel [24]. Regarding modulation, Key sight claims that the uplink modulation is BPSK 1 bits/symbol, QPSK 2 bits/symbol, 8PSK 3 bits/symbol, or optionally 16 QAM 4 bits/symbol. The same options except for 8PSK are available for the downlink modulation.

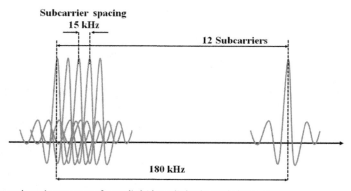

FIGURE 6−7 Frequency-domain structure for uplink/downlink physical channel.

In NB-IoT, the impact of spectrum bandwidth for transmission between UEs and eNB can be analyzed by using Shannon theorem based on the modified number of resource units (RUs) allocated by eNB to the UE and depends on three transmission properties, SNR, bandwidth utilization, and energy per transmitted bit. By considering maximum bandwidth as BW_{max} and repetitions at minimum to $R = 1$, the required SNR for UE can be calculated using Shannon bounds as follows:

$$\text{SNR}_{\text{req}}^{(RU,BW_{max,1})} = 2^{Rb^{(RU,BW_{max,1})}/BW_{max}} - 1 \tag{6.4}$$

where Rb is the data rate of the transmission for bandwidth allocated by the number of RUs, From the data rate of the UE, we can obtain the bandwidth utilization Y of the transmission. Consequently,

$$Y^{(RU,BW_{max,1})} = \frac{Rb^{(RU,BW_{max,1})}}{BW_{max}} \tag{6.5}$$

Let $\frac{Eb}{Nq} = \frac{\text{SNR}_{req}}{Y}$ be the lower bound of the received energy per bit to noise power spectral density ratio, and Eb be the energy per transmitted bit, then,

$$Eb^{(RU,BW_{max,1})} = \frac{Eb^{(RU,BW_{max,1})}}{No} \cdot L \cdot F \cdot No \tag{6.6}$$

where L represents path loss, F is receiver noise figure, and No is thermal noise density. From the above equations, it can be observed that as the number of RUs is greater, the values of the rest of the transmission properties analyzed i.e., SNR_{req}, Y, and Eb, decrease.

6.3.2.2 Operation mode

NB-IoT can be implemented in three operation scenarios: standalone, guard-band, and in-band scenarios. In standalone, the spectrum which is not used for cellular services is utilized. The scenario can also be established by formatting one or more GSMs carrying with a bandwidth of 200 kHz and a guard interval of 10 kHz on both sides of the spectrum to carry NB-IoT traffic, ensuring smooth transition to the LTE massive machine communications [25]. The operation of guard-band scenario with cellular services positioned the NB-IoT traffic in the guard-band of the LTE carriers without allocating LTE resources and avoiding possible interference. In-band scenario will allow the NB-IoT in the LTE carrier sharing the LTE resources, which will efficiently use spectrum resources for LTE or NB-IoT services based on the demand from mobile users or devices and will be more cost-effective [26], and it requires that eNodeB software should be upgraded. Fig. 6−8 shows the process of NB-IoT assigned to a selected carrier from LTE spectrum based on the three operation options.

For standalone option, the initial cost is higher than other operation options because of new hardware required for antenna and RF systems. For both in-band and guard-band options, the initial cost is similar because they use the LTE carrier but spectrum cost is different as in-band option uses coexisting LTE signal while guard-band uses free physical

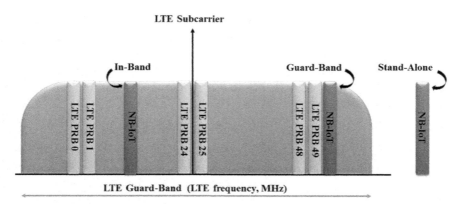

FIGURE 6–8 NB-IoT carrier selection from LTE spectrum.

resource locks. According to the operation options, the deployment of NB-IoT should be transparent to UEs when it is turned on and searches for NB-IoT carrier. It only requires to search for a 100 kHz carrier to facilitating the UE initial synchronization referred as anchor carrier.

6.3.2.3 Transmission mode

The transmission scenarios in NB-IoT for uplink and downlink between the base stations and UE are adaptable based on three coverage enhancement levels, CE level 0 to CE level 2. The CE level 0 corresponds to normal coverage and CE level 2 corresponds to the worst case, where the coverage may be assumed to be very poor [27]. These three different coverage levels are known as normal, robust, and extreme with MCL of 144, 154, and 164 dB, respectively. A list of power thresholds for the received reference signals is broadcasted in the cell for each CE level. The main impact of the different CE levels is that the messages have to be repeated several times. The three coverage enhancement levels govern the operation of uplink and downlink communication depending on several types of channels (see Fig. 6–9). For uplink, there are two physical channels: the Narrowband Physical Uplink Shared Channel (NPUSCH) and the Narrowband Physical Random-Access Channel. The downlink uses three physical channels known as Narrowband Physical Broadcast Channel (NPBCH), Narrowband Physical Downlink Control Channel (NPDCCH), and Narrowband Physical Downlink Shared Channel (NPDSCH). The Physical Downlink Channels are always QPSK-modulated.

NB-IoT supports the operation with either one or two antenna ports using space-frequency block coding and the same transmission scheme is applied to NPBCH, NPDCCH, and NPDSCH [27]. For a transmission mode, all data are sent/received over the NPUSCH and NPDSCH channels. Fig. 6–10 shows the transmission modes for uplink and downlink communication in NB-IoT scenario.

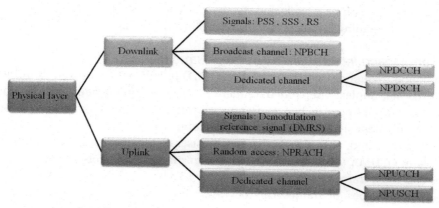

FIGURE 6–9 NB-IoT uplink/downlink channels.

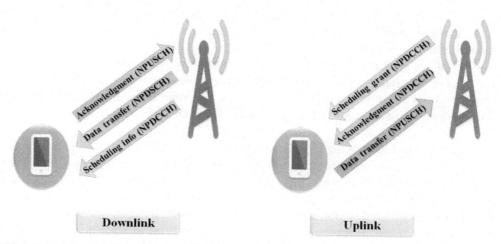

FIGURE 6–10 NB-IoT uplink/downlink transmission mode.

In downlink, communication will take place by a message sending from eNodeB to the device and taking into account a number of measures to conserve battery power by allowing the NB-IoT to configure the process of eDRX and PSM . This allows the device to go into deep sleep mode for a few seconds and it is no longer reachable by the network. For initiating the communication, the device can select downlink TBS on MAC layer from 2 bytes to 85 bytes as approved by 3GPP, and then the selected TBS carries the contents such as data payload and header for IP, UPD, and CoAP [28].

In uplink, device requests to communicate with eNodeB using RACH procedure; when eNodeB receives the request, it will send back a scheduling to the device to indicate for time and frequency allocation for the device, followed by uplink data transfer and ACK/NACK.

In downlink, the TBS on MAC selected by the device is between 2 to 125 bytes and the payload carried depending on higher layer protocol overhead [28].

6.3.2.4 Narrowband-Internet of Things frame structure

NB-IoT supports frequency division duplexing (FDD) in half-duplex as well as in full-duplex mode. In FDD half-duplex, a separated frequency is used for uplink and downlink communication with guard subframe in between and receivers or transmitters do not perform both operations simultaneously [29]. As mentioned before, NB-IoT uses OFDM in downlink and SC-FDMA in uplink and, each OFDM symbol consists of 12 subcarriers with a bandwidth of 180 kHz. Many OFDMA symbols are occupied into on slot. The slots are summed up into subframes and radio frames in the same way as for LTE.

In uplink, two types of frames can be used: single-tone frame and multitone frame (see Fig. 6−11). The single-tone frame is mandatory, which is used to provide capacity in signal-strength-limited scenarios and more dense capacity for one subscriber with a spacing of 15 or 3.75 kHz via random access and slot durations between 0.5 and 2 ms. The multitone frame is used to provide high data rates for devices in normal coverage as an optional capability. The number of subscribers can be 3, 6 or 12 signaled via DIC with a spacing of 15 kHz and a slot duration of 0.5 ms.

In downlink, as shown in Fig. 6−12, the frame structure is the same as for LTE (coexistence with LTE). The bandwidth is 180 kHz for 12 subcarriers separated by 15 kHz. The durations are categorized as one frame for 10 subframes, one subframe equals two slots in 1 ms, one slot requires 0.5 ms duration, which is equal to seven OFDM symbols.

6.3.2.5 Narrowband-Internet of Things networking architecture

NB-IoT networks consist of user equipment (UE) and evolved NodeB (eNodeB) to relay the data to IoT evolved packet core (EPC) which connects to the network-connected application servers. The communication between the NB-IoT UEs and the eNodeB over air interface is based on functions which describe interface access processing and cell management [30]. As shown in Fig. 6−13, the operation of network architecture takes place from UE with the eNodeB that is able to communicate with the IoT EPC through S1-lite interface by sending NAS data to the EPC for initiating processing. The data is forwarded to the application server from IoT platform through the interaction between the IoT EPC with NAS of UE. The IoT platform collects the data from access networks and then collected data are forwards it to the application servers for IoT data processing.

In NB-IoT, the S1-lite interface is considered as an optimized version for control plane based on S1 AP protocol, interface between eNodeB and IoT management core. User plane data is carried by the modified S1 AP messages to support small data handling in an efficient manner and optimized security procedures in CIoT and NB-IoT. The collection of data and voice in NB-IoT is performed by EPC, a framework that unifies voice and data on an IP service architecture as an MME, and voice is treated as just another IP application. This mechanism will allow operators to deploy and operate one-packet network for all cellular networks in addition to WLAN, WIMAX, LTE, and fixed access with the NB-IoT

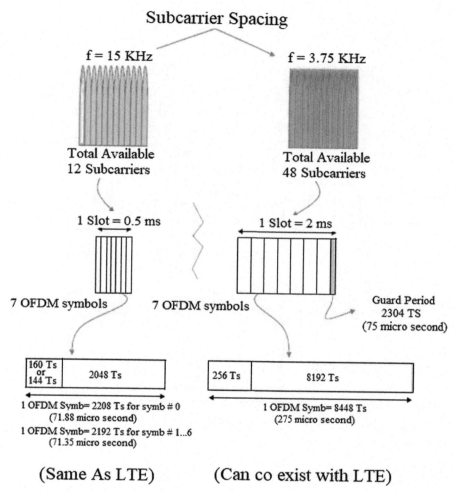

FIGURE 6–11 NB-IoT uplink frame structure.

applications. In NB-IoT, data transmission through network can take two possible options between NB-IoT devices and AS, and these transmissions are based on IP and Non-IP (see Fig. 6–14).

In IP network depending on the capability of radio module, IPv4 and IPv6 with UDP transport protocol are used. On the air interface, TCP is supported for NB-IoT while HTTP and HTTPs over air interface are not implemented because of their overhead on TCP. The non-IP-based data transmission is recommended for NB-IoT because it reduces the transmission data volume. This option can be handled by the network by allowing only one target IP address, which means data can only be sent to one target IP address (server).

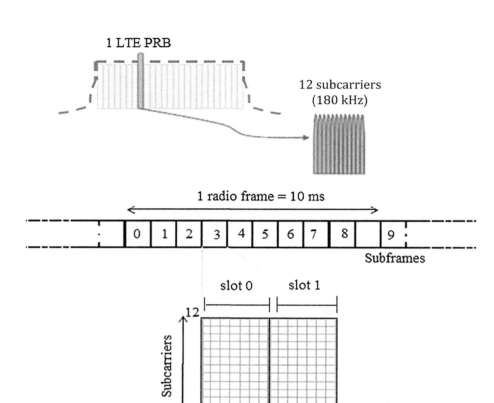

FIGURE 6–12 NB-IoT downlink frame structure.

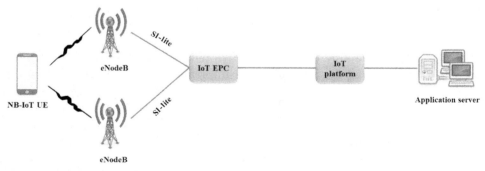

FIGURE 6–13 NB-IoT network architecture.

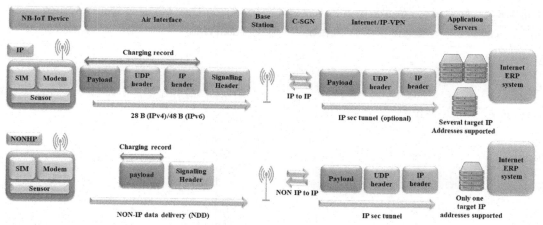

FIGURE 6–14 Data transmission in NB-IoT network.

6.4 Narrowband-Internet of Things-related technologies

In general, NB-IoT has more advantages than other LPWAN technologies, especially in terms of narrowband and peak rate . This means that it has better performance on ultralow power consumption and low data rate services. It is able to efficiently serve in various application cases and to be the main market for LPWAN. Table 6−1 shows the NB-IoT features over other LPWAN technologies as a comparison.

Despite the high performance of NB-IoT compared to other technologies, the integration of some other LPWAN technologies will provide higher performance in some technological aspects for different applications. Sigfox uses efficient bandwidth and low noise levels, resulting in high receiver sensitivity, low-cost antenna design, and ultralow power consumption. Both Sigfox and NB-IoT technologies can be integrated into one network as an integrated NB-IoT/Sigfox's IoT wireless technology. The integrated design solutions are targeting applications such as tracking, wearables, security, agriculture, healthcare, industrial and consumer. The use of integrated NB-IoT and sigfox enables an application with ultra-long battery life tracking devices.

LTE-M can support lower device complexity, extended coverage, and longer battery life. NB-IoT and LTE-M are both presented as standard published by 3GPP, which they operate on licensed spectrum. They have a common property of operation on existing cellular infrastructure. NB-IoT uses the same features of power saving that is used in LTE-M, known as PSM and eDRX. Research on signal processing and artificial intelligence (AI) processors provides a new model for smarter, connected devices that provide low-power cellular IoT connectivity, which allows multimode LTE-M/NB-IoT to ensuring ultralow power consumption and highly efficient performance required for a wide range of cellular IoT use cases. In cellular IoT connectivity, integration of NB-IoT with a module that uses low-power digital signal processing and parallel processing architecture will be allow use of ultra-low power

Table 6–1   Features comparison between NB-IoT and other LPWAN technologies.

Technical metrics	NB-IoT	LTE-M	LoRaWAN	SigFox
Carrier bandwidth	200 kHz	1.4 MHz	125 and 250 kHz	100 Hz
Uplink	SC-FDMA	SC-FDMA	Wide-band linear frequency	Ultra-narrowband
Downlink	OFDMA	OFDMA	Frequency-modulated pulses	(UNB)
Modulation	QPSK	16 QAM	CSS	BPSK
Bands	Licensed LTE	Licensed LTE	915 MHz	<1 GHz
MCL	164 dB	156 dB	155 dB	160 dB
Data rate	250 kbps	1 Mbps	50 kbps	100 bps
Coverage	<25 km	5 km	<20 km	<40 km
Power consumption (battery life)	>10 years	10 years	<10 years	>10 years
Cost	Low	High	Low	Lower than NB-IoT
MAC layer	LTE based	LTE	Aloha based	Aloha based
Connection density (maximum number of devices)	50,000 per cell	>50,000 per cell	100,000 per cell	100,000 per cell
Mobility	No	Yes	Yes	No

concept in IoT . This achieves outstanding power consumption, while retaining the flexibility to support multiple cellular IoT standards.

6.5 Narrowband-Internet of Things applications

Many expected applications of wide-area communications technology do not require high data rates, but instead require a low-cost device that consumes very little power and can be deployed in massive numbers. 3GPP collaborative standards organization has defined a standard for low-power wide-area IoT networking as NB-IoT, which will be able to deploy many kinds of applications such as smart grid, smart cities, smart environment, and smart industry.

6.5.1 Smart grid

NB-IoT provides a secure communications technology for smart grid depending on the network types such as home area network and neighborhood area network or for access core networks [31]. The features of NB-IoT such as excellent coverage and low power consumption will promise to improve the smart grid deployment in metering, real-time monitoring, and electric vehicle (EV) charging. In smart metering, NB-IoT will reduce the complexity and cost of using gateways, allowing the smart meters to directly communicate via cloud, and offer high implementation flexibility because of ability to operate in different kinds of environment. The deployment of NB-IoT in smart metering will provide low device power consumption and low cost of energy used in hardware. In smart metering, automated meter reading services can be integrated with 2G and 4G networks. It will be able to collect real-time power metrics voltage, current, and other information about power consumption. By interfacing the AMI with NB-IoT, meter vendors can provide not only meters but also the complete AMI solution including metering data management and head-end system (HES), via the NB-IoT network. The smart meters data could be received by the utilities data center as shown in Fig. 6−15. Through this way, accurate bills will be sent to the end users via mobile SIM cards or cloud services [32].

NB-IoT can offer more benefits to the water companies for smart water consumption metering and calculation of forecasting future demand. The implementation of NB-IoT achieves the possibility of transferring a set of water information based on metering sensors to the company management system to give an ability to exchange information about real-time meter reading, accumulate measuring, reverse-flow consumption, water temperature, pipe pressure, and tempering alarms. And for smart gas metering, NB-IoT can provide a smart solution business management for gas companies through cloud services.

In smart EV charging, NB-IoT technology will contribute to achieving an intelligent interconnection and interaction between the electrical vehicles, batteries, and charging stations under the control of power supply grid management system. Smart charging provides related information to citizens in real time to enable better-charging management and by using the NB-IoT, millions of devices can be connected with this smart charging service due to massive connectivity in NB-IoT [33]. The use of smart charging system will mitigate the overloading problem in electric grid by switching the information of EVs charge and battery status via bidirectional communication for data exchanging through the NB-IoT network (see Fig. 6−16) with acceptable response times approximated to 2−5 min and data rate between 9.6 and 50 kbps.

6.5.2 Smart cities

The deployment of NB-IoT in a smart cities will open avenues to build a new type of technological cities, by achieving intelligent operation in transportations, tracking systems, environment monitoring, health care, and smart buildings [34]. NB-IoT could connect these technologies together more efficiently and also connect a variety of devices such as monitors or weather sensors, to achieve efficient wireless performance without any constant contention between power consumption and inefficiency. In health care, the integration of wearable appliances with the hospital's management is very important [35]. NB-IoT can provide suitable solutions for such healthcare applications because it supports secure communication over a long range and high energy efficiency that can be used in smart health care to enable the personalization of healthcare allocation to real-time patients and medical professionals under the term of health 4.0 [36]. NB-IoT will provide an efficient way for the integration of wearable appliances with the hospital's managements and with help of edge computing technology, and it will be possible to deal with the requirement of latency in the medical process [35]. By using the massive NB-IoT terminals, cloud computing, and services in health care, the concept of personalization will be achieved. In Health 4.0, improved health care with safe

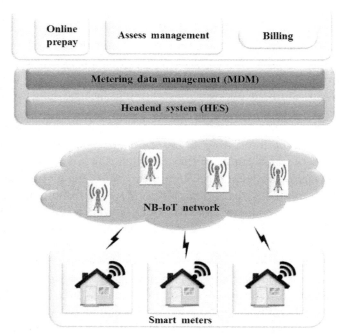

FIGURE 6–15 NB-IoT AMI solution.

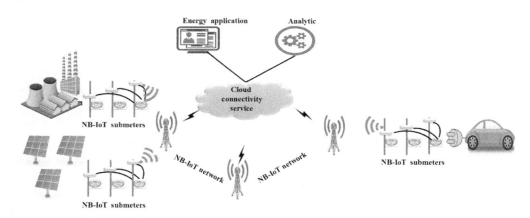

FIGURE 6–16 Smart vehicles charging-based NB-IoT infrastructure.

assembly mechanisms for health services will improve disease treatment strategies and allow for progressive allocation through the disease surveillance repository anywhere, at any time, and for both drug integration and nondrug treatment [37].

In transportation, the solutions to reduce the impact of congestion and improve public transport can be handled by NB-IoT networks for tracking vehicles in real time, collecting information from vehicles or road sensors to the transportation monitoring center for the purpose of traffic analysis and navigation [38]. Another consideration is dealing with road instructions for a drivers group setting, which can be handled through NB-IoT network with vehicle-to-vehicle to inform about changes in transport routes, the nearness or nonappearance of stopping paths, or changes in roadway allowable velocity according to any other different conditions. For street lighting, NB-IoT can help to improve energy utilization and also reduce the difficulty of large-scale street lights management, as well as to provide an intelligent upgrading model for urban street light system. The use of street light controlling system over NB-IoT will allow to get the real-time status of the street lights, energy consumption, and other parameters. The collected information can be used for analysis and for performance improvement. [39]. Through NB-IoT control terminals, the current environmental situation can be analyzed and controlthe street lights by adjusting the light brightness to adaptthe ambient light intensity. Generally, the NB-IoT smart street lighting system may consist of three layers, namely transmission, platform, and application, as shown in Fig. 6−17. The information from NB-IoT sensors as a transmission layer will be routed to the applications in the monitoring center through different types of platforms.

6.5.3 Smart industry

In the smart industries, machines are connected inside the factories with a number of sensors that can be connected to the cloud, allowing for optimal planning and flexibility in manufacturing and maintenance, in addition to being self-monitoring of all related operations. NB-IoT sensors can add special values to the smart industry by predicting the possibilities of machine breakdown in the factories. This will help in scheduling maintenance to minimize factory downtime, by sending information through safe and efficient networks to the monitoring and decision-making center to take a suitable action. The use of AI with an NB-IoT network capable of automatic monitoring of the machine alerts the facility manager when the device needs maintenance by AI algorithms that automate learning and pattern recognition in order to monitor the state of the machine [40]. Industrial automation using NB-IoT can provide a new model for connecting industrial IoT devices to critical industrial systems. It can combine a range of functions such as precise indoor positioning and supporting ethernet via new radio (NR), thereby allowing to access QoS adapters to support the reuse of existing industrial devices and control systems. Fig. 6−18 shows the concept of smart industrial automation through time-sensitive network (TSN) with the interaction of NB-IoT technology.

In smart industrial automation, TSN and industrial control systems typically use ethernet-based transport. The integration of NB-IoT industrial devices with the TSN ethernet domain forms an NR as a technology for supporting a wide variety of services, devices, and deployments. The 5G industry NR across a diverse spectrum will overcome

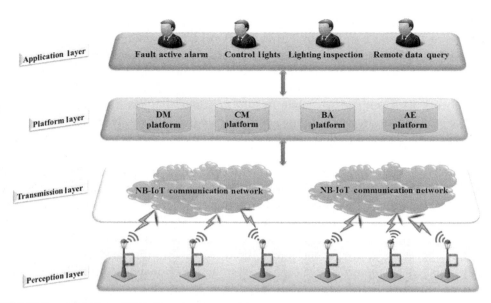

FIGURE 6–17 Smart light-based NB-IoT transmission model.

any deterioration in the performance of industrial automation. It enables more advanced automation, easing flexibility, and enabling greater efficiencies during communication, thereby allowing a greater level of insight in real-time into the state of production and operation [40]. In addition, a 5G industry NR network facilitates the incorporation of technologies such as cloud computing, machine learning, and big data processing. In industry, the NB-IoT in the context of 5G will reduce the IoT connectivity costs. The coexistence of NB-IoT in 5G industry will enhance the mobile broadband in the critical industry communication.

6.6 Narrowband-Internet of Things deployment challenges and solutions

NB-IoT technology is specifically designed to overcome the drawback in many applications. This opens the door to new deployment challenges to deal with rapid growing smart IoT applications related to the requirements of long coverage range, very low power, and very low data rates. Low battery life in many IoT sensor applications is required to avoid high power consumption and costly maintenance by optimizing the life cycle of devices by reducing repetitions to transmit data [41]. In NB-IoT, the maximum battery life is expected in the range of 10 years where small data packets are sent through applications, but one of the biggest challenges for NB-IoT device battery life is that it can be significantly impacted by low coverage [42]. The solution for long battery life lies in that manufacturers should determine

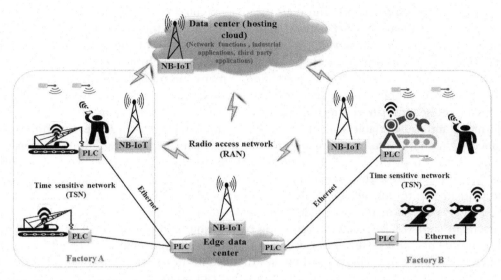

FIGURE 6–18 Industrial automation using NB-IoT technology.

the level of power required to manage devices when controlling the connection process under active, idle, standby, and sleep modes, in addition to the amount of energy required by the devices and management according to applications. From the other point of view, the 3GPP standard specifies PSM to allow devices to enter into deep sleep mode and switch off most of its internal components.

In NB-IoT, extreme coverage for industrial IoT is very important. The target for NB-IoT is to be able to provide sufficient coverage to reach areas in remote location or in difficult access regions. NB-IoT is expected to enable a maximum coverage gain over regular LTE. Depending on the deployment and configuration methods, the gain must be kept at a maximum for all the existing LTE base stations grid [43]. The solution for extreme coverage lies in the fact that the manufacturers must conduct studies on a different number of scenarios for a different radio frequency environment including remote locations. Another challenge in NB-IoT deployment is ensuring the network's ability to cope with the expected volume growth in connected devices. NB-IoT promises several solutions to the industrial IoT; however, the success of the NB-IoT system depends on the manufacturer and service provider solutions to overcome the challenges in battery life, coverage, low cost, and network capabilities by performing field measurement.

Security in NB-IoT is an emerging issue. For NB-IoT deployment, to prevent information from being stolen, the use of cryptographic algorithms such as data encryption, identity authentication, and integrity verification is required. The application layer in NB-IoT occupies a larger amount of data and the requirements are identification and processing of this massive heterogeneous data, integrity, and authentication of data, and access control of data [44]. In NB-IoT network, the data travels encrypted therefore in a secure way. The threat is when data leave NB-IoT network and is sent through the internet from the network operator to the final cloud server. On the other hand, some of the IoT devices may be located in the areas that will probably be attacked by gaining a physical access to device control. The weakness in passwords and encryption in many Internet objects, devices, SIM card, transceivers, or wearables will open a security loopholes for attackers. NB-IoT should provide means of security in connecting various devices of different capabilities in different applications and must enable scalable security solutions management for a large number of IoT devices, in addition to providing end-to-end security measures between IoT devices and Internet host even in the clouds. For these security issues, NB-IoT uses the LTE security levels AS and NAS to secure signaling messages between a UE and MME in the control plane, and to secure the delivery of RRC messages between a UE and eNB in the control plane and IP packets in the user plane. Authentication and trust procedures with identity management should take place to secure the relationship and access between the IoT network components, and ensuring the validation of data analysis engines depending on the implemented application. Threats and attacks that can occur to the devices also require procedural processes to resist some types of attacks that can occur in the IoT.

6.7 Conclusion

NB-IoT and its technical aspects will make the IoT more important in the future. Due to massive increase in the IoT devices and billions of connections in different applications, current communication technologies will face more difficulty in deploying smart applications. The use of NB-IoT will get rid of most of these difficulties such as cost, power consumption, and wide range coverage, in addition to matching different spectrum allocation of operators. It can be used for applications such as smart metering, properties monitoring, agriculture, smart every environment, and logistics management. For all these applications, NB-IoT meets the requirements of rate latency, battery life, and dense wide-area connectivity. In smart cities, NB-IoT with the LTE makes it possible to control street lights, free parking space, and environment conditions. New NB-IoT initiatives will be easily set under any network and services that help to make a very attractive business.

References

[1] P.P. Ray, A survey on Internet of Things architectures, J. King Saud Univ. Comput. Inform. Sci. 30 (2018) 291−319.

[2] G.A. Akpakwu, G.P. Hancke, A survey on 5 generation networks for the internet of things: communication technology and challenges, IEEE Access 6 (2017).

[3] D. Ismail et al., Low-power wide-area networks: opportunities, challenges and directions, in: ICDCN'18 Workshops, Varanasi, India, January 4−7, 2018.

[4] A.-A.A. Boulogeorgos, et al., Low power wide area networks (LPWANs) for Internet of things (IoT) applications: research challenges and future trends, arXiv:1611.07449, 2016.

[5]   M. Chen et al., Narrow band Internet of things, IEEE Special Section on Key Technologies for Smart Factory of Industry 4.0, September 15, 2017, <https://doi.org/10.1109>.

[6]   W. Zhai, Design of narrow band-IoT oriented wireless sensor network in urban smart parking, Int. J. Online Eng. 13(12) (2017), 116−126.

[7]   B. Martinez, et al., Exploring the performance boundaries of NB-IoT, IEEE Commun. Mag. (2019).

[8]   Q.M. Qadir, et al., Low power wide area networks: a survey of enabling technologies, applications and interoperability needs, IEEE Access 6 (2018) 77454−77473.

[9]   R. Ohlan, K. Dalal, Review paper on operational assay of LTE system in stint of OFDMA & SC-FDMA, Int. J. Recent Trends Eng. Res. 2(3) (2016) 499−505.

[10] R.K. Singh, R. Singh, 4G LTE cellular technology: network architecture and mobile standards, Int. J. Emerg. Res. Manag. Technol. 5 (2016).

[11] M.A. Skulysh, O.S. Klimovych, Method of LTE functional units organization with evolved packet core virtualization, Inform. Telecomm. Sci. 6 (2015).

[12] Y. Mehmood, Mobile M2M communication architectures, upcoming challenges, applications, and future directions, EURASIP J. Wirel. Commun. Netw. (2015).

[13] C.M.M. Taher, M. Amer, Hybrid Automatic Repeat Request in LTE, Rochester Institute of Technology, Dubai, 2013.

[14] S. Delbruel, et al., Federation support for long range low power Internet of things protocols, in: FLIP Conference'17, Washington, DC, July 2017.

[15] A. Elnashar, M.A. El-Saidny, Extending the battery life of smartphones and tablets: a practical approach to optimizing the LTE network, IEEE Veh. Technol. Mag. 9(2) (2014) 38−49.

[16] M. Chafii, F. Bader, J. Palicot, Enhancing coverage in narrow band-IoT using machine learning, in: 2018 IEEE Wireless Communications and Networking Conference (WCNC), 15−18 April 2018, IEEE, 2018.

[17] L. Alberto, B. Pacheco, et al., Device-based security to improve user privacy in the Internet of things, Sensors (Basel) (2018). Available from: https://doi.org/10.3390/s18082664.

[18] A. Rakić, Key Aspects of narrow band IoT communication technology driving future IoT applications, in: 2017 25th Telecommunication Forum (TELFOR), 21−22 November 2017, IEEE, 2017, <https://doi.org/10.1109/TELFOR.2017.82493271>.

[19] A. Larmo, A. Ratilainen, J. Saarinen, Article impact of CoAP and MQTT on NB-IoT system performance, Sensors (Basel) 19 (2019) 7. Available from: https://doi.org/10.3390/s19010007.

[20] M. Chafii, F. Bader, J. Palicot, Enhancing coverage in narrow band-IoT using machine learning, HAL, 9 February 2018.

[21] N. Li, C. Cao, C. Wang, Dynamic resource allocation and access class barring scheme for delay-sensitive devices in machine to machine (M2M) communications, Sensors (2017). Available from: https://doi.org/10.3390/s17061407.

[22] L. Wan, Z. Zhang, J. Wang, Demonstrability of Narrowband Internet of Things Technology in Advanced Metering Infrastructure., Springer, 2019.

[23] P. Niemelä, Narrowband LTE in Machine to Machine Satellite Communication, Aalto University, 2018.

[24] P. Andres-Maldonado, et al., Narrow band IoT data transmission procedures for massive machine type communications, IEEE Network 31(6) (2017) 8−15. <https://doi.org/10.1109/MNET.2017.1700081>.

[25] Y.-P. Eric Wang, A primer on 3GPP narrowband Internet of things (NB-IoT), IEEE Commun. Mag. 55(3) (2017).

[26] S.K. Sharma, X. Wang, Towards massive machine type communications in ultra-dense cellular IoT networks: current issues and machine learning-assisted solutions, IEEE Commun. Surv. Tutor. (2018).

[27] A. Rico-Alvariño, et al., An overview of 3GPP enhancements on machine to machine communications, IEEE Commun. Mag. 54(6) (2016) 14–21.

[28] Y.-P.E. Wang, et al., A primer on 3GPP narrowband Internet of things (NB-IoT), arXiv, 2016.

[29] A. Morgado et al., A survey of 5G technologies: regulatory, standardization and industrial perspectives, Digit. Commun. Netw. 4 (2018).

[30] I. Afolabi, End-to-End Mobile Network Slicing, Aalto University, 2017.

[31] S. Javadi, S. Seifvand, NB-IoT applications in smart grid: survey and research challenges, Int. J. Commun. 11 (2017).

[32] M. Faheem, et al., Smart grid communication and information technologies in the perspective of Industry 4.0: opportunities and challenges, Comput. Sci. Rev. 30 (2018).

[33] K. Shuaib, et al., A Secure Discharging Protocol for Plug in Electric Vehicle (SDP-V2G) in Smart Grid, Springer, 2017.

[34] T. van der Vorst, et al., The Wireless Internet of Things: Spectrum Utilization and Monitoring, Radiocommunications Agency Netherlands, 2016.

[35] S.B. Baker, et al., Internet of things for smart healthcare: technologies, challenges, and opportunities, IEEE Access 5 (2017) 26521–26544.

[36] V. Sharma, et al., Security privacy and trust for smart mobile-Internet of things (M-IoT): a survey, arXiv, 2019.

[37] V.V. Estrela, et al., Health 4.0 as an application of industry 4.0 in healthcare services and management, Med. Technol. J., 2(4) (2018).

[38] P. Fraga-Lamas, T.M. Fernández-Caramés, L. Castedo, Towards the Internet of Smart trains: a review on industrial IoT-connected railways, Sensors (Basel) (2017).

[39] J. Guerrero-Ibáñez, et al., Sensor technologies for intelligent transportation systems, Sensors (Basel) 18 (4) (2018).

[40] Shahid Mumtaz et al, Massive Internet of things for industrial applications, IEEE Ind. Electron. Mag. 8 (2017).

[41] J. Lee, J. Lee, Prediction-based energy saving mechanism in 3GPP NB-IoT networks, Sensors (Basel) 17 (2017).

[42] M. El Soussi, et al., Evaluating the performance of eMTC and NB-IoT for smart city applications, arxiv, 2017. <https://doi.org/10.1109/ICC.2018.8422799>.

[43] T.T. Tesfay, et al., Energy saving and capacity gain of micro sites in regular LTE networks: downlink traffic layer analysis, ACM Special Interest Group on Simulation and Modeling New York, 2011.

[44] S. Chacko, Mr. Deepu Job, Security mechanisms and vulnerabilities in LPWAN, IOP Conf. Ser. Mater. Sci. Eng. 396(1) (2018).

7

Long-term evolution for machines (LTE-M)

Suresh R. Borkar

DEPARTMENT OF ELECTRICAL AND COMPUTER ENGINEERING (ECE), ILLINOIS INSTITUTE OF TECHNOLOGY, CHICAGO, IL, UNITED STATES

7.1 Introduction

In the domain of standards-based low-power wide-area networks (LPWAN) operating in licensed bands, narrowband IoT (NB-IoT) and long-term evolution (LTE) for machines (LTE-M) are two interrelated and complementary solutions. Both are based on 4G LTE as the foundation. The two are collectively referred to as LTE IoT and are designed to coexist with existing LTE infrastructure, spectrum, and devices. At the same time, LTE-M also forms the *segue* to and complements future 5G-based IoT networks. Chapter 6, NB-IoT: concepts, applications, and deployment challenges, has covered NB-IoT solution extensively. This chapter builds upon the LTE foundation from Chapter 6, NB-IoT: concepts, applications, and deployment challenges, and provides corresponding coverage for LTE-M.

The next section summarizes the major applications addressed by LTE-M and the primary objectives LTE-M is expected to meet. Subsequent sections cover the architecture and operational aspects of LTE-M. This is followed by discussion on interrelationships from LTE and LTE-M and future coexistence between LTE-M and 5G networks. LTE-M evolution from 3rd Generation Partnership Project (3GPP) Release 13—15 is summarized. The chapter concludes with a discussion on applications and selected use cases for LTE-M.

7.2 LTE-M as low-power wide-area network solution

As indicated in Chapter 1, Introduction to low-power wide-area networks, LPWAN is a wide-area network for meeting requirements for smart and intelligent applications and services. LTE-M is a versatile and powerful standards-based LPWAN solution.

7.2.1 LTE-M introduction

As an LPWAN technology, LTE-M supports a wide array of IoT applications. It is primarily targeted to IoT applications that require comparatively high data rates, low-power

LPWAN Technologies for IoT and M2M Applications. DOI: https://doi.org/10.1016/B978-0-12-818880-4.00007-7

consumption, low latency, mobility, and wide coverage. Voice over LTE (VoLTE) capability is also supported. LTE-M benefits from all the security and privacy features of 3GPP solutions, such as support for user equipment (UE) identity, confidentiality, entity authentication, and data integrity. Key application candidates which require such attributes are smart transportation, critical time-sensitive health services, wearables which monitor vital measurements, and industrial applications among others. LTE-M also covers IoT applications requiring deep coverage where latency, voice capability, mobility, and data speed requirements are less stringent. Such applications include tracking devices, smart grid, smart city, and home automation. Because of its range of performance and coverage as well as use of LTE standard as a base, LTE-M is an LPWAN technology which has applicability to possibly highest number of IoT use cases.

In terms of terminology, eMTC and LTE-M are used synonymously in this discussion notwithstanding the minor differences between these terms [1]. The essentials are the same.

7.2.2 LTE-M objectives

To address these applications, LTE-M is architected to meet the following major objectives [1,2].

- Variable data rates to support a variety of applications and use cases,
- Support for a massive number of devices. 100,000 or more devices per access station can be supported. This is particularly relevant in situations where devices have very low data−throughput requirements,
- As compared to LTE, extended and deeper indoor coverage to overcome path loss and attenuation through walls and floors,
- Support for very low-power operation with 10-year battery lifetime. This is specifically needed for devices at remote locations that are hard to reach or where commercial power supply is not available. The 5-watt-hour AA battery is a primary candidate for such use,
- Selective support for both real-time and mission-oriented applications needing few millisecond latency as well as deferred traffic applications with latency in few seconds range,
- Low device cost comparable to operation in the LTE environment,
- Voice to be supported as part of standard LTE functionality but not necessarily in extended coverage scenarios,
- Compatibility with LTE networks and connection for MTC traffic without requiring a gateway,
- Deployment of LTE-M in-band within an LTE carrier for compatibility reasons, and
- Software upgradable from LTE as a seamless path toward 5G MTC solution.

7.3 LTE-M architecture

LTE-M and NB-IoT have very similar architectures based on LTE constructs [3]. As part of NB-IoT architecture discussion, the main attributes of base LTE network were covered in Chapter 6, NB-IoT: concepts, applications, and deployment challenges. Some of the LTE-M relevant aspects of LTE architecture are reproduced here for the sake of completion.

LTE is a comparatively flat Internet protocol (IP)-based standard to significantly enhance 3G performance in terms of multimedia applications (see Fig. 7−1).

The radio access network (RAN) termed eUTRAN/eNodeB uses single-carrier frequency-division access in uplink (UL) and higher-performance orthogonal frequency-division multiple access in downlink (DL) [4,5]. The packet core is designated as enhanced packet core (EPC). The integrated base station/controller eNodeB provides radio access and management functions and the radio bearer to the LTE devices. EPC manages and handles varying types of user and device traffic, supports and introduces new equipment and applications, and provides support for seamless mobility and service portability across wireless IP networks.

The UL and DL channels are indicated in Fig. 7−2 [6].

Data channels in LTE are formulated using resource blocks (RBs) standardized with a frequency band of 180 kHz. The 1 millisecond transmission time interval (TTI) is the smallest unit of time in which eNodeB is capable of scheduling any user for UL or DL transmission [5]. If a user is receiving DL data, then during each 1 millisecond, eNodeB will assign

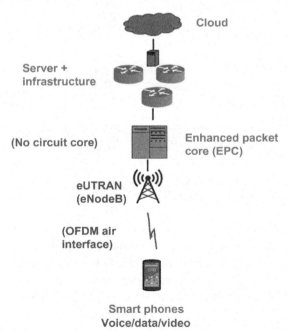

FIGURE 7−1 4G LTE basic architecture.

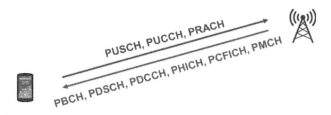

PUSCH: Physical uplink shared channel; data transfer
PUCCH: Physical uplink control channel; control signaling requirements
PRACH: Physical random access channel; random access functions

PBCH: Physical broadcast channel; system information
PDSCH: Physical downlink shared channel; data transfer
PDCCH: Physical downlink control channel; scheduling information
PHICH: Physical channel hybrid ARQ indicator channel; hybrid ARQ status
PCFICH: Physical control format indicator channel; format of signaling
       being received
PMCH: Physical multicast channel; MBMS control information and traffic

FIGURE 7–2 Uplink and downlink physical channels in LTE.

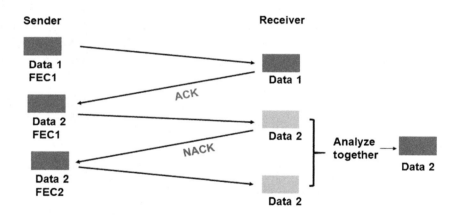

FEC: Forward error correction

FIGURE 7–3 Incremental redundancy HARQ process.

resources and inform user through physical downlink control channel (PDCCH) where to look for its DL data. In case of errors in received packets, incremental redundancy hybrid automatic repeat request (IR-HARQ) process as depicted in Fig. 7–3 is utilized [7].

Based on LTE, LTE-M architecture is developed as a progressive set of 3GPP Releases 13–15. Release 12 was the first attempt in 3GPP to provide basic support for MTC IoT applications using LTE as the foundation [8,9]. Release 12 is introduced in the following to provide the context for the LTE-M releases.

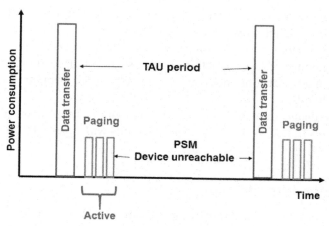

FIGURE 7–4 Power saving mode (PSM).

3GPP Release 12

Consistent with IoT requirements, the main focus in Release 12 for MTC support is to provide higher device power utilization efficiency and simpler and cheaper solution as compared to standard LTE. It defines a new Cat-0 device operation. These include single receive antenna, reduced soft buffer size used for HARQ packet combining, reduced peak data rate (1 Mb/s), and half-duplex (HD) operation.

For reducing power consumption in the device and to enable 10 + -year operation for the battery, the primary assist is power saving mode (PSM) [10]. In standard LTE operations, one way for the UE to conserve its battery is to turn itself off for short periods. However, when it needs to communicate with the network again, it needs to reattach to the network which consumes power. For the PSM feature instead, the UE does not turn itself off but stops checking for paging. It is still registered with the network and maintains connection configurations. During this period, the UE is not reachable. It remains in this hibernate or deep sleep state until a device-originated transaction occurs (see Fig. 7–4).

The extended tracking area updates period results in overall lower battery consumption.

Using single receive antenna and single receive chain operation, the multiple receiver chain for MIMO is eliminated. Reducing soft buffer size allows less overhead. Cat-0 operates up to standard 20 MHz LTE band to be compatible with LTE operations and supports reduced data rate. Lower data-rate operation reduces the complexity and cost for both processing power and memory significantly. HD approach in frequency division duplex (FDD) operation makes it possible to avoid the duplex filter.

7.3.1 Modifications for LTE-M in 3GPP Release 13

Release 13 is considered as the first standards release for both LTE-M and NB-IoT [1,11–13]. The enhancements for LTE-M and NB-IoT over LTE Release 12 are primarily in the areas of coverage extension and significant additional power savings [14]. LTE-M defines a Cat-M1 device operation for these requirements. The physical layer provides basic capabilities of extended coverage and higher power efficiency while at the same time it maintains compatibility with LTE. New media access control (MAC) and higher-layer procedures include mechanisms to augment the layer 1 battery power saving. LTE-M operates in the smallest LTE 1.4 MHz band in a standalone mode and it can also be provisioned as part of larger 3, 5, 10, or 20 MHz LTE bands.

The key LTE-M-related modifications are summarized in the following. Several of these are common between LTE-M and NB-IoT. Modifications which are LTE-M-specific are mentioned as well.

7.3.2 Features for extended coverage

Coverage enhancement (CE) is achieved using a combination of techniques including power boosting of data and reference signals, repetition/retransmission for both control and data channels, and relaxing performance requirements. In enhanced coverage mode for LTE-M, power spectral density boosting and additional repetition are used to reach devices in poor coverage. LTE-M also provides the option of reducing output power to 3 dB for lower implementation cost while still maintaining the coverage.

The typical measure for coverage is maximum coupling loss (MCL) [15]. It is the limiting value of the coupling loss at which a service can be delivered and therefore defines the coverage of the service. MCL is a good measure of the design as it is independent of frequency and environmental factors [15]. Without CE, legacy Release 12 or earlier LTE systems operate up to MCL of approximately 144 dB. In most cases for outdoor urban or suburban environments, the LTE network provides adequate signal strength to satisfy this MCL. However, indoor coverage is more difficult because of in-building environment where penetration loss can be very high. For example, if a device is underground or deep inside a building, the external wall penetration loss and in-building penetration loss can be significant.

Repetitive transmissions: Repeatedly sending the same data over a period of time or transmitting the same transport block multiple times in consecutive subframes using TTI bundling can significantly increase the probability for the receiver device to correctly decode transmitted messages. Repeating in time almost every channel beyond one subframe (1 millisecond) allows enough energy to accumulate for decoding purposes. It improves coverage at cell edge or in poor radio conditions. Also, in TTI bundling, instead of retransmitting the erroneous data with a new set of coded bits, the sender sends redundancy versions of the same set of bits in consecutive TTI based on the packet recovery technique of HARQ (see Fig. 7−5) [7].

This avoids delay and reduces control plane overhead at MAC layer. TTI bundling is utilized both in DL and UL.

Lower-order modulation: By utilizing quadrature phase-shift keying (QPSK) instead of 16 quadrature amplitude modulation (16 QAM), the probability of successful detecting bits increases. However, lower modulation implies lower spectral efficiency (bps/Hz). This results in lower bandwidth but extends the coverage.

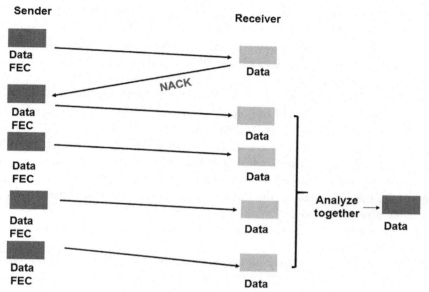

FIGURE 7–5 HARQ with TTI bundling.

Repetitive transmission and lower modulation rate allow LTE-M MCL to increase to 156 dB. This is 12 dB higher than the values for LTE Cat-1 in LTE Release 8 or Cat-0 in LTE Release 12. This results in roughly seven times more coverage.

LTE-M standard supports two CE modes: CE Mode A and CE Mode B. CE Mode A can use up to 32 repetitions whereas CE Mode B can utilize up to 2048 repetitions for the data channels. CE Mode A is the default mode of operation for LTE-M networks, providing efficient operation in coverage scenarios where moderate CE is needed with full mobility support. In this case, channel state information (CSI) feedback is supported. It is designed to maintain the LTE-M advantages of higher data rates, voice call possibility, and mobility support. CE Mode B, which extends the coverage area compared to Mode A, is an optional extension providing even further CE at the expense of throughput and latency. Limited/no mobility support and minimal CSI are available. It is mainly designed to provide coverage deep within buildings. Mode B is intended for applications with stationary or pedestrian speeds that require limited data rates and limited volumes of data over time.

7.3.3 Power saving and extended battery life

Power saving is a key LPWAN requirement to assist IoT devices to conserve battery power and potentially achieve and exceed 10-year battery life. In addition to the mechanism of PSM incorporated in Release 12 Cat-0 devices, power saving is facilitated by a variety of techniques in LTE-M including extended discontinuous reception (eDRX), modified attach procedure, use of control plane for data traffic, and connected mobility mode (CMM). The CE facilities of lower 1.4 MHz LTE channel and reduced modulation rates mentioned in the previous section also reduce complexity and battery power consumption.

eDRX: Currently, standard LTE devices use DRX introduced in Release 10 to extend battery life between recharges [10,16]. A DRX cycle consists of an On duration during which the device checks for paging and a DRX period during which the device is in sleep mode. A device can avoid monitoring the control channel for both sleep and connected modes, thus enabling further device power savings [10]. By momentarily switching off the receive section of the radio module for a fraction of a second, the device is able to save power. The device cannot be contacted by the network while it is not listening, but if the time period is kept small, the device will not experience a noticeable degradation of service. For example, when being accessed by the base station, the device might simply respond a fraction of a second later than if DRX was not enabled.

eDRX allows the time interval during which a device is not listening to the network to be greatly extended (see Fig. 7−6).

For many IoT applications, it may be quite acceptable for the device to not be reachable for a few seconds or longer. For LTE, DRX limit is 2.56 seconds. For LTE-M in sleep mode, the maximum possible DRX cycle length is extended to 43.69 minutes, while for connected mode, the maximum DRX cycle is extended up to 10.24 seconds.

eDRX can be used along with or without PSM for power savings. While not providing the same levels of power reduction as PSM, eDRX may provide a good compromise between device reachability and power consumption for some applications.

Modified attach process: When PSM or eDRX is enabled, the system may lose communication with the device for DL traffic. In LTE-M, a modified attach procedure is implemented, which allows the device to set up an ad hoc connection not scheduled by PSM or eDRX. This feature triggers a packet data network connection establishment via short message service mechanism [13].

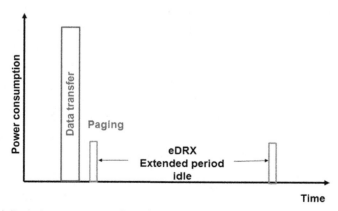

FIGURE 7–6 Extended discontinuous reception (eDRX).

The use of control plane to carry user data traffic is very effective for services that occasionally transmit reasonably small amounts of data. Utilization of control plane for user-traffic results in reduction of amount of signaling required and in reduction in the time needed for setting up data bearer, thus optimizing power consumption. This is particularly relevant for scenarios involving mobility.

CMM is the default mode used by LTE-M wherein the network performs cell reselection to retain the session during Handovers. Standard LTE deployments rely on idle mode mobility wherein the device performs the cell reselection. This is clearly applicable only in the case of nonstationary IoT devices.

7.3.4 Narrowband operation

An LTE-M device follows narrowband operation for the transmission and reception of physical channels and signals. The maximum channel bandwidth is reduced to 1.08 MHz, with predefined set of six contiguous RBs (see Fig. 7−7).

The six 180-kHz RBs along with the guard bands constitute 1.4 MHz, the smallest LTE band. To handle low-bandwidth applications from 10s of kbps to 1 Mbps, the 1.4 MHz LTE band is utilized for both control and data signaling. This implies 375 kbps upload and download speeds in HD and 1 Mbps speed in full-duplex (FD) modes. LTE-M devices follow the same cell search and random access procedures as used by legacy UE.

One implication of narrowband operation is limited frequency, spatial, and time diversity. This results in inability of managing effects of fading and outages. To address this, frequency hopping is introduced among different narrowbands using RF retuning [17]. This hopping is applied to different UL and DL physical channels when repetition is enabled. The utilization of 1.4 MHz channel is distinctively different from the NB-IoT utilization of 180 kHz.

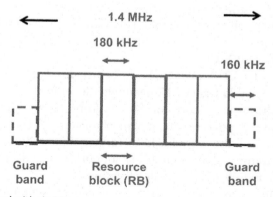

FIGURE 7–7 LTE 1.4 MHz band with six RBs.

7.3.5 Low cost and simplified operation

In addition to modifications associated with key LPWAN specific features like enhanced coverage, power reduction, and narrowband operation, there are several other areas 3GPP release 13 has been enhanced. These include reduction in complexity, simplified operations, and cost reduction. More effective and efficient resource utilization is enabled by simplifying the LTE channel structure via elimination of channels associated with non-MTC type applications. Cat-M1 operation also introduces new control and data channels that are more efficient for narrowband MTC operations.

MTC-oriented channel enhancements: For managing transmission of DL control information in wideband LTE, legacy PDCCH is able to use the first orthogonal frequency-division multiplexing (OFDM) symbols in a subframe. This implies that control and data are multiplexed in the time domain within the same subframe. A narrowband LTE-M device is not able to monitor these channels. As a result, this functionality is replaced by a new control channel called MTC physical downlink control channel (MPDCCH) [4,18]. It is similar to LTE PDCCH with additions of repetitions and frequency hopping support. MPDCCH is used for bandwidth-reduced operation and carries common and device-specific signaling. This new control channel spans up to six RBs in the frequency domain and one subframe in the time domain.

The LTE physical downlink shared channel and physical uplink shared channel are augmented with addition of repetitions and frequency hopping to create the enhanced MTC physical downlink shared channel and MTC physical uplink shared channel. These MTC-oriented channels along with corresponding control channels mentioned above support IoT MTC applications.

Handling legacy control region: In legacy LTE, the number of OFDM symbols which constitute the size of the control region is indicated in physical control format indicator channel (PCFICH) and can potentially change every subframe. In LTE-M, this information is semistatically signaled in the system information block. This eliminates the need for LTE-M devices to decode PCFICH.

HARQ feedback for UL transmissions: In legacy LTE, this information is contained in physical channel hybrid ARQ indicator channel (PHICH). PHICH is eliminated in LTE-M. Retransmissions are designed to be adaptive, asynchronous, and based on new scheduling assignment received in an MPDCCH.

The applicable UL and DL LTE-M channels are depicted in Fig. 7−8. Clearly, there is considerable simplicity compared to legacy LTE channels shown in Fig. 7−2.

Cost reduction: Significant cost reduction results from various modifications associated with narrowband operations, power reduction mechanisms, and simplification of physical layer. Additional features are also introduced to further reduce the cost of LTE-M devices. These include reduced transmission mode (TM) support, reduced number of blind decodings for control channel, and no simultaneous reception.

TMs in LTE pertain to differing approaches of diversity configurations, multiplexing, precoding, etc. in multiantenna situations [19]. In LTE, eight TMs are supported. For LTE-M,

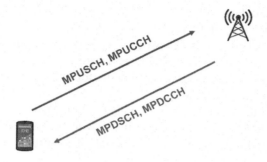

MPUSCH: MTC physical uplink shared channel; acknowledgement, data transfer
MPUCCH: MTC physical uplink control channel; acknowledgement
MPDSCH: MTC physical downlink shared channel; data transfer
MPDCCH: MTC physical uplink control channel; scheduling information,
          data transfer

FIGURE 7–8 LTE-M physical channels.

this is whittled down to only four—TM modes 1, 2, 6, and 9. Blind decoding pertains to decoding of a set of candidates to identify which transmissions are directed to the device itself. No simultaneous reception implies that a device is not required to decode unicast and broadcast data simultaneously.

More efficient signaling: New access control mechanisms such as extended access barring prevent devices from generating access requests when the network is congested, thus eliminating unnecessary signaling. The network can also utilize group-based paging and messaging to more efficiently communicate with multiple devices.

Enhanced resource management: LTE-M allows a large set of devices to share the same subscription. Resources and device management can be consolidated. For example, a group of water meters in a smart city can be collectively provisioned, controlled, and billed.

Emphasis on HD operation: The LTE-M standard supports FDD and time division duplex (TDD) operation for LTE-M deployments in both FD and HD modes in paired and unpaired bands, respectively. HD operation is the primary choice in most LPWAN applications. It is less complex and less costly. This allows the device to implement a simpler radio frequency switch instead of a full duplexer. However, HD operation results in a lower peak rate compared to devices that support FD operation.

7.3.6 Use of LTE priority structure for LTE-M applications

One of the distinctive features of LTE is the very versatile and robust priority assignment structure for user-traffic bearers. There are 13 priority levels ranging from 0.5 (highest) to 9 (lowest) supporting guaranteed bit–rate (GBR) and non-GBR (N-GBR) bearers [20]. The priority levels are characterized by varying combinations of packet delay and packet error rate (PER). Packet delays range from 50 to 300 milliseconds and a range of 10^{-2} to 10^{-6} is available for PER. The standards provide suggested mapping of various application classes to these priority levels but the network operators can apply their own discretion to assign the priority levels for their services. This is a strong and effective platform for managing the varying quality of service (QoS) requirements for LTE-M-based LPWAN services and provides a unique advantage to LTE-M.

Several classes of applications can be mapped for LTE-M usage. A representative mapping is suggested in Table 7–1. The best available PER of 10^{-6} is used since data integrity is very important for MTC type applications. This is paired with three options for packet delay —60, 100, and 300 milliseconds.

MTC1 are applications, which may be delay-tolerant, for example utility meters. MTC 3 type applications may be delay-sensitive, for example remote robotics surgery. MTC 2 can be applied to applications such as smart city which fall in between these two extreme attributes.

7.4 Optimizing long-term evolution core network

So far, the facilities provided by RAN have been discussed. The modified LTE EPC also plays a key role in supporting LTE-M based IoT applications.

The EPC is enhanced to provide efficient signaling and resource management to handle significantly larger amount of signaling and control traffic generated in MTC type applications. This is necessitated by the need to support massive number of connected devices in IoT applications. The MTC traffic puts very challenging requirements as compared to multimedia voice, data, and video broadband applications. Most MTC IoT devices have characteristic of small data and sporadic nature of data transfers. Hence capacity per se is not a major issue for LTE-M operations.

Another approach relates to consolidation of the various entities constituting the EPC. These include consolidating mobility management entity (MME), signaling gateway (SGW), and packet gateway (PGW) into a single core called EPC-lite. The choice of EPC-lite or EPC is driven by considerations of cost versus performance.

Table 7–1 Proposed LTE-M priority structure.

3GPP suggested service	Priority level	Est. packet delay (milliseconds)	Est. packet error rate	Suggested LTE-M application
Nonconversational video	5 (GBR)	300	10^{-6}	MTC 1
IMS signaling	1 (N-GBR)	100	10^{-6}	MTC 2
Mission-critical delay-sensitive signaling	0.5 (N-GBR)	60	10^{-6}	MTC 3

7.5 LTE-M release sequence 13 ≥ 14 ≥ 15

The primary goal in the release sequence 13 ≥ 14 ≥ 15 is to continue to improve battery life, message latency, and other aspects of performance. Both LTE-M and NB-IoT have continued to evolve in 3GPP Releases 14 and 15, delivering more capabilities and better efficiencies for massive IoT applications [21,22]. Facilities such as single-cell multicast, device positioning, and higher data rates are applicable to both LTE-M and NB-IoT. The modifications in areas of enhancing VoLTE are applicable only to LTE-M. The cell size extension feature is applied only to NB-IoT since extended coverage is of more relevance to NB-IoT.

Release 14 enhancements

Release 14 enhancements include increased data throughput, a new Cat-M2 device, multicast support, positioning enhancements, voice optimizations, improved mobility support, and reduction in power usage [21]. For Cat-M2 CE class A devices, the band is increased to 5 MHz whereas Cat-M2 CE class B retains 1.4 MHz band.

Release 14, also called *feMTC*, adds bands 25 and 40. Additional features to increase data rates and decrease latency include larger transport block size and more HARQ processes. Single-cell multicast is used for easy over-the-air firmware upgrades and device positioning for asset location tracking. Observed time difference of arrival (OTDA) enhancement enables capability of real-time location services for IoT use cases such as asset tracking or eCall. eCall is a feature which automatically calls emergency services without human intervention if the situation warrants it. Voice optimization for VoLTE enables devices such as wearables to more efficiently handle voice traffic in HD mode. Support for interfrequency measurements provides better handover execution. For power efficiency, the time the device spends in the connected mode is further reduced. Release assistance information capability is introduced, which allows the device to request that it be released from the connected state after it has completed all its communications.

Release 15 enhancements

Release 15 focuses on improving latency, spectral efficiency, and reduced power consumption [21].

Lower latency is achieved by reducing system acquisition time and supporting early data transmission. The spectral efficiency, and hence the system capacity, is improved in the DL by the introduction of higher-order modulation (64 QAM) and in the UL by the introduction of finer-granularity resource allocation.

To reduce power consumption, additional features such as wake-up signals, new synchronization signals, and improved HARQ feedback are introduced. In addition, a new low-power wake-up radio design, relaxed cell reselection monitoring, semipersistent scheduling, quicker radio resource control release, and lower transmit power classes (e.g., 14 dBm) are introduced.

Further enhancements are also made to load control with level-based access class barring.

TDD spectrum support is emphasized to enable further deployment flexibilities. Operation is extended to 200 km/h velocity.

The Releases 14 and 15 enhancements are optional and are fully backward compatible with Release 13.

7.6 LTE-M compatibility and migration

Because of it being an integral member of the 3GPP standards, LTE-M provides a seamless software upgrade opportunity from standard 4G LTE systems, has a complementary relationship with NB-IoT, and provides an efficient migration path to 5G network. Another dimension of LTE-M evolution is toward proprietary application in private networks.

7.6.1 Migration from LTEto LTE-M

Existing infrastructure for 4G LTE can be leveraged for LTE-M operation. Deployment of LTE-M can be done in-band within a normal LTE carrier and it can also operate in a shared spectrum with an existing LTE network. Existing antennas, radio, or other hardware can be reused without modifications.

As defined in Release 13, LTE-M supports bands 1, 2, 3, 4, 5, 7, 8, 11, 12, 13, 18, 19, 20, 26, 27, 28, 31, 39, and 41, which is a subset of the LTE bands [23]. An LTE-M device can coexist with other LTE UEs on the same eNodeB. A software upgrade to an existing eNodeB is needed to support Cat-M1 devices. The coexistence of LTE-M is facilitated by the fact that it can occupy 1.08 MHz range in any of the LTE frequency bands from 1.4 to 20 MHz. LTE network supporting Cat-M1 devices can utilize multiple narrowband regions with frequency retuning to enable scalable resource allocation and use frequency hopping for diversity across the entire LTE band [24,25]

7.6.2 Coexistence of LTE-M and NB-IoT

Both LTE-M and NB-IoT are optimized for comparatively low complexity and power, deep coverage, and high device density as compared to LTE. There is a good compatible and complementary relationship between LTE-M and NB-IoT since both are based on the LTE platform [26]. The primary difference is their operational bands. LTE-M operates at 1.4 MHz in-band with LTE, whereas NB-IoT can operate at 180 kHz in-band, in guard band, or in a standalone mode [27]. This allows NB-IoT to repurpose GSM frequency structure as well. Retuning of the antenna and using and provisioning the frequencies in the eNodeB can support both technologies.

Major areas in which LTE-M and NB-IoT differ include bandwidth support, latency, power consumption, and device cost. LTE-M has higher throughput with lower latency and battery use is optimized accordingly. The battery life of a Cat-M1 device can be lower than that of Cat-NB1 and power consumption is more at comparatively high data rates as opposed to NB-IoT which consumes less power at low data rates. LTE-M1 can also carry voice for applications such as residential security systems.

NB-IoT is designed for lower data rates, where comparatively larger delays are acceptable. NB-IoT has advantages of support for wider coverage, larger number of devices supported, cheaper devices, and flexibility of using in-band, in guard band, and standalone frequency bands. The typical cost for a Cat-NB1 device is roughly 30% lower than that for Cat-M1 device.

Cat-M1 devices can support HD FDD in addition to TDD, while cat-NB1 devices support only HD FDD. Overall, LTE-M can handle a superset of applications supported by NB-IoT.

Major similarities and differences between LTE-M and NB-IoT as well as with LTE Cat-1 and Cat-0 devices are summarized in Table 7—2 [26,28].

7.6.3 Migration from long-term evolution machine to 5G

5G is primarily targeted for IoT applications via use of mmWave frequency operation [29,30]. LTE-M has a natural relationship and migration path toward 5G network [23,31]. Whereas LTE-M focuses on continued evolution toward support for massive number of devices, 5G is targeted toward enhanced mobile broadband and mission-critical services. LTE-M meets the 5G battery life and message latency requirements for operations in the sub-6GHz frequency range. A smooth operator migration path to and coexistence for LTE-M with 5G new radio (NR) frequency bands are possible while preserving LTE-M deployments. 5G NR can accommodate LTE compatible attributes, for example frame structure. Alignment of 5G NR and LTE-M subcarrier grids is effected via 5G NR duplex frequency configuration. Also, a 5G NR device can rate match around radio resources taken by nondynamically scheduled LTE-M signals.

7.6.4 Private LTE-M networks

Currently, MulteFire 1.0 supports mobile broadband and high-performance IoT applications [32]. The MulteFire Alliance is adapting LTE-M and NB-IoT to operate in the unlicensed

Table 7–2   Device comparisons.

Attribute	LTE Cat-1 Rel 8	LTE Cat-0 Rel 12	LTE-M Cat-M1 Rel 13	NB-IoT Cat-NB1 Rel 13
Max. data rate	10 Mbps (DL) 5 Mbps (UL)	1 Mbps (DL/UL)	1 Mbps (DL/UL)	170 kbps (DL) 250 kbps (UL)
Carrier bandwidth	1.4, 3, 5, 10, 15, 20 MHz	20 MHz	1.4 MHz	180 kHz
Duplex mode	Full	Half	Full/Half	Half
Device transmit power	23 dBm	23 dBm	20/23 dBm	20/23 dBm
Maximum coupling loss	144 dB	144 dB	156 dB	164 dB
Est. modem complexity	100% (ref)	75%	25%	20%

spectrum and enable LPWAN use cases. This will in turn bring new opportunities for private LTE networks.

The roadmap includes application of LTE-M in both 400/800/900 MHz as well as in 1.9/2.4/3.5 GHz frequency bands [32]. The intent is also to have a smooth migration when MulteFire moves to the 5G platform.

7.7 LTE-M use cases

LTE-M has several attributes consistent with the QoS requirements, coverage options, and low-power requirements for a large number of LPWAN applications. Regarding attributes of coverage, battery life, and number of devices, all the major LPWAN options including LTE-M are comparatively similar. LTE-M has major advantages in significant number of use cases because of its comparatively high bandwidth (1 Mbps) and low latency (10–15 milliseconds). Roughly speaking, these attributes allow it to cover a significant number of use cases resulting in ability to handle a superset of use cases compared to other major LPWAN options. Also, like the NB-IoT option, it operates in licensed bands and has the advantage of standardized cellular LTE infrastructure. This provides strong security as compared to proprietary options such as long-range wide-area network (LoRaWAN) and Sigfox. The key advantage with respect to NB-IoT is mobility support. One issue for LTE-M is its somewhat higher estimated device cost of $10 with respect to other major LPWAN options which range from $3 to $7.

The suggested anchor points for LTE-M use cases are shown in Fig. 7–9.

LTE-M-based solution has high suitability for medical applications compared to proprietary solutions such as Sigfox and LoRaWAN. Major advantages are in areas of a wide set of attributes supported and the sensors can be mobile and wearable. In this section, two interrelated case studies relating to e-Health application based on LTE-M are summarized. The primary usage is in real-time monitoring, early diagnosis, preventive care, management of chronic diseases, and medical emergencies. Two representative scenarios are considered—one for basic and the other for advanced remote health monitoring and management. For basic health monitoring support, the need is to collect basic vital data such as pulse, respiratory rate, and body temperature. For advanced monitoring and immediate attention, additional vitals such as blood pressure, electrocardiograms, glucose level, and oxygen saturation are required. There are several architectural solutions available in the literature for such applications [33].

7.7.1 Basic remote health monitoring

For basic monitoring purposes, a health sensor with LTE-M interface capabilities can directly communicate with an LTE-M base station. Fig. 7–10 provides major components which come into play.

The data collected by wearables may in fact be in unformatted form with minimal processing in the wearable to keep the cost, size, and operational life manageable. Wearables are expected to provide mobility support. Wearables need to possess an LTE-M interface which sends the data to the health cloud.

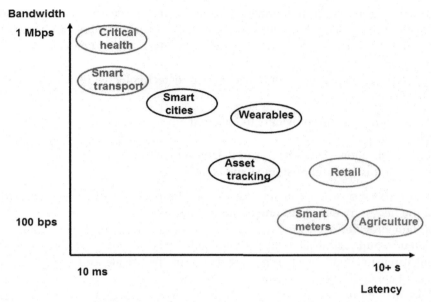

FIGURE 7–9 Illustrative LTE-M use cases.

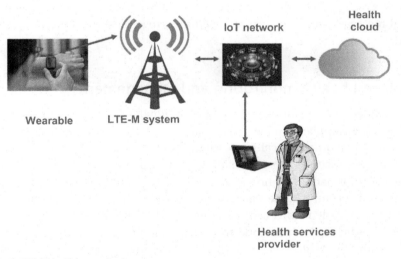

FIGURE 7–10 Basic LTE-M health monitoring.

The health cloud analyzes the data, provides comparative information, and creates the report for analysis and decision-making by the health service provider. Secure health cloud stores the medical information obtained from wearables for analysis purposes. Machine learning offers the potential of identifying trends in medical data, provide treatment plans and diagnostics, and give recommendations to health care professionals that are specific to individual patients. As such, cloud storage architecture should include the implementation of artificial intelligence-based machine learning on big data sets and be capable of extensive processing.

The primary services that can be provided by cloud technologies in health care environments are listed below.

- Applications to health care providers to enable them to work with health data or perform other relevant tasks;
- Tools for virtualization, networking, and database management; and
- The physical infrastructure for storage and servers.

The health service provider can securely access the data and the report from the health cloud, take the necessary decisions, and communicate to the patient.

The solution provides security, error-correcting capabilities, robustness against interference, low latency, and high availability. As with short-range communications, strong security is important to ensure that sensitive patient data remains private and cannot be altered or imitated. Low latency is again important in time-critical applications, such as emergencies, where delays in communication could have detrimental effects on patients. High-quality error-correcting capabilities and significant robustness against interference are essential to maintain the integrity and correctness of the received message. This is important in all health care applications, but particularly in emergency situations. Lastly, high availability is essential to ensure that messages will be delivered at all times, regardless of where the patient is physically located.

7.7.2 Advanced health monitoring and management

An alternate architecture provides consolidation of individual sensor data locally via a central node or a smartphone application (see Fig. 7–11).

For local consolidation and local processing purposes, a central node receives data from the sensor nodes. It processes this information, may implement some decision-making, and then forward the information to the cloud via the LTE-M network. Similar function as the one provided by the central node can also be provided by an application on an LTE-M-enabled smartphone. A dedicated central node is preferable to a smartphone as it can be a very low-power device and can be optimized for health-oriented functions. For sensors to communicate with the central node or the smartphone, a short-range communications method, for example Bluetooth or Zigbee can be utilized.

The rest of the entities and activities including the cloud facility are similar to the first case study discussed in Section 7.7.1.

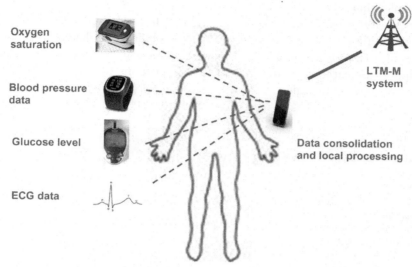

Oxygen
saturation

Blood pressure
data

Glucose level

ECG data

LTM-M
system

Data consolidation
and local processing

FIGURE 7–11 Local consolidation and processing.

7.8 Concluding remarks

LTE-M is derived from the 3GPP 4G LTE standard. It leverages the extensive LTE advantages in terms of efficient flat IP architecture, mobility, priority handling mechanisms, security, and a globally established base. It has strong ties with legacy 3GPP 2G and 3G networks as well as a comparatively seamless migration to the powerful 5G solution [31]. LTE-M can be efficiently utilized for a wide range of applications with varying range of bandwidths and latencies. It has considerable similarity to NB-IoT but has advantages in the terms of a range of bandwidth, latency, and mobility. LTE-M fits very well with standard 4G LTE and 5G networks in terms of compatibility and migration. Its disadvantages include complexity and cost for the end entities such as sensors and devices.

Simulation results on a specific configuration show that LTE-M meets and exceeds some of the targets specified in the 3GPP standards [34]. It needs to be emphasized that the performance targets and the measurements are highly sensitive to the parameters and facilities used. For example, for lower bandwidth operations and in the absence of mobility, the targets as well as measurements increase significantly. The reported simulation results pertain to low data rates of 1400 bps in DL and 250 bps in UL. For these conditions, the results show a coverage gain of 21 dB as compared to the corresponding 3GPP target of 18 dB with respect to LTE operations. This would be particularly relevant for applications that can tolerate low data rates and longer latencies.

An important feature of NB-IoT and LTE-M is that they share the same numerology as LTE. This allows spectrum to be shared between the two systems without causing mutual interference.

For IoT applications requiring higher data rates, low latency, full mobility, and voice in typical coverage situations, LTE-M is the best LPWAN technology choice [34]. And for IoT applications requiring deep coverage where latency, mobility, and data speed requirements are less stringent, LTE-M is a strong LPWAN contender as well. Overall, this versatility allows LTE-M to support an extremely wide array of IoT applications which helps to increase volume and drive economies of scale.

Acronyms

3GPP	3rd generation partnership project
CE	coverage enhancement
CMM	connected mobility mode
CSI	channel state information
EAB	extended access barring
eDRX	extended discontinuous reception
EPC	enhanced packet core
ESB	extended access barring
FD	full duplex
FDD	frequency division duplex
FDMA	frequency division multiple access
GBR	guaranteed bit rate
HD	half duplex
HARQ	hybrid automatic repeat request
IMM	idle mode mobility
IoT	internet of things
IP	internet protocol
LPWAN	low-power wide-area network
LTE	long-term evolution
LTE-M	LTE for machines
M2M	machine to machine
MAC	media access control
MCL	maximum coupling loss
MME	mobility management entity
MPDCCH	MTC physical downlink control channel
MPDSCH	MTC physical downlink shared channel
MPUCCH	MTC physical uplink control channel
MPUSCH	MTC physical uplink shared channel
MTC	machine-type communication
NB-IoT	narrowband IoT
OFDMA	orthogonal frequency-division multiple access
N-GBR	non-GBR
OTDA	observed time difference of arrival
PBCH	physical broadcast channel
PCFICH	physical control format indicator channel
PDCCH	physical downlink control channel
PDN	public data network
PDSCH	physical downlink shared channel

PGW	packet gateway
PHICH	physical channel hybrid ARQ indicator channel
PMCH	physical multicast channel
PRACH	physical random access channel
PSD	power spectral density
PSM	power saving mode
PUCCH	physical uplink control channel
PUSCH	physical uplink shared channel
QAM	quadrature amplitude modulation
QCI	QoS class identifier
QoS	quality of service
QPSK	quadrature phase-shift keying
RAN	radio access network
RB	resource block
RNC	radio network controller
RRC	radio resource control
SC-FDMA	single-carrier FDMA
SGW	signaling gateway
SINR	signal to interference plus noise ratio
SMS	short message service
TAU	tracking area updates
TBS	transport block size
TDD	time division duplex
TTI	transmission time interval
UE	user equipment
VoLTE	voice over LTE

References

[1] LTE-M deployment guide to basic feature set requirements, <https://www.gsma.com/newsroom/wp-content/uploads//CLP.29-v2.0.pdf>.

[2] D. Maidment, et al., LTE Cat-M—A Cellular Standard for IoT, ARM limited, 2016.

[3] 3GPP, UTRA-UTRAN long-term evolution (LTE) and 3GPP system architecture evolution (SAE), <ftp://ftp.3gpp.org/Inbox/2008_web_files/LTA_Paper.pdf>.

[4] LTE Access Network (E-UTRAN); 3GPP TS 36.300 version 14.2.0 Release 14, <https://www.etsi.org/deliver/etsi_ts/136300_136399/136300/14.02.00_60/ts_136300v140200p.pdf>.

[5] 3GPP, LTE, <https://www.3gpp.org/technologies/keywords-acronyms/98-lte>.

[6] 3GPP, LTE: physical channels and modulation, TS 36.211 evolved universal terrestrial radio access, (E-UTRA); Physical channels and modulation; <https://www.etsi.org/deliver/etsi_ts/136200_136299/136211/14.02.00_60/ts_136211v140200p.pdf>.

[7] C.M.M. Taher, M. Amer, Hybrid Automatic Repeat Request in LTE, Rochester Institute of Technology, Dubai, 2013. <https://pdfs.semanticscholar.org/6b24/754829400d44e637c138e9dd1da7d323cd1f.pdf>.

[8] Dino Flore, 3GPP RAN—Release 12 and beyond, <https://www.3gpp.org/news-events/1579-ran_rel12_and_beyond>.

[9] Y. Mehmood, Mobile M2M communication architectures, upcoming challenges, applications, and future directions, EURASIP J. Wirel. Commun. Netw. (2015).

[10] 3GPP, TS 24.301, Non-access protocol for evolved packet system, 2015, <https://www.arib.or.jp/english/html/overview/doc/STD-T63v11_00/5_Appendix/Rel12/24/24301-c80.pdf>.

[11] 3GPP, Release 13 analytical view version September 9th 2015; RP-151569 (replace by TR), <https://www.3gpp.org/ftp/Information/WORK_PLAN/Description_Releases/>.

[12] A. Rico-Alvarino, M. Vajapeyam, H. Xu, X. Wang, Y. Blankenship, J. Bergman, et al., An overview of 3GPP enhancements on machine to machine communications, IEEE Commun. Mag. 54 (6) (2016) 14−21.

[13] GSMA, NB-IoT deployment guide to basic feature set requirements, 2017, <www.gsma.com/>.

[14] Nokia, LTE evolution for IoT connectivity, 2017, <https://onestore.nokia.com/asset/200178>.

[15] Altair, Coverage analysis of LTE-M category-M1, Jan 2017, <https://altair-semi.com/wp-content/uploads/2017/02/Coverage-Analysis-of-LTE-CAT-M1-White-Paper.pdf>.

[16] 3GPP, eUTRAN—UE procedures in idle mode, TS 36.304 version 10.3.0 Release 10, <https://www.etsi.org/deliver/etsi_ts/136300_136399/136304/10.03.00_60/ts_136304v100300p.pdf>.

[17] PUSCH Frequency Hopping, <https://www.sharetechnote.com/html/Handbook_LTE_FrequencyHopping_PUSCH.html>.

[18] LTE-BL/CE (LTE-M), <https://www.sharetechnote.com/html/Handbook_LTE_BL_CE.html>.

[19] J. Wannstrom, LTE-advanced, <https://www.3gpp.org/technologies/keywords-acronyms/97-lte-advanced>.

[20] 3GPP, Policy and charging control architecture, ETSI TS 123 203, <https://www.etsi.org/deliver/etsi_ts/123200_123299/123203/13.07.00_60/ts_123203v130700p.pdf>.

[21] 3GPP, Release 14 description; TR 21.914, <https://www.3gpp.org/DynaReport/21-series.htm>.

[22] 3GPP, Release 15 description; TR 21.915, <https://www.3gpp.org/DynaReport/21-series.htm>.

[23] User equipment (UE) radio access capabilities, <https://www.3gpp.org/ftp/Specs/archive/36_series/36.306/>.

[24] Qualcomm, Leading the LTE IoT evolution to connect the massive Internet of things, July 2017, <https://www.qualcomm.com/media/documents/files/whitepaper-leading-the-lte-iot-evolution-to-connect-the-massive-internet-of-things.pdf>.

[25] Mathworks, LTE-M downlink waveform generation, <https://www.mathworks.com/help/lte/examples/ltem-downlink-waveform-generation.html>.

[26] M. Soussi, et al., Evaluating the performance of eMTC and NB-IoT for smart city applications, Conference paper, arXiv:1711.07268v1 [cs.IT], 20 Nov 2017.

[27] P. Reininge, 3GPP standards for the Internet of things, <https://www.3gpp.org/Information/presentations/presentations_2016/2016_11_3gpp_Standards_for_IoT.pdf>.

[28] S. Chalapati, Comparison of LPWA technologies and realizable use cases, SCTE-ISBE-NCTA-CableLabs Fall 2018 Technical Forum, <https://www.nctatechnicalpapers.com/Paper/2018>.

[29] Release 15, <https://www.3gpp.org/release-15>.

[30] G.A. Akpakwu, G.P. Hancke, A survey on 5 generation networks for the Internet of things: communication technologies and challenges, <https://www.researchgate.net/publication/321513584_A_Survey_on_5G_Networks_for_the_Internet_of_Things_Communication_Technologies_and_Challenges>.

[31] GSMA, Mobile IoT in the 5G future—NB-IoT and LTE-M in the context of 5G, April 2018, <www.gsma.com/MobileIoT>.

[32] MultiFire, LTE-based technology for operating in unlicensed & shared spectrum, <https://www.multefire.org/>.

[33] S.B. Baker, W. Xiang, I. Atkinson, Internet of things for smart healthcare: technologies, challenges, and opportunities, IEEE Access 5 (2017). <https://ieeexplore.ieee.org/document/8124196>.

[34] Altair, Evolution of LTE-M towards 5G IoT requirements, <https://altair-semi.com/wp-content/uploads/2018/03/LTE-M-Performance-Towards-5G-White-Paper-V1.1-Final.pdf>.

8

TV white spaces for low-power wide-area networks

Anjali Askhedkar[1], Bharat Chaudhari[1], Marco Zennaro[2], Ermanno Pietrosemoli[2]

[1]SCHOOL OF ELECTRONICS AND COMMUNICATION ENGINEERING, MIT WORLD PEACE UNIVERSITY, PUNE, INDIA [2]T/ICT4D LABORATORY, THE ABDUS SALAM INTERNATIONAL CENTRE FOR THEORETICAL PHYSICS, TRIESTE, ITALY

8.1 Introduction

With the growth in wireless services and higher data rate applications, the demand for spectrum is also increasing. Spectrum is a limited resource and static frequency allocation schemes cannot accommodate this ever increasing demand. Studies suggest that the licensed spectrum is underutilized. It can be made available by what is termed as opportunistic usage of the frequency bands. This novel spectrum management technique is aptly supported by cognitive radio (CR) technology [1]. The primary users are the licensed users for a specific part of the spectrum in a certain geographical area. Secondary users are unlicensed users that make use of the spectrum without causing harmful interference to the primary users. CR senses the spectrum and finds the band of frequencies void of information carrying signals at a given time and space in the spectrum. This spectrum is called the white space and the devices that use this white space are called white space devices (WSDs). The spectrum allocated to various licensed services such as TV broadcasting and even cellular is underutilized for a significant amount of time, especially in sparsely populated areas. The free spectrum in the TV bands constitutes the TV white spaces (TVWS) as shown in Fig. 8−1.

TVWS have excellent propagation characteristics and have attracted worldwide attention for wireless communications. Various countries such as United States, United Kingdom, Canada, Singapore, and Mozambique have already formulated regulations for the usage of TVWS for wireless communications (fixed and mobile), and many others such as Japan and Hong Kong are actively considering to do the same.

According to the CEPT (European Conference of Postal and Telecommunications Administrations) Report 24 from 2008, "White Space" is a label indicating a part of the spectrum, which is available for a radio communication application (service, system) at a given time in a given geographical area on a noninterfering/nonprotected basis with regard to other services with a higher priority on a national basis. The TVWS gets its name from the

LPWAN Technologies for IoT and M2M Applications. DOI: https://doi.org/10.1016/B978-0-12-818880-4.00009-0

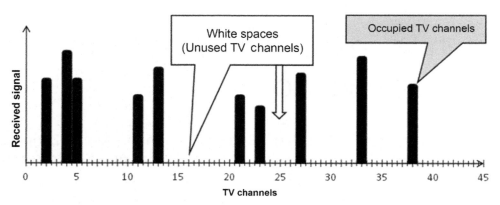

FIGURE 8–1 White spaces in the TV spectrum band.

fact that it falls in the spectrum range of 470–890 MHz (international telecommunication union-region 3 frequency allocations) which encompasses the band allocated mainly for TV broadcasting as the primary service. The amount of available TVWS spectrum varies across different geographical locations depending on various factors such as television channels utilization, actual TV coverage, level of tolerable interference to the primary TV broadcasting service, and geographical features. The TVWS availability mainly depends on:

- *Frequency*: Some channels in the TV band plan are purposely left unused in some locations such as guard bands, to avoid interference.
- *Height of TV transmitter and receiver antennas*: An indoor antenna might not be able to detect a distant TV signal and presume the existence of an unoccupied channel, when a roof antenna at the same location would reveal a busy one. This situation is heavily dependent on the TV broadcasting antenna height.
- *Space*: Some geographical areas may be outside the current TV coverage. Also, some geographical separation areas may be already present as planned between locations using the same TV channels.
- *Time*: A licensed broadcasting transmitter may not use the assigned frequency channel during a specific period of time; thus TVWS becomes available for use on noninterference basis [2].

Benefits of TVWS
The use of TVWS for wireless applications brings with itself a number of benefits as below:

- *Long range*: TV transmission frequencies being in the very high frequency and lower UHF range have excellent propagation characteristics over long distances. Therefore, TVWS are suitable for applications that require long transmission range. The lower frequency ranges of TVWS facilitate obstacle penetration, which also make them less vulnerable to multipath and fading, thus enabling applications such as connected home and consumer applications.
- *Bandwidth*: TVWS offer different bandwidth choices for applications with large bandwidth and high bit-rate applications. Multiple adjacent channels in TVWS can be used concurrently for bandwidth hungry applications [3].

Table 8–1 Comparison of TVWS with other wireless technologies.

Feature	TVWS	Cellular	Wi-Fi	ZigBee	LoRa
Spectrum	Licensed	Licensed	Unlicensed	Unlicensed	Unlicensed
Coverage	Large	Large	Small	Small	Large
Power Consumption	Medium	High	High	Low	Low
Data Rate/Bandwidth	Medium	High	High	Low	Low
Latency	Low	High	Low	Medium	High
Cost	Low	Recurrent	Low	Low	Low

- *Availability of TV spectrum*: Ample TVWS may be widely available as compared to industrial, scientific, and medical (ISM) bands, more so in the rural and suburban areas where there are fewer TV stations, as compared to urban areas. The flexible nature of TVWS technology means that more spectrum can be taken advantage of in the least serviced areas.
- *Low-risk regulation*: Because TVWS is a secondary spectrum–use technology, there is no need to reallocate spectrum in order to regulate its use. It does not commit the regulator to giving away a spectrum band for years to come. Whether TVWS succeeds is a risk for the market, not for the regulator.
- *A great rural technology*: While mobile technology has been an access boon for the developing world, mobile operators still struggle to deploy access in rural areas where low incomes and sparse populations do not make it a viable economic plan for the establishment and maintenance of mobile base stations. TVWS has specific advantages that make it well suited to being a complementary access technology. First, TVWS use of the ultra-high frequency (UHF) spectrum band offers better propagation characteristics than other technologies higher up in the spectrum band. This means that individual base stations can reach further, thereby lowering the total number of base stations required for a given areas. Second, UHF spectrum doesn't require strict line-of-sight between radios. This will also lower the cost of deployment, thereby reducing the need for high towers and more complex network design [4].

Table 8–1 gives a comparative analysis of TVWS with other wireless technologies emphasizing its advantages.

8.2 Architecture

8.2.1 Identification of TV white spaces

For TVWS to be utilized according to the CR concept, the available spectrum needs to be identified. There are three possible approaches to determine the TVWS, either used individually or in a combination [2].

The first approach is based on centrally maintaining geolocation database. It is currently the most common model to identify available TV spectrum based on frequency allocation and network planning information provided by TV operators. The secondary user/device needs to identify its location by means of a GPS or any other accurate geolocation determination technique. Then it sends geolocation information to a central geolocation database, using an alternate communication channel. In turn, the device receives information from the database about free channels are available at its location, if any and at which power level, it is allowed to transmit. Only after that, the device can start its transmission on one of the assigned channels. This requires the availability of a communication channel for the database query. Since it does not make sense to provide this capability in the user equipment (need to have another radio channel devoted to the database query), this approach is applied only at the TVWS base station. Normally the base station is connected to the Internet by fiber optic, wired Ethernet, or an alternative wireless technology such as Wi-Fi, cellular, or satellite. The base station uses this access for the database query and after being authorized, it can start communicating with the user devices within its range using the assigned TVWS frequencies.

The second approach is based on the use of pilot (beacon) channel. A dedicated channel is used to broadcast the information about the current spectrum usage and free channels available. However, it is difficult to find a common worldwide (or region-wide) frequency allocation for such special pilot channels due to different frequency allocations in different countries. Additionally, interference between transmitted beacons of the different regions has to be mitigated. Such a solution has a limited success, since allotting a separate channel is an additional overhead and a complex task, whereas the other options, especially geolocation databases, seem to be more promising [5].

The third approach for TVWS identification is based on spectrum sensing. It requires the secondary user to sense its environment to determine the available TVWS channels at its location. The device conducts measurements on a desired channel to determine the presence of any primary licensed user and also on adjacent channels to decide transmission power restraints. Energy detection, matched filter detection, and cyclostationary feature detection are some of the commonly used spectrum sensing methods. Spectrum sensing does not depend on connection to a local database or beacon and it is an added advantage in far-flung or rural areas. Spectrum sensing can be implemented in the user devices and in the base station. In fact, the reliability of this approach can be further enhanced by performing the spectrum sensing simultaneously with the help of number of end-user devices, which in turn can convey their results to the base station. In such a scenario, the base stations can perform a better assessment of spectrum availability.

Geolocation database or spectrum sensing methods alone may not be enough to make an accurate estimation. A hybrid method that involves combined spectrum sensing and database can be employed with rewarding outcomes [6].

8.2.2 Architecture based on geolocation database

The geolocation database approach for opportunistic spectrum access is a currently viable option for TVWS technology and has been adopted by both FCC and European Telecommunications Standards Institute (ESTI). The technical requirements of the TV white space device (TVWSD) are specified in ETSI EN 301 598 for WSDs, wireless access systems operating in 470−790 MHz band [2].

TVWSD is controlled by TV white space database (TVWSDB) and operates in TVWS. ETSI EN defines two types of TVWSDs, namely Equipment Type A with integral, dedicated, or external antennas and is intended for fixed use only; and Equipment Type B with an integral or a dedicated antenna and is not intended for fixed use. The devices are also categorizes as master TVWSD which has geolocation capability and communicates with a TVWSDB for operation in the TVWS, whereas the slave TVWSD does not have geolocation capability and communicates with the master and other slave devices while in operation.

As per the ETSI EN 301 598, TVWSDB is the database system approved by the relevant national regulatory authority, which can communicate with TVWSDs and provide information on TVWS availability. The TVWSDB formulates and updates a set of operational parameters such as channel frequency, bandwidth, time validity, and operational power and communicates them to the master TVWSD. The TVWSD master passes these parameters to its slave TVWSDs. The slaves configure themselves according to received parameters before start operating in the TVWS frequencies to avoid any kind of interference to the primary service.

Fig. 8−2 shows a typical architecture for the deployment of TVWS networks. The TVWSD communicates with a data repository to get the listing of the certified TVWSDBs. In this case, the TVWSD uses a web-listing hosted on a server to get the information about the TVWSDBs and also how to communicate with them. The TVWSDB stores information relating to the terrestrial frequencies in a geographical area, interference management tools, etc. Before transmitting in the TVWS band, the TVWSDs communicate with the TVWSDB and declare

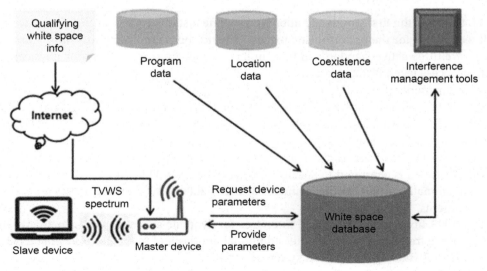

FIGURE 8–2 Architecture of geolocation database-based TVWS network.

their location parameters to the TVWSDB, in accordance with which the TVWSDB allots a certain set of operational parameters to the TVWSD for commencing operation. The slave devices get their operational parameters through the master device.

8.3 TV white spaces regulations and standards

Most countries have their own set of regulations for the use of TVWS either as licensed or unlicensed operation. Countries such as United States, United Kingdom, Canada, Singapore, and Mozambique have already formulated regulations for the usage of TVWS for wireless communications (fixed and mobile), and many others such as Japan and Hong Kong are actively considering to do the same. Licensed access to TVWS is mainly enabled for wireless broadband and mobile data services in the United States [2−4]. Unlicensed use of TVWS is promising for the developing nations. For unlicensed access, the federal communications commission (FCC) classifies devices as fixed or portable and these devices have to conform to the specified regulations for unlicensed operation. The FCC also allows accessing TVWS with hybrid licensing models. Many international organizations work on creating a wide variety of standards and specifications for the use of TV band white spaces. Some of them are described below:

IEEE 802 LAN/MAN Standards Committee produces standards for wireless networking devices, including wireless local area networks (WLAN), wireless personal area networks (WPAN), wireless metropolitan area networks (Wireless MAN), and wireless regional area networks (WRAN). Some of the IEEE 802 standards that are relevant to TV broadcast bands include IEEE 802.11af (WLAN), IEEE 802.15.4m (WPAN), IEEE Std. 802.22-2011 (WRAN), and IEEE 802.19.1 (coexistence).

- IEEE 802.22 is the first worldwide attempt to define a standardized air interface based on CR techniques for the opportunistic use of TV bands on a noninterfering basis. Currently, there are three active standards in this family officially approved: 802.22 for physical and media access layer, 802.22.1 for interference protection, and 802.22.2 covering installation and deployment.
- IEEE 802.11af is a standard that adapts 802.11 for TV band operation so as to leverage the success of Wi-Fi, while addressing the issues derived from extended range and unlicensed spectrum congestion, implementing wireless broadband networks in the bandwidth allocated to TV broadcasts stations, and has been called superWiFi and also White-Fi.
- IEEE 802.15.4m is chartered to specify a physical layer for 802.15.4 and to enhance and add functionality to the existing standard 802.15.4-2006 MAC, meeting TVWS regulatory requirements.
- IEEE P1900.4a for architecture and interfaces for dynamic spectrum access networks in white-space frequency bands defines additional entities and interfaces to enable efficient operation of white-space wireless systems.
- IEEE P1900.7 is a result of the Dynamic Spectrum Access Networks (DySPAN) Standards Committee Working Group, the successor of IEEE P1900 Standards Committee, with a

focus on improved spectrum usage addressing the radio interface for white space dynamic spectrum access radio systems supporting fixed and mobile operation.

Internet Engineering Task Force (IETF) protocol to access white spaces (PAWS) is a standard for interfaces and access to the white space database by a WSD operating in the television broadcast band.

White-Space Alliance has created interoperability tests and certification procedures for the IEEE 802.22 systems, which is called Wi-FAR™. It has also created interoperability tests and certification procedures for the IETF PAWS protocol, which is called as the WSAConnect Specification.

ECMA-392: ECMA International is the successor of the European Computer Manufacturers Association. ECMA-392 is a standard that specifies a physical layer and a medium access sublayer for wireless devices that operate in TV frequency bands. It is mainly directed at personal and portable wireless devices.

European Telecommunications Standards Institute - European Standard, Telecommunications Series (ETSI EN) 301 598: This European standard for WSDs and wireless access systems operating in the 470−790 MHz frequency band, applies to TVWS devices controlled by a geo-location database.

8.4 TV white spaces protocols and technologies

Many protocols have been developed for the use of TVWS. Some of the important protocols are discussed in this section.

8.4.1 TV white spaces identification protocols

These protocols are mainly designed for determining the availability of TVWS channels using the spectrum sensing and geolocation database hybrid methods. For outdoor TVWS detection, important protocols are Waldo, SenseLess, and V-Scope. Whereas for indoor TVWS detection, WISER or FIWEX can be used.

White Space Local DetectOr system (Waldo): Waldo [7] is a system that enables low-cost, local white space detection by taking advantages of the signal features along with location to model white space availability using low-cost sensors. Waldo can be deployed on an Android phone to efficiently detect white spaces without draining the phones resources.

SenseLess: SenseLess [8] is another approach toward building a white spaces network, where WSDs mainly rely on a combination of an up-to-date database, sophisticated signal propagation modeling, and an efficient content distribution mechanism to ensure efficient, scalable, and safe white space network operation.

V-Scope: It is a vehicular sensing framework aimed to collect wide area spectrum measurements for evaluating the accuracy of spectrum occupancy databases [9]. A V-Scope utilizes spectrum sensors mounted on public vehicles for collecting and reporting measurements from the road (opportunistic wardriving).

Wideband Software Extensible Radio (WiSER): WiSER [10] is a wideband open-source SDR platform supporting new experimental research in the fields of dynamic spectrum and

CR networking. The main features are hardware virtualization capable of supporting multiple radios and an open-source software toolkit. WiSER consists of three modules: indoor positioning system, white space database, and real-time sensing module. Indoor positioning system is used to determine users' locations. White space database registers the indoor white space availability. Real-time sensing module collects real-time signal strengths of all TV channels at different locations and reports the results to white space database. WiSER takes users' locations as the inputs and outputs the indoor white space availability at the given indoor locations.

FIWEX (cost-eFficient Indoor White space Exploration): FIWEX [11] is a novel cost-efficient indoor white space exploration method that detects white space exploiting the location dependence and channel dependence of TV channels' signal strength in indoor environments. UHF TV channels in a building are measured and the temporal and spatial features of indoor white spaces are extracted to design an FIWEX mechanism.

8.4.2 TV white spaces network protocols

TVWS differ from the conventional Wi-Fi spectrum in three facets: spectrum fragmentation, spatial variation, and temporal variation. These differences make the network design over TVWS challenging and fundamentally different from Wi-Fi networks [12]. A few examples of protocols that make use of TVWS for a networking plan are Cognitive Radio System (CR-S), Cognitive Network over White Spaces (Kognitiv Networking over White Spaces, KNOWS), White Spaces Indoor Network (WINET) and Sensor Network over White Spaces (SNOW). WhiteFi (White space networking with Wi-Fi-like connectivity) is built on the KNOWS platform and provides wireless broadband access. WhiteNet is similar to WhiteFi but provides larger coverage. TVWS has also been used to provide rural broadband connectivity in developing nations [3].

KNOWS: It is a hardware-software platform that detects TVWS through collaborative sensing. KNOWS includes a spectrum-aware medium access control protocol and algorithm to deal with spectrum fragmentation. It enables dynamic spectrum access and sharing of white spaces by adaptively allocating the spectrum among contending users. Instead of the conventional, static channelization approach, it employs a distributed scheme that dynamically adjusts the operating frequency, the occupancy time, and bandwidth, based on the instantaneously available white spaces, the contention intensity, and the user demand. If there are few users in the system, KNOWS provides each user with a larger chunk of the bandwidth and provides smaller portions to all users if there are more competing nodes [13].

WINET: WINET is a design framework for indoor multi-access-point white space network. Access point placement, its association, and spectrum allocation is optimized. Hence, the spectrum fragmentation, spatial variation, and temporal variation of TVWS are handled effectively with carrier sense multiple access/collision detection based medium access control. WINET uses WiSER to obtain and track the indoor white space availability [12].

SNOW: Wireless sensor networks (WSNs) in wide-area wireless monitoring and control systems require numerous sensors to be connected over long distances. Existing WSN technologies with short range, such as those based on IEEE 802.15.4, form many-hop mesh networks that complicate the network design. SNOW [14] is a scalable sensor network architecture that takes advantage of the TVWS. Many WSN applications need low data rate, low power operation, and scalability in terms of geographic areas and the number of nodes. Scalability and energy efficiency are achieved by splitting channels into narrowband orthogonal subcarriers and enabling packet receptions on the subcarriers in parallel with a single radio. SNOW employs a distributed implementation of orthogonal frequency division multiplexing at the physical layer that enables distinct orthogonal signals from distributed nodes, thus reducing the probability of packets collisions normally associated with the longer range of WSDs as compared with Wi-Fi. Its media access control protocol handles subcarrier allocation among the nodes and transmission scheduling.

8.5 TV white spaces for low-power wide-area network

Low-power wide-area network (LPWAN) is a broad term incorporating various implementations and protocols, both proprietary and open-source that share common characteristics as the name suggests:

- Low power: Operates on small, inexpensive batteries for years.
- Wide area: Has an operating range that is typically more than 2 km in urban settings.

One of the reasons for low power and wide range is transmission at very low data rate. Typical LPWAN technologies can only send less than 1000 bytes of data per day. LPWAN technology works well in situations where devices need to send very short messages over a wide area while maintaining battery life over many years. Most of the existing LPWAN protocols such as LoRa, Sigfox, random phase multiple access, narrowband-Internet of things (NB-IoT), and others use the unlicensed ISM band for communication. With the escalating number of IoT devices, ISM band too is getting crowded. TVWS bands with good coverage are the promising alternatives for LPWAN and a lot of research is focused in this field. Weightless is such a protocol designed for Machine-to-Machine (M2M) applications in white space [15].

Weightless [16] is a LPWAN wireless communication protocol designed to connect smart machines to the Internet, the so-called M2M communications, over distances ranging from a few meters to about 10 km. Originally, there were three published Weightless connectivity standards: Weightless-P, Weightless-N, and Weightless-W, out of which Weightless-W was designed to operate in the TVWS. Patents would only be licensed to those qualifying devices conforming to the defined standards by the Weightless Special Interest Group. Thus, the protocol, whilst open, may be regarded as proprietary. The base station uses a database query approach to determine free TV channels in the area. Each base station can communicate at varying speeds with nearby or far away terminals using a variety of modulation and encoding techniques. Weightless achieves long range with low power by spreading the transmitted signal and sophisticated modulation techniques such as quadrature amplitude modulation. The use of the white space spectrum does not provide guaranteed spectrum to allow for uplink and downlink pairing, so time division duplex operation is essentially practiced. For good interference tolerance, frequency hopping is employed. Weightless-W is the best for use in the smart oil and gas sector, because of abundant available TVWS [4,17].

8.6 Applications

TVWS bands with good coverage capacity can support a wide range of applications such as sensor networks, telehealth, home networking, private networks, wireless video surveillance, extended range Wi-Fi, and as backhaul [2].

Rural broadband: Remote and rural areas can benefit to a great extent by the deployment of TVWS-based broadband connectivity because of long-range characteristics of TV bands, greater availability of TVWS in these areas and reduced number of repeaters as compared with other technologies. Fig. 8–3 illustrates two main types of deployment scenarios. One depicts the master-slave kind of communication between the WSDs, WSD Master, and its slaves, a point to multipoint topology. The slave WSD may be directly incorporated in the user equipment, or it can be fitted with Wi-Fi or any other communication technology

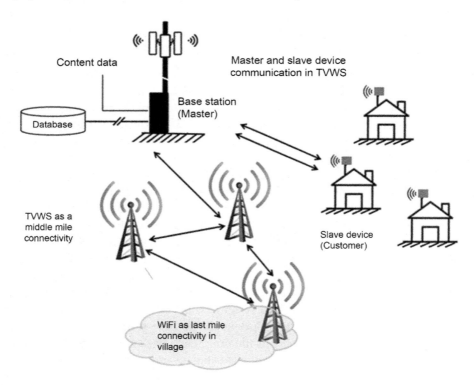

FIGURE 8–3 Rural broadband implementation using TVWS.

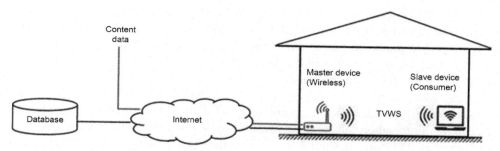

FIGURE 8–4 In-building solution using TVWS.

attuned to the user equipment. The other scenario portrays the TV spectrum being used to provide middle mile connectivity to the base stations and repeaters to reach the distant rural areas, which are then served by Wi-Fi in the last mile. Other deployment scenarios applicable to the rural areas can also be envisioned.

Hotspot coverage/in-building solutions for multimedia and broadband: TVWS due to their intrinsic ability to penetrate deep inside buildings seem useful for providing hotspot coverage and in-building solutions both for broadband applications and for NB-IoT applications. TVWSDs operating in the UHF band have greater coverage as compared to the devices in the ISM bands and hence a TVWS device can serve a larger coverage area than one in the ISM band. Fig. 8–4 below shows a typical organization for in-building solutions in TVWS.

Smart grids: TVWS can be used to provide middle mile connectivity to facilitate smart metering and other smart grid solutions. The communication between the control center and the smart meter equipment at the customer premises can be done using TVWS. Fewer TVWS channels are needed as these applications need less bandwidth.

M2M communication, sensor networks, and smart cities: M2M communication consists of very low-power radio transmitters used for low-data rate industrial and commercial applications such as monitoring, tracking, metering, and control, leading toward "smart machines" and "smart cities." M2M solutions have been mostly developed using a number of proprietary technologies but also with cellular technologies and solutions such as LoRa and SigFox. TVWS bands with their higher coverage ranges are apt for these use cases. Weightless is a specification for low-rate M2M devices using the white spaces in TV bands under a noninterference basis.

8.7 Challenges and opportunities

There are certain challenges to the widespread use of TVWS technology such as licensing, standardization, geolocation database updating, TVWS fragmentation and availability, interference and coexistence, security, and mobility.

Several countries have opened up the licensed TV bands for other users. These can be utilized via dynamic spectrum access by secondary users. Regulations for secondary usage of underutilized TV spectrum need to be in place for large-scale TVWS proliferation. Many standards have come up in the context of TVWS communication systems and some are still being developed. National regulatory authorities and TV operators need to share relevant information freely for the building of the geolocation database. It is essential that this database is regularly updated with the necessary parameters before any secondary user utilizes the spectrum which is an additional burden. TVWS availability is higher in suburban and rural areas than in urban areas. A combination of TVWS and other wireless technology options can be put together and applications need to be implemented accordingly [3].

The secondary TVWS users should not cause interference to the primary licensed users. For this, they need to operate at much lower powers which may not support long range. Research in this domain and development of protocols then becomes vital. Coexistence is an issue with homogeneous as well as heterogeneous networks operating in the TVWS that needs to be tackled with the growing application scenarios of TVWS in IoT and smart cities. TVWS are frequently discontinuous in the frequency spectrum, which poses a challenge for seamless connectivity and mobility. Protocols that can handle such fragmented spectrum need to be developed for bandwidth-intensive applications. Hardware limitations for fragmented spectrum bands in terms of antenna design also impact the scalability of such TVWS networks, which should be able to accommodate a much wider frequency range relatively to the carrier frequency as compared to Wi-Fi.

TVWS in the lower frequency range require larger antennas. For certain applications such as WSNs, such large antennas are impractical. Research on design of smaller devices and antenna is a must for facilitating TVWS networks. Short-range technologies often neglect to implement security, relying on the inherent protection offered by the vanishing of the signal beyond a certain range. Nevertheless, these devices can be used to attack other networks and must therefore be properly protected. As a variety of M2M, IoT, and long-range applications are envisaged with TVWS technology, the development of security protocols becomes important. TVWS availability is location- and time-dependent. Therefore, a lot of work needs to be done to enable TVWS in application areas such as vehicular communication where mobility is indispensable.

TVWS have been explored for several applications including wireless broadband Internet access. There is a huge scope for the use of TVWS for many applications including remote sensing, monitoring and control, asset tracking, IoT, smart utility networks, smart transportation, smart factories, smart agriculture, smart cities, and so on. It can support a variety of applications such as smart metering, vehicle tracking, smart cars—vehicle diagnostics and upgrades, health monitoring, traffic sensors, smart appliances, smart e-payment infrastructure, industrial machine monitoring, and many more.

TVWS is undoubtedly an attractive alternative to the growing spectrum demand. It uses UHF bands with better propagation characteristics, allowing large coverage and deeper in-building penetration. It is particularly well suited to provide Internet access in sparsely populated areas, especially in developing countries, where there are a plenty of unused TV channels. It is also emerging as a facilitator for smart cities and indoor applications in developed countries. Unfortunately, the current high prices of available equipment have prevented its wide-scale adoption, and the predicted mass adoption has not yet materialized.

References

[1] J. Mitola, G.Q. Maguire, Cognitive radio: making software radios more personal, IEEE Personal Communications 6 (4) (1999) 13–18.

[2] Broadband deployment through TV-White Space.

[3] <http://tec.gov.in/pdf/Studypaper/TVWS_Final.pdf>.

[4] M. Rahman, A. Saifullah, A comprehensive survey on networking over TV white spaces, Preprint submitted to Pervasive and Mobile Computing, August 2019.

[5] E. Pietrosemoli, M. Zennaro, TV white spaces, a pragmatic approach, December 2013. <http://wireless.ictp.it/tvws/book/tvws.pdf>.

[6] P. Sharma, TVWS-enabler for smart cities. <http://www.whizpace.com>.

[7] A. Arteaga, A. Navarro, Availability of TV white spaces using spectrum occupancy information and coverage maps, IEEE Lat. Am. Trans. 14 (6) (2016).

[8] A. Saeed, K.A. Harras, E. Zegura, M. Ammar, Local and Low-Cost White Space Detection. <https://www.cc.gatech.edu/~amsmti3/files/waldo_icdcs17.pdf?>.

[9] R. Murty, R. Chandra, T. Moscibroda, P. Bahl, SenseLess: a database-driven white spaces network, in: 2011 IEEE International Symposium on Dynamic Spectrum Access Networks (DySPAN), Aachen, 2011, pp. 10–21.

[10] Z. Tan, S. Banerjee, V-scope: an opportunistic wardriving approach to augmenting TV whitespace databases, Proceedings of the 19th Annual International Conference on Mobile Computing & Networking, ACM, 2013, pp. 251–254.

[11] <http://www.winlab.rutgers.edu/projects/wiser/Index.html>.

[12] D. Liu, Z. Wu, F. Wu, Y. Zhang, G. Chen, FIWEX: compressive sensing based cost-efficient indoor white space exploration, Proceedings of the 16th ACM International Symposium on Mobile Ad Hoc Networking and Computing, ACM, 2015, pp. 17–26.

[13] J. Zhang, W. Zhang, M. Chen, Z. Wang, WINET: indoor white space network design, in: 2015 IEEE Conference on Computer Communications (INFOCOM), Kowloon, 2015, pp. 630–638.

[14] Y. Yuan, et al., KNOWS: cognitive radio networks over white spaces, in: 2007 2nd IEEE International Symposium on New Frontiers in Dynamic Spectrum Access Networks, Dublin, 2007, pp. 416–427.

[15] A. Saifullah, M. Rahman, D. Ismail, C. Lu, J. Liu, R. Chandra, Low-power wide-area network over white spaces, IEEE/ACM Trans. Netw. 26 (4) (2018) 1893–1906.

[16] <https://www.iotforall.com/lpwan-benefits-vs-iot-connectivity-options/>.

[17] <www.weightless.org>.

9

Performance of LoRa technology: link-level and cell-level performance

Daniele Croce, Michele Gucciardo, Giuseppe Santaromita, Stefano Mangione, Ilenia Tinnirello

ENGINEERING DEPARTMENT, UNIVERSITY OF PALERMO, PALERMO, ITALY

9.1 Introduction

In recent years, we have assisted to an impressive proliferation of wireless technologies and mobile-generated traffic, which is now the highest portion of the total internet traffic and will continue to grow with the emergence of Internet-of-things (IoT) applications [1]. Such a proliferation has been characterized by a high-density deployment of base stations (based on heterogeneous technologies, such as 4G cellular base stations and Wi-Fi access points), as well as by high-density wireless devices, not limited to traditional user terminals. Indeed, with the advent of IoT applications, many smart objects, such as domestic appliances, cameras, and monitoring sensors, are equipped with a wireless technology.

In this chapter, we consider the emerging LoRa technology, which represents a critical example of wireless technology working in high-density scenarios. Indeed, LoRa technology has been conceived for low-power wide-area networks (WANs), characterized by low data rate requirements per single device, large cells, and heterogeneous application domains, which may lead to extremely high numbers of devices coexisting in the same cell. For this reason, LoRa provides different possibilities to orthogonalize transmissions as much as possible—carrier frequency, spreading factor (SF), bandwidth (BW), coding rate (CR)—and provide simultaneous collision-free communications. However, despite the robustness of the LoRa physical layer (PHY) [2] patented by Semtech, in WAN scenarios where multiple gateways can be installed, the scalability of this technology is still under investigation [3]. Current studies are mostly based on the assumption that the utilization of multiple transmission channels and SFs lead to a system that can be considered as the simple superposition of independent (single channel, single SF) subsystems [4]. This is actually a strong simplification, especially because the SFs adopted by LoRa are pseudo-orthogonal [5] and therefore, in near-far conditions, collisions can prevent the correct reception of the overlapping transmissions using different SFs.

For characterizing these phenomena, in this chapter we provide two main contributions: a link-level characterization of LoRa modulation (based on our previous work [6]) and then,

LPWAN Technologies for IoT and M2M Applications. DOI: https://doi.org/10.1016/B978-0-12-818880-4.00010-7

exploiting such link-level properties, we provide a complete cell-level model study of multi-link LoRa systems. Regarding the first aspect, we characterize LoRa modulation showing that collisions between packets of different SFs can indeed cause packet loss. Modifying the software transceiver presented in [7], we quantify the power difference for which capture effects and packet loss occur, for all combinations of SFs, showing that the cochannel rejection thresholds can be very different (on average 10 dB—an order of magnitude) lower than the theoretical ones presented in [8], with values as low as −8 dB. Such power difference between two radio signals can easily appear in common LoRa application scenarios, contradicting the common belief that different SFs can be considered as orthogonal in practice.

Second, regarding the analysis at the cell level, we developed simple yet accurate models of LoRa performance, deriving the aggregated capacity and data extraction rate of a LoRa cell with one or multiple gateways. The models take into account heterogeneous transmission times among the contending stations and heterogeneous probabilities of collisions. We validate our models with the LoRaSim simulator [9] and with a custom Matlab simulator. Our thorough analysis demonstrates that channel captures and inter-SF collisions can significantly impact performance. Finally, we quantify the performance achievable by using multiple gateways and we show that it might be better to distribute them at the edge of the cell rather than on a regular grid.

The rest of the chapter is organized as follows: after a brief literature review in Section 9.2, we provide a background description of LoRa modulation and a link-level characterization of its main features in Section 9.3. The analysis of cell capacity is presented in Section 9.4, where we model and quantify the impact of channel captures and inter-SF collisions. In Section 9.5, numerical results built on top of link-level and cell-level models are presented. We analyze the capacity improvements achievable with multiple gateways and the performance impact of topology in Section 9.6 and finally conclude our chapter in Section 9.7.

9.2 Related work

Since LoRa is a fairly new technology, relatively few works exist in the literature. Link-level studies are mainly based on the experimental characterization of the coverage and on the rejection of interference properties of Semtech's patented LoRa PHY [10]. The paper in [8] quantifies the signal-to-interference-ratio (SIR) power thresholds needed to reject interference caused by LoRa signals modulated with different SFs. However, the presented theoretical results are very different from our experimental ones. In [5], the performance of LoRa is compared to ultra-narrowband technologies (such as Sigfox), where it is shown that ultra-narrowband has a greater coverage although LoRa networks are less sensitive to interference.

Studies at the cell level are based on the characterization of the link behavior. In [3], after demonstrating the capture phenomena between LoRa frames, authors quantify the capacity of a cell in simulation [9]. The simulator assumes that a 6 dB power ratio between the collision packets is needed for channel captures and that different SFs can be considered completely orthogonal [4]. This last hypothesis is a strong simplification, since LoRa SFs are pseudo-orthogonal [5] and inter-SF collisions can appear in near-far scenarios. In [11], the authors propose a solution to improve LoRa performance in high-density scenario, but the details of the model used are not provided.

Capture effects can significantly increase the performance of wireless systems, because the strongest received signal might be correctly demodulated even in the case of collision. Several approaches to model this phenomenon have been presented, estimating interference power, collision times, channel fading, etc. Given the number of interfering frames and their power, it is possible to estimate channel captures when the SIR is greater than the capture threshold [12]. Alternatively, the highest possible interference level can be mapped to a vulnerability range from which interfering signals do not affect packet reception [13]. Despite the simplicity of this approach, this model provides good results when the network operates in stable conditions. Thus, in this chapter, we generalize the concept of vulnerability in both intra-SF and inter-SF collisions.

9.3 LoRa link-level behavior

In this section, we provide a characterization of LoRa link-level performance, which will then be exploited in the next sections to develop our cell-level model of LoRa systems.

9.3.1 LoRa modulation and demodulation

LoRa modulation is derived from *chirp spread spectrum (CSS)*, which makes use of *chirp signals*, that is, frequency-modulated signals obtained when the modulating signal varies linearly in the range [f_0, f_1] (upchirp) or [f_1, f_0] (downchirp) in a symbol time T. Binary modulations, mapping 0/1 information bits in upchirps/downchirps, have been demonstrated to be very robust against in-band or out-band interference [2]. LoRa employs an M-ary modulation scheme based on chirps, in which symbols are obtained by considering different circular shifts of the basic upchirp signal. The temporal shifts, characterizing each symbol, are slotted into multiples of time $T_{chip} = 1/BW$, called chip, being $BW = f_1 - f_0$ the BW of the signal. It results that the modulating signal for a generic nth LoRa symbol can be expressed as:

$$f(t) = \begin{cases} f_1 + k(t - n\, T_{chip}) & \text{for } 0 \le t \le n\, T_{chip} \\ f_0 + k(t - n\, T_{chip}) & \text{for } n\, T_{chip} < t \le T \end{cases}$$

where $k = (f_1 - f_0)/T$ is the slope of the frequency variations. The total number of symbols (coding i information bits) is chosen equal to 2^i, where i is called SF. The symbol duration T required for representing any possible shift is $2^i\, T_{chip} = 2^i/BW$. It follows that, for a fixed BW, the symbol period and the temporal occupancy of the signal increase with larger SFs. Fig. 9−1 shows the modulating signal used for a basic upchirp and three examples of circular

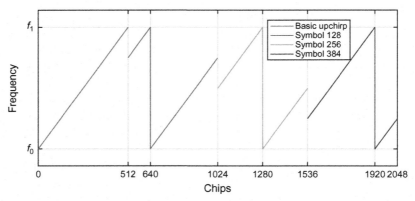

FIGURE 9–1 Modulating signal with *SF* = 9 for one basic upchirp and three symbols: 128, 256, and 384.

shifts obtained for an SF equal to 9: the symbol time is 512 T_{chip}, while the three exemplary shifts code the symbols 128, 256, and 384.

The preamble of any LoRa frame is obtained by sending a sequence of at least eight upchirps followed by two coded upchirps, used for network identification (sync word), and two and a quarter base downchirps. Payload data are then sent by using the M-ary modulation symbols. LoRa provides three *BW* settings (125, 250, or 500 kHz) and seven different SF values (from 6 to 12). In general, a larger BW translates in data rate increase and receiver sensitivity deterioration. Conversely, higher SFs can be used for improving the link robustness at the cost of a lower data rate.

An interesting feature of LoRa modulation is the pseudo-orthogonality of signals modulated under different SFs, which can be exploited for enabling multiple concurrent transmissions. Indeed, the cross-energy between two signals modulated with different SFs is almost zero, regardless of the starting of the symbol times.

An example of signal separation is illustrated in Fig. 9−2, where one signal modulated with SF equal to 9 (blue line) is overlapped to two symbols modulated with SF equal to 8 (red line). Generally, in the absence of interference, a single-user receiver will synchronously multiply the received signal to the base downchirp. This results in a signal comprising only two frequencies: $f_n = -kn \cdot T_{chip}$ and $f_n - BW = -(f_1 - f_0) - kn \cdot T_{chip}$. Both frequencies will be aliased to the same frequency f_n by downsampling at the rate *BW*. The estimated symbol index $\hat{n}$ corresponds to the position of the peak at the output of an FFT, as described in [8].

In the case of collisions with other LoRa symbols, we can distinguish two different scenarios, depending on the interfering spreading factor SF_{int}. First, if the SF_{int} is the same as the one the receiver is listening for, the above receiver will observe multiple peaks at the output of the fast fourier transform (FFT). Indeed, assuming that the two transmissions are received at the same power and that the reference signal is perfectly synchronized with the receiver, the FFT will show a maximum peak corresponding to the reference symbol and two smaller peaks corresponding to two partially overlapping interference symbols, as shown in Fig. 9−3. In the example, the two peaks have a lower power than the reference symbol (this depends

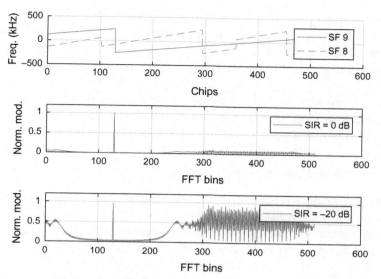

FIGURE 9–2 An example of collision between signals modulated with different SFs. A LoRa symbol modulated with SF equal to 9 (solid line) and two overlapping interfering symbols with SF equal to 8 (dashed line) received at different SIR levels, and iFFT output after multiplication with the base downchirp and downsampling.

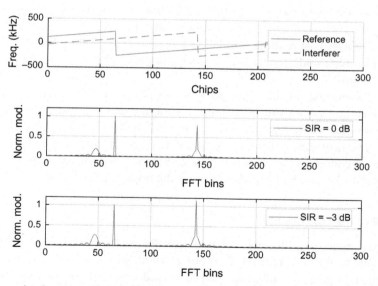

FIGURE 9–3 An example of capture effect within signals modulated with the same SF. A LoRa reference symbol (solid line) and two partially overlapping interfering symbols (dashed line) received at different SIR levels, and iFFT output after multiplication with the base downchirp and downsampling.

on the transmitted symbol and on the offset with the receiver). However, a SIR of just 3 dB is sufficient for the reference symbol to combat the interfering signal and "capture" the channel. This means that LoRa exhibits a very high capture probability with the same SF.

Second, when the SF_{int} is different from the one the receiver is interested in, after multiplication with the base downchirp and downsampling, the interfering signal will still be a chirped waveform, resulting in a wide-band spectrum with low spectral density, as shown in Fig. 9−2. Since the receiver estimates the transmitted symbol by looking for a peak, the cochannel rejection in this scenario results much higher (≈ -20 dB in the figure).

9.3.2 LoRa physical layer coding

Up to now, we have neglected the impact of bit coding schemes. Indeed, the patented LoRa PHY includes several mechanisms to make the system more robust to interference. After transmitting the preamble, both header and payload bits of LoRa frames are mapped to symbols by a pipeline of processing operations, which include: Hamming coding,[1] whitening, shuffling & interleaving, and gray coding. These operations have been specifically designed for increasing robustness toward synchronization errors or narrowband interference, which can be a serious issue for CSS-based modulations. In fact, in case of synchronization errors or narrowband interference, the receiver described in the previous section will most probably mistake the transmitted symbol, mapped to frequency f_n after the inverse FFT (IFFT), for one of the immediately adjacent symbols. Since gray coding ensures that adjacent symbols are mapped to bit patterns differing in one position only, the receiver is able to identify the less reliable bits (at most two bits) of each received symbol. The purpose of the LoRa interleaver is spreading unreliable bits among several codewords, thus enabling even the 4/5 Hamming code (consisting in a simple parity check) in exhibiting a significant channel coding gain.

Although these coding mechanisms are effective against synchronization errors, they do not mitigate decoding errors due to colliding symbols. We confirmed this observation by studying the performance of a software-defined LoRa receiver under controlled received signals [6]. Indeed, in our experiments, the probability of the distance between the transmitted symbol and the decoded one in presence of inter-SF collisions follows approximately a binomial distribution and is not concentrated around the adjacent symbol.

9.3.3 Cochannel rejection

To quantify the impact of collisions and measure the cochannel rejection, we performed a number of experiments on a real point-to-point LoRa link in presence of continuous collisions of packets. Our goal is to identify a SIR threshold under which the demodulation of the received frame is affected by errors. To this purpose, we used a Semtech SX1272 transceiver,

---

[1] The Hamming codes used in LoRa have a coding rate between 4/5 and 4/8, and can reveal or correct at most one error on the AWGN channel.

Table 9–1   SIR thresholds with SX1272 transceiver.

SF_{ref}	7	8	9	10	11	12
7	1	−8	−9	−9	−9	−9
8	−11	1	−11	−12	−13	−13
9	−15	−13	1	−13	−14	−15
10	−19	−18	−17	1	−17	−18
11	−22	−22	−21	−20	1	−20
12	−25	−25	−25	−24	−23	1

SF_{int}

controlled by an Arduino Yun, for characterizing the behavior of a commercial LoRa receiver in presence of inter-SF and intra-SF collisions. We modified the LoRa synthesizer presented in [7] to encode, modulate, and generate I/Q samples of a real LoRa packet, which are then transmitted over the air with a USRP B210 board through GNU radio. With this LoRa synthesizer, we generated two traces (one for the interferer and one for the reference LoRa link) for each combination of SFs, composed of a stream of 500 packets with a 20-byte payload (for the reference SF_{ref}) and adjusting the payload length of the interfering SF_{int} to match the length of the reference trace, that is, with an equivalent time on air. The offset of each interfering packet, overlapped in time to the packets of the reference link, has been randomly selected within a window, which guarantees that the two packets collide for at least one symbol. We filled the payload of all frames with randomly generated bytes, except for the two bytes that specify the destination address and the payload length. In particular, we assigned the destination address of the SX1272 receiver only to the packets of the reference link. This allows the receiver to discard the interfering packets when they are modulated with the same SF of the reference ones (i.e., $SF_{int} = SF_{ref}$). Finally, we scaled the amplitude of the interfering packet stream to achieve the desired SIR, varying from −30 to +6 dB, and added it to the reference stream. The resulting combined stream, which emulates the traffic generated by two different transmitters, was transmitted through the USRP toward the receiving SX1272 module.

The results of the experiments are summarized in Table 9–1, for all SF combinations. The table shows that LoRa cochannel rejection is almost independent of the interfering SF and on average is around −16 dB, about 10 dB—an order of magnitude—higher than the theoretical results of [8], with values as high as −8 dB.[2] This contradicts the common belief that SFs can be considered orthogonal, because in typical near-far conditions, when the interferer is much closer to the LoRa receiver, inter-SF collisions can indeed be an issue.

---

[2] Note that the results of [8] have not been justified in the paper. The values presented in Table 9−1 of [8] follow $10log_{10}(2^{S_{Fref}})$ in the lower triangular part, while on the upper triangular part are similar to $10log_{10}(2^{S_{Fref}}) + 10log_{10}(2^{S_{Fint}} - 2^{S_{Fref}})/(2^{S_{Fint}} - 2^{S_{Fref}}))$. However, these results do not reflect the actual behavior of LoRa, as demonstrated by our experimental results.

9.4 Analysis of cell capacity

The performance of LoRa networks depends on the simple channel access scheme, which is basically a nonslotted Aloha (without carrier sense). Under Poisson arrivals, the performance of nonslotted Aloha systems is well known: for a single cell, if G is the normalized load provided by the cell nodes, then the throughput is $S = G \cdot e^{-2G}$.

For each Spreading Factor $i \in [6,12]$, being n_i the number of nodes, ToA_i the time on air of a frame of P bytes using SFi and s the (uniform) source rate of the nodes in pkt/second, then the (normalized) load can be computed as $G_i = n_i \cdot s \cdot ToA_i$. Since high SFs translate in higher ToA and offered load, in order to keep the same data extraction rate (DER, defined as the quota of frames successfully received at the gateway), the number of nodes must be proportionally reduced. Next, we will generalize this model by describing the effect of channel captures and inter-SF collisions.

9.4.1 Channel captures

First, let us consider a scenario where collisions are between packets, employing the same SF only (i.e., SFs are completely orthogonal). For simplicity, we assume that only a single frame can interfere with the target node at a time, which is a reasonable hypothesis when the network is stable (i.e., offered load < 1). During a collision, the capture effect appears when the SIR is higher than a certain threshold, as reported in Table 9−1 (from our experiments, the SIR threshold for channel captures is 1 dB independently of the SF). Considering a node at distance r from the gateway, ignoring the impact of fading and assuming that attenuation is proportional to $r^{-\eta}$, we can depict this condition as a circle of radius, $\min(\alpha r, R)$, with R the radius of the cell and $\alpha = 10^{S\,IR/10_\eta} > 1$, as shown in Fig. 9−4.

Thus a target station that uses SF i only competes with part of the total load G_i: if α is small, then the competing load is also small and vice versa. Based on the considerations

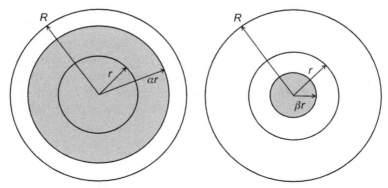

FIGURE 9–4 Distribution of nodes competing with terminals at distance r: channel captures (left) and inter-SF collisions (right).

above, we generalize the basic Aloha model to quantify the performance of LoRa in the presence of channel captures: for simplicity, if packets have the same time on air, the throughput $S_c(i)$ can be computed as:

$$S_c(i) = 2\pi \int_0^{R/\alpha} \delta_i e^{-2\frac{\alpha^2 r^2}{R^2} G_i} r \cdot dr + \delta_i(\pi R^2 - \pi R^2/\alpha^2)e^{-2 \cdot G_i}$$

where $\delta_i = G_i/(\pi R^2)$ is the offered load density of SF i. It results:

$$S_c(i) = \frac{1}{2\alpha^2}(1 - e^{-2G_i}) + G_i\left(1 - \frac{1}{\alpha^2}\right)e^{-2G_i} \tag{9.1}$$

and the DER is $S_c(i)/G_i$.

Although this model ignores collisions involving multiple packets, it can be considered as an upper bound which, in stable network conditions, is usually very tight to real results.

9.4.2 Inter-spreading factor collisions

Due to imperfect orthogonality of the SFs, the target node operating with SF i will compete with both nodes n_i employing the same SF, as well as other terminals n_{-i} using different SFs. Such inter-SF collisions will arise from nodes nearer to the gateway, which frames will be received with power exceeding the cochannel rejection threshold. In our results, we have shown experimentally that this threshold is on average about -16 dB, almost independently of the SF employed by the nodes. Thus, as shown in the right of Fig. 9–4, we map the SIR threshold to a circle of radius $\beta \hat{r}$, with $\beta = 10^{S\ IR/10_\eta} < 1$.

For now, let us ignore channel captures and consider only inter-SF collisions. The probability that a frame of a target device is successfully received using SF i depends on the probability of finding the medium idle at the beginning of the transmission and for the successive ToA_i required for the transmission to be completed. Being $G_{-i} = \sum_{k \neq i} G_k$, the load offered by nodes with different SF than i and $G^*_{-i} = n_{-i}ToA_i s$, we can compute the throughput of SF i with imperfect orthogonality as:

$$S_{io}(i) = e^{-2G_i} \cdot 2\pi \int_0^R \delta_i e^{-\frac{\beta^2 r^2}{R^2}(G_{-i}+G^*_{-i})} r \cdot dr$$

$$= G_i e^{-2 \cdot G_i} \frac{1 - e^{-\beta^2(G_{-i}+G^*_{-i})}}{\beta^2 \cdot (G_{-i} + G^*_{-i})} \tag{9.2}$$

which, compared to the Aloha performance in the orthogonal hypothesis $S_o(i) = G_i e^{-2G_i}$, is clearly smaller.

Following the same approach and considering both capture effects and inter-SF collisions, the throughput can be calculated for each SF i by considering the intra-SF interference quota $\min(\alpha^2 r^2/R^2, 1)$ and the portion of inter-SF collisions $\beta^2 r^2/R^2$:

$$S_{io,c}(i) = \frac{G_i e^{-2G_i}(1 - e^{-2G_i - (G_{-i} + G^*_{-i})\alpha^2/\beta^2})}{2\alpha^2 G_i + \beta^2(G_{-i} + G^*_{-i})} +$$

$$= \frac{G_i e^{-2G_i}}{\beta^2(G_{-i} + G^*_{-i})}(e^{(G_{-i} + G^*_{-i})\alpha^2/\beta^2)} - e^{(G_{-i} + G^*_{-i})\beta^2})$$

(9.3)

Finally, dividing the throughput by the offered load, the average DER can be computed.

9.4.3 Model extension: nonuniform spreading factor allocation

The SF allocation and distribution among the terminals can influence the capacity of a LoRa cell. Indeed, capture effects and inter-SF collisions depend on the power ration of the signals, which is obviously influenced by the relative position of the terminals. For example, let us consider inter-SF collisions between two SFs, without channel captures for the sake of simplicity. Generally, higher SFs are allocated to nodes further away from the gateway. Therefore, suppose that spreading factor $SF_1 < SF_2$ is assigned to users close to the gateway, up to a certain distance d (with $r < d < R$), while SF_2 is assigned to users with distance r greater than d, that is, $d < r < R$.

In this scenario, and ignoring possible fading effects, users with spreading factor SF_1 will never be interfered by nodes with SF_2 because these users are located deterministically at greater distances, and thus the power received at the gateway will always be lower. Therefore, the performance obtained by SF_1 will be $S_1 = G_1 \cdot e^{-2\,G_1}$, as if the SFs were perfectly orthogonal. Vice versa, terminals using spreading factor SF_2 will likely be affected by nodes using SF_1 as these terminals are closer to the gateway, and thus the interference density will be higher than in the previous uniform allocation case. The previous throughput equations can be easily extended with the nonuniform load density functions $\delta_2 = G_2/(\pi R^2 - \pi d^2)$ and $\delta_1 = G_1/(\pi d^2)$.

9.5 Numerical results

To validate our analytical model, we used a custom Matlab simulator that we validate with the LoRaSim simulator [9] used in [3]. Additionally, we included log-normal fading, which was not considered in LoRaSim. To simplify the analysis, we set an average SIR threshold for all SF combinations. Unless otherwise specified, the transmission parameters used by the terminals are as follows: transmit power = 14 dBm, BW = 500 kHz, CR = 4/5, payload length $P = 20$ byte, source rate $s = 1$ pkt/90 second. Finally, for clearness of presentation, in some figures, we deliberately show the performance obtained only with a subset of SFs (e.g., SF6, 9, and 12), although similar results apply for the omitted SFs as well.

9.5.1 Channel capture effects

First, we evaluate the performance improvement of LoRa thanks to capture effects. In these experiments, each SF is used in isolation, that is, when all nodes use the same SF. Fig. 9−5 shows the DER for all SFs (for easiness of comparison) with a conservative capture SIR

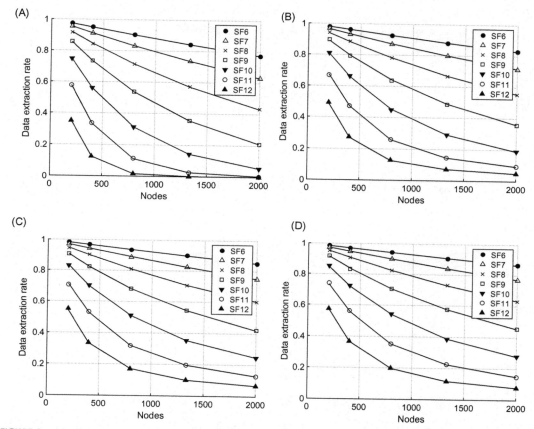

FIGURE 9–5 Simulation (markers) and analytical model (lines) results for channel capture effect with several values of η: (A) Aloha (no capture), $\eta = 4$; (B) capture, $\eta = 2$; (C) capture, $\eta = 4$; and (D) capture, $\eta = 7$.

threshold of 1 dB, attenuation exponent $\eta = 2, 4, 7$ and up to $n_{i=}2000$ nodes. From the figure, it is clear that the analytical model closely follows the simulation results. Fig. 9–5A also shows the performance without channel captures (for comparison with a pure Aloha system), which is obviously lower. The results show that the DER can grow significantly due to channel captures. Additionally, in scenarios with strong attenuation (high values of η), capture effects may help increase capacity further.

9.5.2 Interfering spreading factors

We now evaluate the impact of inter-SF collisions. Fig. 9–6 shows the performance with up to $N = 2000$ nodes, when all SFs are allocated uniformly among nodes (i.e., each SF is used by $n_i \approx N/7$ terminals). Imperfect orthogonality is accounted for SIR values lower than -10dB, while the perfectly orthogonal case is shown for comparison in Fig. 9–6A. Once again, the proposed analytical model closely follows the simulation results.

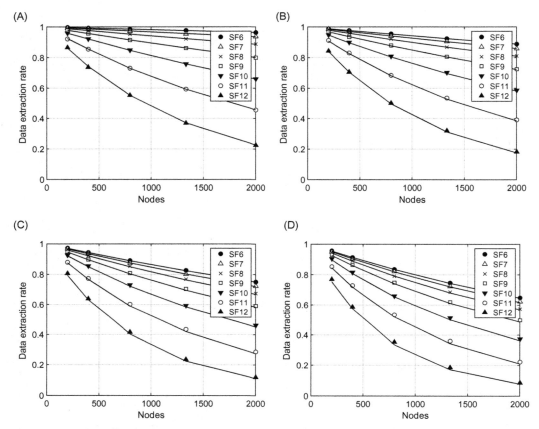

FIGURE 9–6 Simulation (markers) and analytical model (lines) results for interfering SFs with several values of η: (A) Orthogonal SFs and $\eta = 4$; (B) interfering SFs, $\eta = 2$; (C) interfering SFs, $\eta = 4$; (D) interfering SFs, $\eta = 7$.

It is clear from the figure that inter-SF collisions can impact the performance, especially in strong attenuation scenarios (increased η). In absolute terms, the DER is reduced particularly for lower SFs (6–8), which collide with the longer frames of higher SFs. However, in relative terms, the reduction is worse for the highest SFs (10–12): for example, compared to Fig. 9–6A, with $\eta = 4$ and 2000 nodes, Fig. 9–6C shows that the DER of SF12 is reduced by about 50%. This is because the time on air is much higher and thus the vulnerability period is increased.

9.5.3 Impact of fading

We now analyze the impact of fading on LoRa communications. To this purpose, we implemented in the LoRaSim simulator a log-normal fading model with $\sigma = 0$, 3, 6, 10 dB. Fig. 9–7 shows the capacity of LoRa in presence of fading and channel captures for few SFs

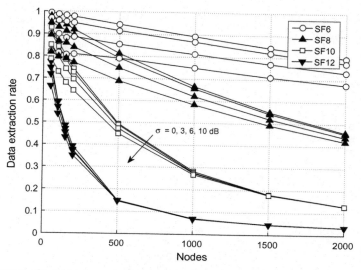

FIGURE 9–7 Performance of LoRa in presence of log-normal fading ($\sigma = 0$, 3, 6, 10 dB), capture effect and $\eta = 4$.

only for clarity of explanation (i.e., ignoring inter-SF interference). It is interesting to note that fading impacts lower SFs (6–9), but higher SFs are almost unaffected. This is because faded frames are not received at the gateway, and therefore the collision probability is reduced, partially compensating the lost frames due to fading. Similar results can be achieved by considering also inter-SF collisions, which are omitted for simplicity.

9.5.4 Nonuniform spreading factor allocation

The above results indicate that using higher SFs for far distance nodes may not improve performance in presence of congestion or fading. Indeed, high SFs are more vulnerable to collisions, while closer nodes are generally received with higher power. For example, Fig. 9–8 compares the performances of SF6 in competition with SF9 or SF12, with uniform SF allocation (dashed line) or when nodes near the gateway use SF7 and faraway terminals use SF9 (or SF12), as described in Section 9.4.3. The figure proves that DER of higher SFs decreases, while SF7 improves its DER because collisions from other SFs are reduced. This means that, even in presence of strong fading channels, using high SFs for far distance, nodes may be counterproductive when the network load is high.

9.6 Capacity with multiple gateways

In a single LoRa cell, deploying more than one gateway is a common strategy to increase capacity. Indeed, when multiple gateways receive the same frame, replicated packets are filtered by the network server. The capacity improvement with multiple gateways depends on the topology of the gateways with respect to the nodes and the interfering areas are not

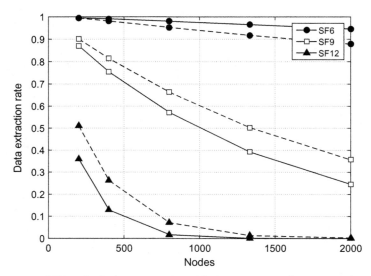

FIGURE 9–8 Performance of SF9 or SF12 when competing with SF6. Comparison between uniform distribution of nodes (dashed lines) and higher SFs allocated to faraway nodes (solid lines). SIR = −10 dB and $\eta = 4$.

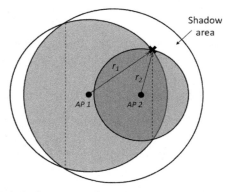

FIGURE 9–9 Competing area with multiple gateways for the node on the X and "shadow area" created by the same node.

simple circles anymore. Fig. 9−9 shows, for example, a scenario with two gateways only: a node on the X experiences interference from the nodes in the gray areas of the image. To model this scenario, we computed numerically the average success probability γ_k in presence of k interfering terminals. This way we decouple the geometric effects (positioning) from the collision probability (offered load).

For evaluating γ_k, we randomly positioned $k + 1$ nodes in the cell, quantifying the number of nodes with distance smaller than $r\alpha$ from at least one gateway, repeated the experiment multiple times, and averaged the results. Obviously, the possible capacity increase reaches a limit that depends on the SIR threshold. For example, Fig. 9−10 shows the

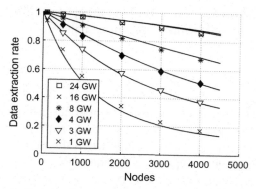

FIGURE 9–10 DER obtained in a multigateway scenario with a regular grid deployment: model (lines) and simulation results (points).

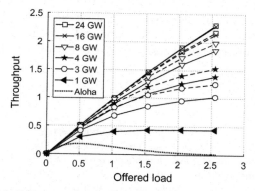

FIGURE 9–11 Grid deployment (solid) versus edge deployment (dashed).

capacity of SF7 with gateways deployed in a regular grid. The figure shows that deploying more than 16 gateways does not improve the DER significantly. The figure also shows that our model matches closely the simulator results, validating the numerical evaluation of the gamma coefficients.

Employing a different gateway topology can significantly impact the DER. In particular, it turns out that with a few gateways (up to a dozen), it is better to deploy them at the edge of the cell instead of in a regular grid. Only with more than 16 gateways, a regular grid is preferable, although the DER increase is only marginal. The reason is shown in Fig. 9–9: in the network topology, the node on the X creates a "shadow area" for other nodes closer to the edge of the cell, interfering with such nodes. With few gateways, the shadow area is statistically larger when the gateway deployment is in a regular grid and smaller if the gateways are on the edge of the cell. Indeed, Fig. 9–11 compares the DER obtained with the regular grid or when gateways are at the edge of the cell. For example, deploying three gateways on the edge (dashed lines) increases the DER by about 25% with respect to a grid deployment (solid line). Conversely, with 16 gateways or more, the grid distribution should be preferred, but in this case the DER increases by only 5% approximately.

9.7 Conclusion

In this chapter, we have shown that, because of imperfect orthogonality between different SFs, a LoRa network cell cannot be studied as a simple superposition of independent networks working on independent channels. Indeed, when the power of the interfering SF significantly overcomes the reference SF, the correct demodulation of the reference SF can be prevented. Based on an accurate link-level analysis, we developed an analytical framework to model the capacity of a LoRa cell. We demonstrated that inter-SF collisions can impact performance and that allocating higher SFs to faraway nodes may be counterproductive. Multiple gateways deployed in the same cell can increase channel captures and mitigate such issues, but the DER increase is marginal after 16 gateways. Finally, we showed that in presence of few gateways, their deployment should be on the edge of the cell to increase diversity as much as possible. The analytical framework presented in this chapter provides important results for network operators and planning of future LoRa networks.

Acknowledgments

This work has been partially funded thanks to the National Research Project DEMAND (DistributEd MANagement logics and Devices for electricity savings in active users installations) supported by the Italian Ministry for the Economic Development MISE—identification code: CCSEB 00093.

We also thank the Institute of Electrical and Electronics Engineers for allowing us to reuse portions of published articles: Copyright © IEEE. All rights reserved. Reprinted, with permission, from D. Croce, M. Gucciardo, S. Mangione, G. Santaromita, I. Tinnirello, Impact of LoRa imperfect orthogonality: analysis of link-level performance," IEEE Commun. Lett. 22 (4) (2018) 796−799. Personal use of this material is permitted. However, permission to reuse this material for any other purpose must be obtained from the IEEE.

References

[1] Worldwide connected devices forecast, <www.statista.com>.

[2] Semtech. LoRa modulation basics. AN1200.22, Revision 2. <www.semtech.com>, May 2015.

[3] M.C. Bor, U. Roedig, T. Voigt, J.M. Alonso, Do LoRa low-power wide-area networks scale? Proc. MSWiM, ACM, New York, 2016, pp. 59−67.

[4] B. Reynders, S. Pollin, Chirp spread spectrum as a modulation technique for long range communication, in: SCVT 2016, Mons, 2016, pp. 1−5.

[5] B. Reynders, W. Meert, S. Pollin, Range and coexistence analysis of long range unlicensed communication, in: ICT 2016, Thessaloniki, 2016, pp. 1−6.

[6] D. Croce, M. Gucciardo, S. Mangione, G. Santaromita, I. Tinnirello, Impact of LoRa imperfect orthogonality: Analysis of link-level performance, IEEE Commun. Lett. 22 (4) (2018) 796−799.

[7]   M. Gucciardo, I. Tinnirello, D. Garlisi, Demo: A cell-level traffic generator for LoRa networks, in: Proceedings of the 23rd Conference on Mobile Computing and Networking (MobiCom'17), October 2017, Snowbird, UT, USA.

[8]   C. Goursaud, J.M. Gorce, Dedicated networks for IoT: PHY/MAC state of the art and challenges, EAI endorsed Trans. Internet Things 1 (1) (2015) e3.

[9]   Available at <http://www.lancaster.ac.uk/scc/sites/lora/>.

[10] A. Augustin, J. Yi, T. Clausen, W. Townsley, A study of LoRa: long range & low power networks for the Internet of things, Senors 16 (9) (2016) 1466.

[11] D. Bankov, E. Khorov, A. Lyakhov, On the limits of LoRaWAN channel access, in: 2016 International Conference on Engineering and Telecommunication (EnT), Moscow, 2016, pp. 10–14.

[12] M. Zorzi, R.R. Rao, Capture and retransmission control in mobile radio, IEEE J. Sel. Areas Commun. 12 (8) (2006) 1289–1298.

[13] D.J. Goodman, A.A.M. Saleh, The near/far effect in local ALOHA radio communications, IEEE Trans. Veh. Technol. 36 (1) (1987) 19–27.

Energy optimization in low-power wide area networks by using heuristic techniques

Zeinab E. Ahmed[1], Rashid A. Saeed[2,3], Amitava Mukherjee[4], Sheetal N. Ghorpade[5]

[1]DEPARTMENT OF COMPUTER ENGINEERING, UNIVERSITY OF GEZIRA, SUDAN [2]DEPARTMENT OF ELECTRONICS ENGINEERING, SUDAN UNIVERSITY OF SCIENCE AND TECHNOLOGY (SUST), KHARTOUM, SUDAN [3]DEPARTMENT OF COMPUTER ENGINEERING, TAIF UNIVERSITY, ALHAWIYA, TAIF, SOUTH AFRICA [4]DEPARTMENT OF COMPUTER SCIENCE AND ENGINEERING, ADAMAS UNIVERSITY, BARASAT, KOLKATA, INDIA [5]RMD SINHGAD SCHOOL OF ENGINEERING, PUNE, INDIA

10.1 Introduction

Energy efficiency (EE) is an open research problem globally due to issues associated with energy consumption specifically in low-power devices. For the networks with billions of sensors and low-power devices, the efficient use of limited energy resources based on multiple tasks of applications is critically important to support these advanced technologies. In addition, energy consumption is due to energy loss or wastage. Energy loss refers to energy supplied to the system that is not directly consumed by computing activities such as power transport and conversion, cooling, and lighting. Whereas energy wastage refers to the energy consumed by other equipment without generating any useful output specifically, the energy used by servers without performing useful work. This chapter discusses in-depth review of existing meta-heuristic optimization techniques used for EE in wireless networks comprised of low-power devices and sensors. The chapter is carried on the basis of classification of various optimization algorithms. It deals with the detailed analysis and comparison of their performances. EE and basic categories of optimization techniques are discussed in Sections 10.2 and 10.4, respectively. The detailed process flow of various meta-heuristic optimization techniques is well explained in Section 10.5. Section 10.6 includes in-depth review of existing meta-heuristic optimization techniques used for energy optimization in wireless sensor networks (WSNs) and the analysis of their

LPWAN Technologies for IoT and M2M Applications. DOI: https://doi.org/10.1016/B978-0-12-818880-4.00011-9

Table 10–1   List of acronyms.

	Acronyms
CO	Combinatorial optimization
GA	Genetic algorithms
PSO	Particle swarm optimization
ABC	Artificial bee colony
ACO	Ant colony optimization
TS	Tabu search
SA	Simulated annealing
MA	Memetic algorithms
EE	Energy efficiency
DE	Differential evolution algorithm
SI	Swarm intelligence
LPWANs	Low-power wide area networks
E3	Economical energy efficiency
LoRa	Long range
RPMA	Random phase multiple access
NB-IoT	Narrowband-Internet of things
GWO	Gray wolf optimization

performances is described in Section 10.7. The list of acronyms that appear in this chapter is given in Table 10–1.

10.2 Energy efficiency

In the last decades, one of the most key challenges faced by modern networks (cloud computing, fog computing, Internet of things (IoT), and networked computing systems) is effective utilization of energy at different domains. The effective utilization of energy means that save energy, minimize the energy consumption, and make the quantity of energy consumed for workload suitable for it. The design of energy-efficient communication networks is of much importance; numerous methods have been developed in the last few decades for designing energy-efficient communication networks but still it is a challenging task [1]. Optimized scheduling of all resources involved in the communication network will reduce energy consumption. Low-power wide area networks (LPWANs) [2] and economical energy efficiency (E3) [3] are the recent approaches which supports low energy consumption.

10.3 Low-power wide area networks

Recently, LPWANs represent a new communication technology, which is designed to establish great distance communications with low energy consumption, long transmission range, low cost, and long battery lifetime [4]. Various applications use LPWAN technology such as leak detection, precision agriculture, street lighting control, environment monitoring, and smart city [5]. LPWAN is suitable for connecting different devices that require sending a small amount of data over a long range. A set of LPWAN technologies have been designed to operate on the medical, scientific, and industrial, including [6–8]:

- Long range (LoRa)
- Random phase multiple access (RPMA)
- Ingenu
- Narrowband-Internet of things (NB-IoT)
- Telensa
- Weightless
- Bluetooth low energy
- Sigfox

LPWAN technologies utilized the unlicensed or licensed frequency bandwidth [9]. There are several issues and challenges facing the LPWAN [2], as follows:

- Scaling networks to massive number of devices
- Interference control and mitigation
- High data-rate modulation techniques
- Interoperability between different LPWA technologies
- Link optimizations and adaptability
- Authentication, security, and privacy
- Mobility and roaming
- Support for service level agreements
- Support for data analytics

10.4 Optimization techniques

Optimization technique is an approach to find the best solution to complex problems by minimizing or maximizing one or more objective functions stand on one or more decision variables that give a value of the objective function [10]. Optimization techniques are widely applied in various fields, namely medicine, commerce, finance, transportation engineering, and any decision-making processes. Combinatorial optimization, nonlinear programming, and linear programming are designed to solve a wide range of optimization problems [11]. Optimization techniques are classified as exact methods and approximation methods (heuristics and meta-heuristics methods). The exact methods are considered one of the techniques used for solving combinational optimization problems, including the Branch and Bound, the Branch and Cut, and the Simplex [12]. The exact methods are classified based on how the graph can be formulated into four major groups: Arc-node formulation, path formulation, arc-based formulation, and spanning tree formulation [13]. Approximate methods can be divided into two types: heuristics algorithms and meta-heuristics algorithms. Heuristic

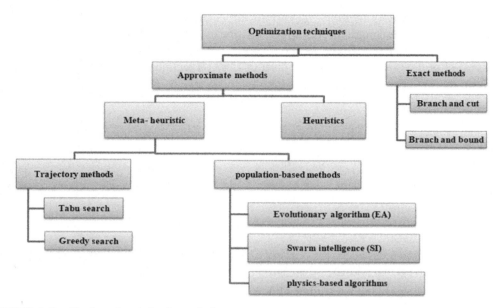

FIGURE 10–1 Classification of optimization techniques.

algorithms find approximate solutions but have acceptable time and space complexity. There are general heuristic strategies that are successfully applied to manifold problems. The heuristics and meta-heuristic techniques include genetic algorithm (GA), ant colony optimization (ACO), simulated annealing (SA), particle swarm optimization (PSO), Gray Wolf Optimization (GWO), and so on [14]. The classification of optimization techniques is shown in Fig. 10−1.

10.4.1 Heuristics methods

For the complex problems, conventional methods may not give the best solution at a suitable time. To overcome this drawback, many heuristic algorithms have been proposed to find far near-optimal solutions within an acceptable time. Heuristics were first introduced by G. Polya [15] and improved later in the 1970s, to solve specific problems in different domains of the real life. Heuristics methods are also suitable to deal with big data [16]. The main objective of heuristics approach is to increase the efficiency of well-known methods and separate not-fit conditions from fit ones by decreased risk exposure, the error rate for prediction, and also increased competitive advantages [17].

10.4.2 Meta-heuristics methods

Meta-heuristics are high-level algorithms that are becoming popular in many fields for solving very complex optimization problems with limited domain knowledge and computation capacity. If the search space is large, process is iterative, and the search is exhaustive, then it is better to use meta-heuristics than heuristics to determine near-optimal solutions with less computational effort. Furthermore, meta-heuristics are used for the optimal utilization of resources [18]. Meta-heuristic algorithms are classified into two categories: population-based methods and trajectory-based/single solution-based methods [19]. The single solution-based methods involve Tabu search (TS) and Greedy search, while the population-based algorithms are classified into three groups: swarm intelligence (SI), evolutionary algorithm (EA), and physics-based algorithms. The EA includes the memetic algorithm (MA), GA, and differential evolution (DE) algorithm. SI such as the artificial bee colony (ABC) algorithm, ACO algorithm, PSO, and GWO [20]. The choice of suitable meta-heuristics for a specific problem is very important. Single solution-based methods are more efficient to determine the local optima, but not suitable for global optima. Population-based algorithms are more suitable to find the best optimal solution [21]. Hybrid meta-heuristic approaches by integrating two or more algorithms are proposed to solve complex multiobjective optimization problems.

10.5 Classification of meta-heuristics methods

The classification of meta-heuristics is based on the characteristic and working mechanism of the meta-heuristics. The classification of meta-heuristics methods will be discussed as follows.

10.5.1 Genetic algorithms

GAs are robust ways that can be used in search and optimization issues based on Darwin's principle of natural selection [22,23]. GA is one of the best optimization algorithms having great potential to deal with various multiobjective problems [13]. The idea behind GA is that the combination of exceptional characteristics from different ancestors generates the better and optimized offsprings, that is, having an improved fitness function than the ancestors [24]. Implementing this mechanism iteratively, the offsprings gets more optimized, resulting in higher sustainability in the environment. The parameter set of the optimization problem is required to be coded as a finite-length string or chromosome [17]. Population in GA is a collection of strings or a chromosome [25,26]. Adaptation of chromosome to the environment is evaluated using objective function. Hence, the objective function of GA is called as fitness function. The basic GA is composed of three operators [27]:

- *Selection*: Forms population by selecting parents to reproduce chromosomes at first stage, then the chromosomes generated in first stage are selected to generate population for the next stage.
- *Crossover*: This operator requires two parents to generate several offsprings by the combination of genes.
- *Mutation*: This operator is created by constrained random modifications of one or more genes of chromosomes.

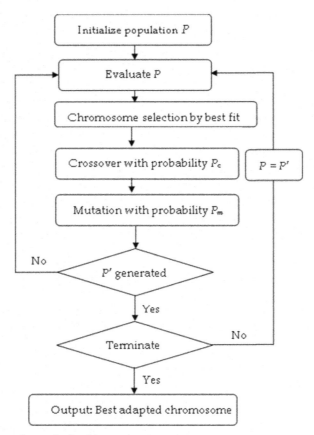

FIGURE 10–2 Process flow of genetic algorithm.

Mutation is used to explore the solution space and crossover to reach to the local optima [28]. In GA population evolves iteratively, it starts with a randomly generated initial population; a new population is generated from the existing population by selection, crossover, and mutation. The process flow of GA is shown in Fig. 10–2. Randomly select an initial population P of n chromosomes. At every iteration, a new population P' is generated by choosing two parents from P with the probability of selection proportional to their fitness. Generate new offsprings by crossover from selected parents with probability P_c, and then by random mutation recombine these chromosomes with some probability $P_m(0.001 \leq P_m \leq 0.01)$. Newly generated population replaces the existing population.

10.5.2 Particle swarm optimization

Kennedy and Eberhart [29] developed SI model inspired by birds flocking behavior called as PSO algorithm. The PSO has particles determined from natural swarms by combining

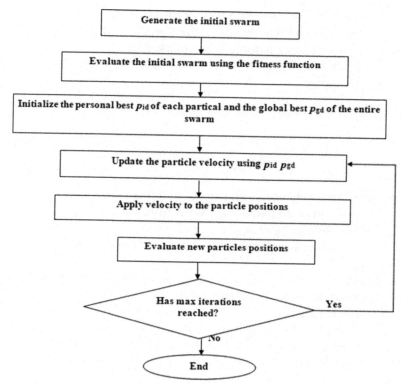

FIGURE 10–3 Process flow of particle swarm optimization.

self-experiences with social experiences using communications based on iterative computations. In PSO algorithm, a candidate solution is presented as a particle. To search out for
a global optimum, it uses a collection of flying particles (changing solutions) in a search
area (current and possible solutions) as well as the movement toward a promising area.
PSO optimizes a problem by having a population of candidate solutions and moving these
particles around in the search space according to simple mathematical formulae over the
particle's position and velocity. Each particle's movement is influenced by its local best-
known position and is also guided toward the best-known positions in the search space,
accordingly each particle updates its position to the better one than the previous position.
This is expected to move the swarm toward the best solutions [30]. The process flow for
PSO is shown in Fig. 10–3 In PSO, each single solution is a "bird" in the search space
called as "particle." All the particles have fitness values that are evaluated using fitness
function to be optimized and have velocities that direct the flying of the particles. The particles (solutions) fly through the problem space by following the recent optimum particles.
PSO is initialized with a group of random particles and then it searches for optima by
updating generations. In each iteration, every particle is updated by following two "best"

values. The first value is the best solution (fitness) which it has achieved so far. This value is called p_{best}. Another "best" value that is tracked by the particle swarm optimizer is the best value, obtained so far by any particle in the population. This best value is a global best and called g_{best}. After finding the two best values, the particle updates its velocity and positions with following equations:

$$V_{id}(t+1) = V_{id}(t) + c_1 R_1 (p_{id}(t) - x_{id}(t)) + c_2 R_2 (p_{gd}(t) - x_{id}(t)) \tag{10.1}$$

$$x_{id}(t+1) = x_{id}(t) + v_{id}(t+1) \tag{10.2}$$

where

V_{id} is the rate of position change of ith particle in dth dimension and t denotes iteration count, x_{id} is the position of ith particle, p_{id} is the historically best position of particle, and p_{gd} is the position of swarm's global best particle.

R_1 and R_2 are two n-dimensional vectors with random numbers uniformly selected between [0,1].

c_1 and c_2 are position constant weighting parameters called as cognitive and social parameters, respectively. Memory update is done by updating p_{id} and p_{gd} when the following condition is met.

$p_{id} = p_i$, if $f(p_i) > f(p_{id})$ and $p_{gd} = g_i$ if $f(g_i) > f(p_{gd})$, where $f(x)$ is the objective function subject to maximization. Once terminated, the algorithm reports the values of p_{gd} and $f(p_{gd})$ as its solution.

10.5.3 Ant colony optimization algorithm

ACO is inspired by the foraging behavior of ants [31]. At the core of this behavior is the indirect communication between the ants with the help of chemical pheromone trails, which enables them to find short paths between their nest and food sources. Blum [32] exploited this characteristic of real ant colonies in ACO algorithms to solve global optimization problems. Dorigo [33] developed the first ACO algorithm and since then numerous improvements of the ant system have been proposed. ACO algorithm has strong robustness as well as good dispersed calculative mechanism. ACO can be combined easily with other methods; it shows well performance in resolving the complex optimization problem. ACO optimizes a problem by having an updated pheromone trail and moving these ants around in the search space according to simple mathematical formulae over the transition probability and total pheromone in the region. At each iteration, ACO generates global ants and calculates their fitness. Update pheromone and edge of weak regions. If fitness is improved, then move local ants to better regions, otherwise select new random search direction. Update ant pheromone and evaporate ant pheromone. The continuous ACO is based on both local and global search. Local ants have the capability to move toward latent region with best solution with respect to transition probability of region k,

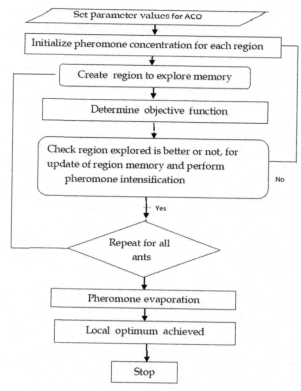

FIGURE 10–4 Process flow of ant colony optimization.

$$P_k(t) = \frac{t_k(t)}{\sum\limits_{j=1}^{n} t_j(t)} \qquad (10.3)$$

where $t_k(t)$ total pheromone at region k and n is number of global ants.

Pheromone is updated using the following equation:

$$t_i(t+1) = (1-r)t_i(t) \qquad (10.4)$$

where r is the pheromone evaporation rate.

The probability of selection of region for local ants is proportional to pheromone trail. The process flow for ACO is shown in Fig. 10–4.

10.5.4 Tabu search

TS is a local heuristic method based on neighborhood , it explores the solution space by constantly replacing recent solution with best non visited neighboring solution, new solution

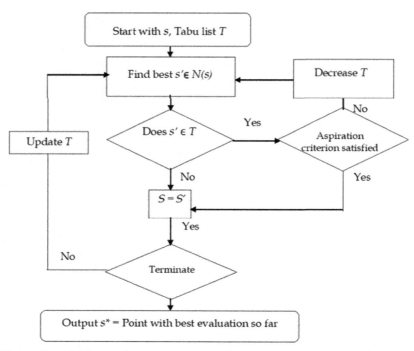

FIGURE 10–5 Process flow of Tabu search.

may be less efficient[34]. It explores the solution space by constantly replacing recent solution with best nonvisited neighboring solution, new solution may be less efficient. A fundamental concept in TS is that the intelligent search must be based on learning; the usage of flexible memory explores beyond optimality and exploits the earlier state of the search to influence its future states [35]. The process flow of TS is as shown in Fig. 10−5. TS begins iteratively with local or neighborhood search from one solution to another still selected termination criteria satisfied. Each solution s has neighborhood $N(s) \subset S$ and solution $s' \in N(s)$ is generated from s by move. The objective function of TS is to minimize $f(s)$. TS method permits moves which improve the current objective function value and ends when no improving solution can be established. The algorithm starts by selection of $s \in S$, then find $s' \in N(s)$ such that $f(s') < f(s)$. If no such s' found, then s is the local optimum and algorithm stops. Otherwise, designate s' to new s and repeat the process.

10.5.5 Simulated annealing algorithm

SA is a standard probabilistic meta-heuristic for the global optimization problem of a given function in a large search space for locating a good approximation to the global optimum. It is frequently used when the search space is discrete. The main advantage of SA is its capability of moving to states of higher energies. Kirkpatrick et al. [36] introduced SA to solve combinatorial optimization problem. The algorithm of SA is based on two loops called as

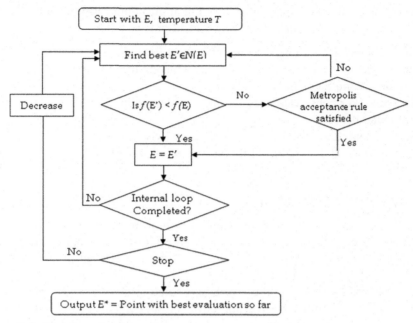

FIGURE 10–6 Process flow of simulated annealing.

internal loop and external loop. Iterations in an internal loop continue, until the system becomes stable. Whereas as external loop reduces the temperature to simulate annealing of stable systems. The internal loop generates new state by basic alterations in previous one and then applies it to the Metropolis acceptance rule. The best state generated by the algorithm is preserved and updated successively by internal loop. The process flow of SA is shown in Fig. 10−6.

Point E' is generated within a state space from the existing point E at each step in the algorithm. Point E' is accepted unconditionally if it has a lower cost function than E. But if it has a higher cost, then it is accepted using the metropolis criterion. The SA program is ended if an acceptable solution is originated or if a designated final temperature is reached. SA is successful in a wide range of non -deterministic optimization problem (NP)-hard optimization problems.

10.5.6 Artificial bee colony optimization

The ABC optimization was proposed by Dervis Karaboga to solve real-world and numerical problem in [39], which is inspired by the intelligent behaviour of honey bee swarms to getting a source for their food. ABC algorithm models consist of three groups of bees: scout bees discover all food source positions randomly based on the dances, employed bee exploiting a source of food which come from scouts bees, and onlooker bees to evaluate food quality [38]. ABC has been widely used in several applications in many different fields such as training neural networks, signal processing applications, and machine learning community [37]. The general structure of the ABC algorithm is as follows:

Step 1: initialization of solutions (the population of food sources).
Step 2: Employed bees search new food sources having more nectar within the neighborhoods of the food source.
Step 3: The onlooker bees evaluates food quality depending on the information provided by the employed bees.
Step 4: Start to search for new solutions randomly by scout bees.
Step 5: Repeat steps 2, 3, and 4 until the best solution is achieved.

10.5.7 Gray wolf optimization

GWO [40] is one of the most recent bio-inspired optimization methods. It mimics hunting procedure of a pack of gray wolves. It has effective imitation more than hunting in the pack and can be used in network optimization. In GWO, there are three prime solutions: alpha (α), beta (β), and delta (δ). Solution α is derived from α wolves and is the best solution, while β and δ solutions are from β and δ wolves, treated as second and third best solutions, respectively. All other solutions are considered to be omega (ω) solutions that are evolved from ω wolves. Hunting in the pack is directed by α, β, δ, and ω trails these three candidate solutions. The first step in the hunting process is encircling the prey. Gray wolves have the ability to recognize the location of prey and encircle them. The hunt is usually guided by the alpha. The beta and delta might also participate in hunting occasionally. However, in an abstract search space we have no idea about the location of the optimum (prey). In order to mathematically simulate the hunting behavior of gray wolves, it is assumed that the alpha (best candidate solution), beta, and delta have better knowledge about the potential location of prey. Therefore, the first three best solutions obtained so far are saved and oblige the other search agents (including the omegas) to update their positions according to the position of the best search agent. As mentioned above, the gray wolves finish the hunt by attacking the prey when it stops moving. Gray wolves mostly search according to the position of the alpha, beta, and delta. They diverge from each other to search for prey and converge to attack prey. Gray wolves diverge from each other during exploration and converge during the exploitation process. Process flow of GWO is as in Fig. 10−7.

The objective function of GWO mainly focuses on finding the optimal solution, say, x in the particular search space as represented by

$$\text{minimize } f(x), x = (x_1, x_2, x_3, \ldots, x_n) \in R^n \tag{10.5}$$

where n is the number of dimensions contained in a solution. $x \in F \in S$, where F is the feasible region in the search space S, which defines a n-dimensional rectangle R. The domain size for rectangle R is $l_b(i) \leq x(i) \leq u_b(i).l_b$ and u_b are lower and upper bounds, respectively. Constraints in the feasible region can be given as

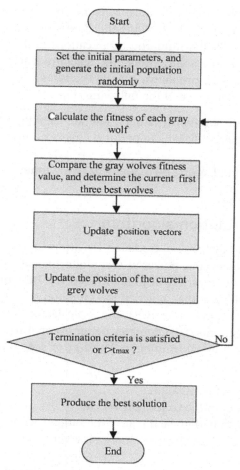

FIGURE 10–7 Process flow of gray wolf optimization.

$$g_j(x) \leq 0, \quad for \ j = 1, 2, \ldots, r \tag{10.6}$$

$$h_j(x) = 0, \quad for \ j = r + 1, \ldots, m \tag{10.7}$$

If any solution x satisfies the constraint g_j or h_j in region F, then g_j is considered to be an active constraint at x.

10.5.8 Memetic algorithms

MAs are developed by combining local search (LS) with EA to solve several complex optimization problems [41]. MAs are derived by considering the concept of meme which is nothing but a part of the information encoded in computational representations for the capabilities of local refinements [42]. On the other hands, MAs are the extensions of evolutionary algorithms that use a separate LS process to refine the fitness of an individual by hill-climbing. MAs [43] have been utilized in the numerous applications such as telecommunication and networking, scheduling and timetabling problems, machine learning and robotics, and NP-hard combinatorial optimization problems. MAs [44] address the problem by specific algorithmic structure through a sequence of the following steps:

- Selection of the candidate solutions (parents) that are used to create new solutions.
- The combination of existing solutions to perform crossover and mutation for offspring generation.
- Improve the quality of an offspring by refinement of the fitness of an individual.
- Update of the population.

10.5.9 Differential evolution algorithm

DE is a population-based stochastic introduced by Storn and Price [45]. DE algorithm is more simple, efficient, robust, and powerful technique for solving optimization problems, which performs well on a wide variety of problems [46,47]. DE works similar to GA, but the performance of DE is sensitive to the choice of the mutation strategy and settings of the control parameters such as crossover rate (CR), the scale factor (F), and the population size (NP). Due to all these constraints, DE generates better optimal solution, for the optimization problems which includes a time-consuming trial-and-error search [48]. The steps involved in DE algorithm are as follows [49]:

- Starts initializing a random population.
- The mutation process to produce a mutant vector.
- Crossover operation is applied to each pair of the target vector.
- Identify whether new offspring obtained after crossover is better than the current one. If it is better then select it.

10.6 Adaptation of meta-heuristics techniques for energy optimization

In this section, various approaches based on meta-heuristic techniques for EE in communication networks are reviewed and analyzed.

10.6.1 Genetic algorithm

Sensor self-clustering method for dynamically organizing heterogeneous WSN by using GA that optimizes the network life is proposed in Ref. [50]. The proposed method provides a framework to integrate clustering factors and multiple heterogeneities that employ network locality, expected energy expenditure, remaining energy, and distance to the base station in search for an optimal, dynamic network structure for heterogeneous WSN.

Hussain et al. [51] has proposed intelligent energy-efficient hierarchical clustering protocol by using GA to increase the network lifetime and give more energy-efficient clusters for routing in WSNs. The simulation results show that the proposed method performs better than the traditional cluster-based protocols in terms of EE and lifetime.

In Ref. [52], the authors presented a novel method energy-efficient coverage control algorithm (ECCA) based on multiobjective genetic algorithms (MOGAs) for application-specific data gathering in a WSN. The basic goal of ECCA is to minimize the number of working nodes while maintaining full coverage. The proposed method extends the network lifetime by saving energy while meeting the coverage requirement. The simulation results show that the proposed algorithm can achieve balanced performance on different types of detection sensor models while maintaining high coverage rate.

An energy-efficient delay-constrained based on GA mechanism to resolve quality of service (QoS) multicast routing problem in mobile ad hoc network (MANET) is proposed in [53]. The proposed algorithm aims to find the bounded end-to-end delay and reduces the total energy consumption cost of the multicast tree. Crossover and mutation operations are applied directly on trees, which simplify the coding operation and omit the coding/decoding process. The simulation results have proven that the proposed algorithm is effective and efficient.

Tang and Pan [54], introduced a new method a hybrid genetic algorithm (HGA) for the energy-efficient virtual machine placement problem that considers the energy consumption in both the communication network and physical machines (PM) in a data center. For the set of randomly generated test problems, the performance of HGA and the original GA is compared in [54] and the energy consumption in the data cetnre is also determined by both the approaches. Results for proposed method showed that the HGA generated a significantly better quality of solutions than the original GA in terms of performance and efficiency.

GA-based energy-efficient multicast routing problem with multiple QoS constraints in MANETs is addressed in [55]. Source tree–based routing algorithm is used for generating the shortest path multicast tree to decrease delay time. In addition, improved genetic sequence and topology encoding are used to prolong the lifetime of mobile nodes that calculate the residual battery energy of all nodes in a multicast tree. The simulation results explain that the proposed method is an efficient and robust algorithm for multicast route selection.

10.6.2 Simulated annealing algorithm

Marotta et al. [56] proposed a novel mixed integer linear programming model for the virtual machines' consolidation based on SA to minimize the energy consumption wastefulness in cloud computing. The proposed method aims to evaluate the attractiveness of the possible virtual machine (VM) migrations and to increase the overall cost-efficiency by reducing the number of active nodes.

In Ref. [57], the authors proposed a SA algorithm to solve energy resources, scheduling based on virtual power player (VPP) operating in a smart grid. The results of the SA approach are effective to solve the envisaged problem and proved to be able to obtain good solutions in low execution times, providing VPPs with suitable decision support for which the huge amount of distributed resources will lead to combinatorial explosion making deterministic approaches useless in practice.

10.6.3 Particle swarm optimization

The authors [58] have proposed a novel usage of PSO in the cluster head (CH) selection, which makes it a semidistributed method known as PSO semidistributed (PSO-SD). The PSO-SD applies the fitness function to reduce the intracluster distance from the sensor nodes to the cluster head. The function optimized the location of the cluster head, which influences the expected number of packet retransmissions along the path, which helps in optimizing overall energy consumption in the network. The simulation results illustrate that our PSO-SD delivers better performance in terms of lifetime, energy consumption, and an average number of packets communicated to the base station (BS).

An energy-efficient CHs selection algorithm based on PSO known as PSO-ECHS is proposed in Ref. [59]. The algorithm is developed by using an efficient particle representation and fitness function. In the proposed method, the authors measure the performance by using several metrics such as energy consumption, network lifetime, and packets receiving. The proposed approach is better in performance measurement in terms of energy consumption since PSO-ECHS deals with the fitness function, which takes into account the sink distance, intracluster distance, and the residual energy of nodes, by reducing the distance between sensor nodes and their CHs to find their optimal locations.

Multiobjective MOPSO algorithm is used to manage the resources of the network by finding the optimal number of clusters in MANETs [60]. The optimal number of clusters can provide an energy-efficient solution and reduce the network traffic when managed intercluster and intracluster traffic by the CHs. Also, It reduces the routing cost of a packet, so that the CHs can do the job of routing network packets within the cluster or to the nodes of other clusters because the least number of nodes is involved in routing a packet. The simulation results illustrate that proposed MOPSO-based methods outperform these two algorithms [weighted clustering algorithm (WCA) and comprehensive learning particle swarm optimization (CLPSO)-based clustering] in finding the optimal number of clusters as well as providing multiple options for the user.

Wang et al. [61] proposed a novel energy-aware VM placement optimization approach for a heterogeneous virtualized data center by using the improved PSO. This approach aims to develop optimal VM replacement scheme with the lowest total energy consumption while satisfying the resource requirements of the cloud services.

A linear programming and nonlinear programming formulation for energy-efficient clustering of the WSN, which is based on PSO, is proposed in [62]. The algorithm is developed by using an efficient particle encoding scheme and fitness function to take care of energy consumption of the varying number of sensor nodes and the gateways (GWs). Also, the algorithm builds a trade-off between the number of data forwards and transmission distance through load balancing clustering.

In Ref. [63], the authors proposed a novel method based on PSO algorithm for energy-aware cluster-based protocol to extend the sensor network lifetime. The main idea of the proposed protocol is the optimization of energy management of the network by minimizing the intracluster distance between a CH and the cluster member. Also, centralized control algorithm is developed to implement all operations at the BS and compared the performance of proposed protocol low-energy adaptive clustering hierarchy (LEACH) and LEACH-C (improved version of LEACH). Results from the simulations demonstrate that the proposed protocol delivers more data and a higher network lifetime to the BS compared to LEACH and LEACH-C.

10.6.4 Ant colony optimization algorithm

In Ref. [64], the authors proposed a new method for energy-efficient routing in energy-constraint WSNs based on an ant colony algorithm known as data aggregation. The data aggregation for energy efficient (EE) depends on the number of sources. The simulation results proved that improve EE up to 45% when using optimal aggregation compared to approximate aggregation schemes in a moderate number of sources, whereas 20% EE in a large number of source nodes.

Lin et al. [65] proposed a family of energy-efficient data aggregation ant colony algorithms to achieve the global optimum in saving energy and prolonging network lifetime. To increase the probability of selecting global optima, ACO is modified in following categories:(1) elitist strategy-based DAACA; (2) max−min-based DAACA, (3) the basic algorithm of DAACA, and (4) ant colony system-based DAACA. In addition, the authors applied a number of simulations to evaluate and compare the performances between the DAACA family and other data aggregation algorithms. The results proved that the DAACA have higher superiority on EE and prolonging the network lifetime.

Energy-efficient ant-based routing algorithm is proposed in Ref. [66]. ACO helps for energy savings with maximized network lifetime and minimizes communication by choosing the shortest path between the sensor nodes and a sink node with less communication hops. The experimental results showed that the method leads to very good results in several WSN scenarios.

Gao et al. [67] proposed a multiobjective ant colony system algorithm for the VM placement problem in a cloud computing environment aiming to minimize power consumption and total resource wastage. The performance of the proposed algorithm is compared with a MOGA, a well-known bin-packing algorithm and an max-min ant system algorithm. Computational experiments on benchmark problems are carried out. The results demonstrate that the proposed algorithm can compete efficiently with two single objective approaches to the problem.

An ant colony-based scheduling algorithm (ACB-SA) is proposed to solve the efficient-energy coverage (EEC) of WSN problem [68]. The ACB-SA, unlike the traditional ACO algorithm, improves the performance for real-number space by applying the new initialization method based on the random selection of parameters for the probabilistic sensor detection model and the modified construction graph. The simulations are carried out to check the effectiveness of the ACB-SA in the EEC problem and compared with conventional ACO algorithm, PSO, and the three pheromones ACO (TPACO) algorithm.

In Ref. [69], the authors presented an efficient routing algorithm by using the ACO algorithm for large-scale cluster-based WSNs. The main objective is to find the optimal route from the CHs to the BS and leads to a shorter convergence time and less routing overhead. The results show a higher system lifetime, lower power consumption, and more loads, balancing for the proposed routing algorithm when compared the method with cluster-based routing without optimization and an ACO-based routing algorithm without clustering.

10.6.5 Artificial bee colony

Kruekaew et al. [70] proposed a new method to optimize the scheduling of VM on cloud computing by using ABC. This approach enhances the performance of cloud task scheduling due to the use of improved ABC algorithm for resources utilization and reduces the makespan of data processing time. The hybrid approach that uses ABC algorithm, the Longest Job First scheduling algorithm, and scheduling based on the size of tasks outperforms in balancing workload and changing environment.

Saleem et al. [71] proposed a bee-inspired power-aware routing protocol called as BeeSensor, which utilizes a simple bee agent model and requires less processing time and network resources. The experiment results demonstrate that BeeSensor delivers better performance in a dynamic WSNs scenario as compared to a WSN optimized version of the AODV protocol; also the computational complexity, control overhead, and bandwidth requirements are significantly smaller. In addition, lifetime achieved by BeeSensor is also larger than AODV.

A novel energy-efficient clustering mechanism based on an ABC algorithm is proposed in Ref. [72]. The routing algorithm which uses fitness functions of energy levels, energy consumption, and QoS is named as an improved version of Cluster-based WSN routings using an ABC algorithm considering Quality of service (ICWAQ). The proposed method selects optimum CH nodes to prolong the network lifetime and minimize energy consumption. Also, it employs a service quality mechanism by considering delays between the signals received from the clusters. The performance of the proposed approach is compared with protocols based on LEACH and PSO. The results of simulation results prove that the ICWAQ routing protocol can effectively maximize the network lifetime, minimize transfer delays comparing the other techniques, and successfully be applied to WSN routing protocols.

Honey bee mating algorithm for ad hoc routing is proposed in Ref. [73], which is applied for data clustering, scheduling and resource allocation, and optimization problems. This proposed approach has restructured the ABC algorithm from the initialization phase to the implementation phase. Selection of less number of parameters in this approach leads to less energy consumption, less traffic in the network, and consequent improvement in the lifetime of the battery as well as the network efficiency. The result shows that the proposed system provides optimal value in all aspects such as response time and throughput.

In Ref. [74], authors have proposed a new routing algorithm for energy-efficient routing in MANETs called BeeAdHoc. BeeAdHoc, a reactive source routing algorithm, achieves similar/better performance to that of DSR, AODV, and DSDV but consumes significantly less energy as compared to these state-of-the-art algorithms because it utilizes less control packets to do routing. The results of simulations show that the proposed algorithm achieves the objectives by sending less control packets and distributing data packets on multiple paths, and as a result, the algorithm is energy-efficient.

10.6.6 Gray wolf optimization

Diwan and Khan [75] has proposed the smart cluster forming solution based on fuzzy logic and gray wolf optimization (GWO)-based CH selection scheme to achieve EE in WSNs. In this method, while forming clusters each node evaluates the parameters such as residual energy, distance to BS, and distance to CH; based on these parameters, probability of being connected to any available CH is determined. In the next step, GWO algorithm is used to select CH by comparing the position of node in the cluster with the gray wolf hierarchy. Network lifetime and throughput of proposed scheme are compared with that of conventional LEACH-based scheme in both stable and unstable period of network operation. Fuzzy logic and GWO-based scheme outperforms conventional scheme.

A three-level hybrid clustering routing protocol algorithm (MLHP) based on the Gray wolf optimizer (GWO) for WSNs is proposed in Ref. [76]. A centralized selection is proposed for Level One, in which the BS plays a great role in selecting CHs. In Level Two, a GWO routing for data transfer is proposed, where nodes select the best route to the BS to save more energy. And, a distributed clustering based on a cost function is proposed for Level Three. The algorithm was evaluated through tests of a network's EE, lifetime, and stability period. Comparisons were made with the best-known routing protocols to measure the performance of the proposed algorithm. The results showed improved performance of the proposed algorithm in terms of longer network lifetime, longer stability period, and more residual energy when compared with the other algorithms.

Sharawi et al. [76] have introduced a CH selection optimization model in WSNs based on GWO. Fitness function used in this approach ensures coverage of the WSN and is fed to the GWO to find its optimum. Results of the introduced model are compared with the LEACH routing protocol. Lifetime, residual energy, and network throughput performance indicators are examined; the introduced system outperforms the LEACH in almost all topologies [77].

10.7 Performance analysis

This section provides the analysis and evolution of meta-heuristic techniques for energy optimization. The analysis is based on the impact of meta-heuristics techniques on the energy optimization for many approaches such as clustering mechanism, routing mechanism, LPWAN, and VM mechanism.

10.7.1 Energy optimization with clustering mechanism

Clustering is one of the most efficient techniques that save energy through extended lifetime of the whole network.Lot of research has been reported in the literature for the hybrid approach based on clustering and meta heuristics for energy optimization. In Refs. [47,48], the researchers proposed using GA for energy-efficient clusters in WSN. The simulation results explain that using GA with clusters increases the network life time and provides more energy efficiency than the traditional cluster-based protocols.

Energy Efficient clustering approach using PSO is proposed in [58,59] to optimize the energy consumption of the network while managing the network traffic. All proposed algorithm takes into consideration a variety of parameters, such as intracluster distance, transmission power, sink distance, degree of nodes, and residual energy, to select CH for each cluster. The studies show that PSO-ECHS has greater EE and more power than no CHs [75].

10.7.2 Energy optimization with routing mechanism

Several techniques have been proposed by researchers to resolve QoS multicast routing problem in MANET. In Ref. [50], the authors proposed that GA depends on bounded end-to-end delay and minimum energy cost of the multicast tree. While authors [55] designed a source tree–based routing algorithm is proposed in [55] and used the small population size in the GA to built the shortest path multicast tree. In [53], BeeAdHoc is presented, a new routing algorithm for energy-efficient routing in MANETs. The results of simulation experiments show that BeeAdHoc consumes significantly less energy than dynamic source routing protocol (DSR), ad hoc on-demand distance vector routing protocol (AODV), and destination sequenced distance vector routing protocol (DSDV).

In Ref. [71] authors proposed a new bee-inspired power-aware routing protocol for WSNs called BeeSensor and also compared BeeSensor with AODV. The results of experiments show that BeeSensor delivers better performance than AODV protocol. In Ref. [69], the authors proposed a routing algorithm for the cluster-based large-scale WSNs using the ACO. The simulation results showed a higher system lifetime and load balancing for the proposed routing algorithm compared to other routing algorithms. A three-level hybrid clustering routing protocol algorithm (MLHP) based on the gray wolf optimizer (GWO) for WSNs [76] performs very well in terms of longer network lifetime, longer stability period, and more residual energy when compared with the other algorithms.

10.7.3 Energy optimization with virtual machine

VM placement is one of the important approaches for enhancing EE and resource utilization in both PM and the communication network in a data center. HGA outperforms the original GA [54]. An approach based improved PSO reduces the total energy consumption in a heterogeneous virtualized data center. It is proven that proposed approach significantly outperforms other approaches [61].

Multiobjective ant colony system algorithm is better choice for the VM placement problem to minimize total resource wastage and power consumption. Also, ABC algorithm is suitable for optimizing the scheduling of VM on cloud computing due to its ability to effectively utilize the reduce makespan and increased system resources [70].

10.7.4 Energy optimization in low-power wide area network

In Ref. [5], authors addressed the maximum lifetime coverage problem by a greedy algorithm that aims at determining a sleep/active schedule for sensors in order to maximize the time span when all the targets are continuously covered by using LoRaWAN. Results explain that the greedy algorithm performs very well whether the energy consumed in sleep mode is negligible or not and no matter the dispersion of the sensors' energies, which is contrary to existing algorithms.

Rady et al. [78] designed the optimal deployment of LPWAN IoT GWs by using a computational method. The proposed method examined network-agnostic (using grid method and spatial algorithm), which showed competitive results compared to a network-aware (using K-means clustering). Results showed that the proposed approach decreases the costs of the network and increases its sustainability.

In Ref. [79], the authors analyzed the feasibility of providing over-the-air software updates for three LPWAN technologies (IEEE-802.15.4g, LoRa, and Sigfox) and discussed the best suited update method and compared for three different scenarios: network stack update, application updates, and full system updates. The study prove that a single applications may have less energy between 6 to 38 times when it compared with firmware update.

10.8 Conclusion

Numerous scheduling and resources utilization techniques for EE in communication networks have been developed in the last few decades, but still it is a challenging task. A scheduling and resources utilization technique for EE developed may perform well for one scenario but not for the other. Hence, neither the single scheduling nor resources utilization method is applicable to all types of networks nor do all the methods perform well for one specific scenario. In this chapter, we have presented the in-depth review of recent scheduling and resources utilization techniques for EE in communication networks and their deviations. These methods are studied analytically and comparison is carried out on the basis of distinct parameters. The state-of-the-art review will be useful in the selection of the appropriate method. Such study and evaluation are also essential for refining the performance of existing algorithms and for developing new powerful algorithms. Their performances can be enhanced by the use of hybrid approach and correct optimization.

References

[1] H. Rong, H. Zhang, S. Xiao, C. Li, C. Hu, Optimizing energy consumption for data centers, Renew. Sustain. Energy Rev. 58 (2016) 674−691.

[2] U. Raza, P. Kulkarni, M. Sooriyabandara, Low power wide area networks: an overview, IEEE Commun. Surv. Tutor. 19 (2) (2017) 855−873.

[3] Z. Yan, M. Peng, C. Wang, Economical energy efficiency: an advanced performance metric for 5G systems, IEEE Wirel. Commun. 24 (1) (2017) 32−37.

[4] K. Mekki, E. Bajic, F. Chaxel, F. Meyer, Overview of cellular LPWAN technologies for IoT deployment: Sigfox, LoRaWAN, and NB-IoT, 2018 IEEE International Conference on Pervasive Computing and Communications Workshops (PerCom Workshops), IEEE, 2018, pp. 197–202.

[5] D.T. Chakonté, E. Simeu, M. Tchuente, Lifetime optimization of wireless sensor networks with sleep mode energy consumption of sensor nodes, Wireless Netw. (2018) 1–10.

[6] P. Thubert, A. Pelov, S. Krishnan, Low-power wide-area networks at the IETF, IEEE Commun. Stand. Mag. 1 (1) (2017) 76–79.

[7] Q.M. Qadir, T.A. Rashid, N.K. Al-Salihi, B. Ismael, A.A. Kist, Z. Zhang, Low power wide area networks: a survey of enabling technologies, applications and interoperability needs, IEEE Access 6 (2018) 77454–77473.

[8] Z. Qin, F.Y. Li, G.Y. Li, J.A. McCann, Q. Ni, Low-power wide-area networks for sustainable IoT, IEEE Wirel. Commun. 26 (2019) 140–145.

[9] N.I. Osman, E.B. Abbas, Simulation and modelling of LoRa and Sigfox low power wide area network technologies, 2018 International Conference on Computer, Control, Electrical, and Electronics Engineering (ICCCEEE) (, IEEE, 2018, pp. 1–5.

[10] Gavrilas, M., 2010. Heuristic and metaheuristic optimization techniques with application to power systems, in: Proceedings of the 12th WSEAS international conference on Mathematical Methods and Computational Techniques in Electrical Engineering, Romania.

[11] R. Franz, Optimization Problems, Chapter 1, Springer, 2011.

[12] R. Martí, G. Reinelt, The Linear Ordering Problem: Exact and Heuristic Methods in Combinatorial Optimization, vol. 175, Springer Science & Business Media, 2011.

[13] A. Dixit, A. Mishra, A. Shukla, Vehicle Routing Problem With Time Windows Using Meta-Heuristic Algorithms: A Survey. In Harmony Search and Nature Inspired Optimization Algorithms, Springer, Singapore, 2019, pp. 539–546.

[14] X. Yu, M. Gen, Introduction to Evolutionary Algorithms, Springer Science & Business Media, 2010.

[15] G. Pólya, How to Solve It, Princeton University, Princeton, NJ, 1945.

[16] M. Khamees, A. Albakry, K. Shaker, Multi-objective feature selection: hybrid of salp swarm and simulated annealing approach, International Conference on New Trends in Information and Communications Technology Applications, Springer, Cham, 2018, pp. 129–142.

[17] H. Altinbas, G.C. Akkaya, Improving the performance of statistical learning methods with a combined meta-heuristic for consumer credit risk assessment, Risk Manag. 19 (4) (2017) 255–280.

[18] S. Memeti, S. Pllana, A. Binotto, J. Kołodziej, I. Brandic, Using meta-heuristics and machine learning for software optimization of parallel computing systems: a systematic literature review, Computing 101 (2018) 893–936.

[19] C. Blum, A. Roli, Metaheuristics in combinatorial optimization: overview and conceptual comparison, ACM Comput. Surv. (CSUR) 35 (3) (2003) 268–308.

[20] K.K.H. Ng, C.K.M. Lee, F.T. Chan, Y. Lv, Review on meta-heuristics approaches for airside operation research, Appl. Soft Comput. 66 (2018) 104–133.

[21] M. Mohammadi, M. Nastaran, A. Sahebgharani, Development, application, and comparison of hybrid meta-heuristics for urban land-use allocation optimization: Tabu search, genetic, GRASP, and simulated annealing algorithms, Comput. Environ. Urban Syst. 60 (2016) 23–36.

[22] J.H. Holland, Adaptation in Natural and Artificial Systems: An Introductory Analysis with Applications to Biology, Control, and Artificial Intelligence, MIT Press, 1992.

[23] M. Gendreau, J.Y. Potvin (Eds.), Handbook of Metaheuristics, vol. 2, Springer, New York, 2010.

[24] S.N. Sivanandam, S.N. Deepa, Introduction to Genetic Algorithms, Springer Science & Business Media, 2007.

[25] K.F. Man, K.S. Tang, S. Kwong, Genetic algorithms: concepts and applications [in engineering design], IEEE Trans. Ind. Electron. 43 (5) (1996) 519–534.

[26] O. Kramer, Genetic Algorithm Essentials, vol. 679, Springer, 2017.

[27] A. Konak, D.W. Coit, A.E. Smith, Multi-objective optimization using genetic algorithms: a tutorial, Reliability Engineering & System Safety 91 (9) (2006) 992–1007.

[28] G. Syswerda, Simulated Crossover in Genetic Algorithms. In Foundations of Genetic Algorithms, vol. 2, Elsevier, 1993, pp. 239–255.

[29] J. Kennedy, R. Eberhart, Particle swarm optimization, Proceedings of IEEE International Conference on Neural Networks, 4, IEEE Press, 1995, pp. 1942–1948. 2.

[30] P. Singh, M. Dutta, N. Aggarwal, A review of task scheduling based on meta-heuristics approach in cloud computing, Knowl. Inf. Syst. 52 (1) (2017) 1–51.

[31] M. Dorigo, G. Di Caro, Ant colony optimization: a new meta-heuristic, Proceedings of the 1999 Congress on Evolutionary Computation-CEC99 (Cat. No. 99TH8406), Vol. 2, IEEE, 1999, pp. 1470–1477.

[32] M. Dorigo, T. Stützle, Ant colony optimization: overview and recent advances, Handbook of Metaheuristics, Springer, Cham, 2019, pp. 311–351.

[33] M. Dorigo, M. Birattari, T. Stutzle (2006). Ant colony optimization: artificial ant as a computational intelligence technique. University libre de Bruxelles. IRIDIA Technical report Series, Belgium.

[34] F. Glover, Tabu search—part I, ORSA J. Comput. 1 (3) (1989) 190–206.

[35] M. Gendreau, J.Y. Potvin, Tabu search, Search Methodologies, Springer, Boston, MA, 2005, pp. 165–186.

[36] S. Kirkpatrick, C.D. Gelatt, M.P. Vecchi, Optimization by simulated annealing, Science 220 (4598) (1983) 671–680.

[37] P. Bujok, Cooperative model for nature-inspired algorithms in solving real-world optimization problems, International Conference on Bioinspired Methods and Their Applications, Springer, Cham, 2018, pp. 50–61.

[38] S. Bitam, M. Batouche, E.G. Talbi, A survey on bee colony algorithms, 2010 IEEE International Symposium on Parallel & Distributed Processing, Workshops and Phd Forum (IPDPSW), IEEE, 2010, pp. 1–8.

[39] D. Karaboga, B. Gorkemli, C. Ozturk, N. Karaboga, A comprehensive survey: artificial bee colony (ABC) algorithm and applications, Artif. Intell. Rev. 42 (1) (2014) 21–57.

[40] S. Mirjalili, S.M. Mirjalili, A. Lewis, Grey wolf optimizer, Adv. Eng. Softw. 69 (2014) 46–61.

[41] N. Krasnogor, J. Smith, A tutorial for competent memetic algorithms: model, taxonomy, and design issues, IEEE Trans. Evolut. Comput. 9 (5) (2005) 474–488.

[42] Y.S. Ong, M.H. Lim, N. Zhu, K.W. Wong, Classification of adaptive memetic algorithms: a comparative study, IEEE Trans. Syst. Man Cybern. B Cybern. 36 (1) (2006) 141–152.

[43] P. Moscato, C. Cotta, A. Mendes, Memetic algorithms, New Optimization Techniques in Engineering, Springer, Berlin, Heidelberg, 2004, pp. 53–85.

[44] F. Neri, C. Cotta, Memetic algorithms and memetic computing optimization: a literature review, Swarm Evolut. Comput. 2 (2012) 1–14.

[45] X.S. Yang, X. He, Swarm intelligence and evolutionary computation: overview and analysis, Recent Advances in Swarm Intelligence and Evolutionary Computation, Springer, Cham, 2015, pp. 1–23.

[46] J. Brest, V. Zumer, M.S. Maucec, Self-adaptive differential evolution algorithm in constrained real-parameter optimization, 2006 IEEE International Conference on Evolutionary Computation, IEEE, 2006, pp. 215–222.

[47] S. Dey, S. Bhattacharyya, U. Maulik, Efficient quantum inspired meta-heuristics for multi-level true colour image thresholding, Appl. Soft Comput. 56 (2017) 472−513.

[48] R. Mallipeddi, P.N. Suganthan, Q.K. Pan, M.F. Tasgetiren, Differential evolution algorithm with ensemble of parameters and mutation strategies, Appl. Soft Comput. 11 (2) (2011) 1679−1696.

[49] H.P. Borges, O.A.C. Cortes, D. Vieira, An adaptive metaheuristic for unconstrained multimodal numerical optimization, International Conference on Bioinspired Methods and Their Applications, Springer, Cham, 2018, pp. 26−37.

[50] M. Elhoseny, X. Yuan, Z. Yu, C. Mao, H.K. El-Minir, A.M. Riad, Balancing energy consumption in heterogeneous wireless sensor networks using genetic algorithm, IEEE Commun. Lett. 19 (12) (2015) 2194−2197.

[51] S. Hussain, A.W. Matin, O. Islam, Genetic algorithm for energy efficient clusters in wireless sensor networks, Fourth International Conference on Information Technology (ITNG'07), IEEE, 2007, pp. 147−154.

[52] J. Jia, J. Chen, G. Chang, Z. Tan, Energy efficient coverage control in wireless sensor networks based on multi-objective genetic algorithm, Comput. Math. Appl. 57 (11−12) (2009) 1756−1766.

[53] T. Lu, J. Zhu, Genetic algorithm for energy-efficient QoS multicast routing, IEEE Commun. Lett. 17 (1) (2013) 31−34.

[54] M. Tang, S. Pan, A hybrid genetic algorithm for the energy-efficient virtual machine placement problem in data centers, Neural Process. Lett. 41 (2) (2015) 211−221.

[55] Y.S. Yen, Y.K. Chan, H.C. Chao, J.H. Park, A genetic algorithm for energy-efficient based multicast routing on MANETs, Comput. Commun. 31 (10) (2008) 2632−2641.

[56] A. Marotta, S. Avallone, A simulated annealing based approach for power efficient virtual machines consolidation, 2015 IEEE 8th International Conference on Cloud Computing, IEEE, 2015, pp. 445−452.

[57] T. Sousa, H. Morais, Z. Vale, P. Faria, J. Soares, Intelligent energy resource management considering vehicle-to-grid: a simulated annealing approach, IEEE Trans. Smart Grid 3 (1) (2012) 535−542.

[58] B. Singh, D.K. Lobiyal, A novel energy-aware cluster head selection based on particle swarm optimization for wireless sensor networks, Hum. Cent. Comput. Inf. Sci. 2 (1) (2012) 13.

[59] P.S. Rao, P.K. Jana, H. Banka, A particle swarm optimization based energy efficient cluster head selection algorithm for wireless sensor networks, Wirel. Netw. 23 (7) (2017) 2005−2020.

[60] H. Ali, W. Shahzad, F.A. Khan, Energy-efficient clustering in mobile ad-hoc networks using multi-objective particle swarm optimization, Appl. Soft Comput. 12 (7) (2012) 1913−1928.

[61] S. Wang, Z. Liu, Z. Zheng, Q. Sun, F. Yang, Particle swarm optimization for energy-aware virtual machine placement optimization in virtualized data centers, 2013 International Conference on Parallel and Distributed Systems, IEEE, 2013, pp. 102−109.

[62] P. Kuila, P.K. Jana, Energy efficient clustering and routing algorithms for wireless sensor networks: particle swarm optimization approach, Eng. Appl. Artif. Intell. 33 (2014) 127−140.

[63] N.A. Latiff, C.C. Tsimenidis, B.S. Sharif, Energy-aware clustering for wireless sensor networks using particle swarm optimization, 2007 IEEE 18th International Symposium on Personal, Indoor and Mobile Radio Communications, IEEE, 2007, pp. 1−5.

[64] R. Misra, C. Mandal, Ant-aggregation: ant colony algorithm for optimal data aggregation in wireless sensor networks, 2006 IFIP International Conference on Wireless and Optical Communications Networks, IEEE, 2006. p. 5.

[65] C. Lin, G. Wu, F. Xia, M. Li, L. Yao, Z. Pei, Energy efficient ant colony algorithms for data aggregation in wireless sensor networks, J. Comput. Syst. Sci. 78 (6) (2012) 1686−1702.

[66] T. Camilo, C. Carreto, J.S. Silva, F. Boavida, An energy-efficient ant-based routing algorithm for wireless sensor networks, International Workshop on Ant Colony Optimization and Swarm Intelligence, Springer, Berlin, Heidelberg, 2006, pp. 49−59.

[67] Y. Gao, H. Guan, Z. Qi, Y. Hou, L. Liu, A multi-objective ant colony system algorithm for virtual machine placement in cloud computing, J. Comput. Syst. Sci. 79 (8) (2013) 1230−1242.

[68] J.W. Lee, J.J. Lee, Ant-colony-based scheduling algorithm for energy-efficient coverage of WSN, IEEE Sens. J. 12 (10) (2012) 3036−3046.

[69] A.A. Salehpour, B. Mirmobin, A. Afzali-Kusha, S. Mohammadi, An energy efficient routing protocol for cluster-based wireless sensor networks using ant colony optimization, 2008 International Conference on Innovations in Information Technology, IEEE, 2008, pp. 455−459.

[70] B. Kruekaew, W. Kimpan, 2014. Virtual machine scheduling management on cloud computing using artificial bee colony, in: Proceedings of the International Multi-Conference of Engineers and Computer Scientists, vol. 1, pp. 12−14.

[71] M. Saleem, M. Farooq, BeeSensor: a bee-inspired power aware routing protocol for wireless sensor networks, Workshops on Applications of Evolutionary Computation, Springer, Berlin, Heidelberg, 2007, pp. 81−90.

[72] D. Karaboga, S. Okdem, C. Ozturk, Cluster based wireless sensor network routing using artificial bee colony algorithm, Wirel. Netw. 18 (7) (2012) 847−860.

[73] B.C. Mohan, R. Baskaran, Energy aware and energy efficient routing protocol for adhoc network using restructured artificial bee colony system, International Conference on High Performance Architecture and Grid Computing, Springer, Berlin, Heidelberg, 2011, pp. 473−484.

[74] H.F. Wedde, M. Farooq, T. Pannenbaecker, B. Vogel, C. Mueller, J. Meth, et al., BeeAdHoc: an energy efficient routing algorithm for mobile ad hoc networks inspired by bee behavior, Proceedings of the 7th Annual Conference on Genetic and Evolutionary Computation, ACM, 2005, pp. 153−160.

[75] P. Diwan, M. Khan, Energy efficient communication for WSNs using Grey-Wolf optimization algorithm, Int. J. Eng. Comput. Sci. 5 (12) (2016) 19793−19805.

[76] M. Sharawi, E. Emary, 2017. Impact of grey wolf optimization on WSN cluster formation and lifetime expansion, in: Ninth International Conference on Advanced Computational Intelligence (ICACI), Feb. 2017.

[77] l-Aboody, H.S. Al-raweshidy, Grey wolf optimization-based energy-efficient routing protocol for heterogeneous wireless sensor networks, in: 2016 4th International Symposium on Computational and Business Intelligence (ISCBI), 2016, pp. 101−107. 10.1109/ISCBI.2016.7743266.

[78] M. Rady, M. Hafeez, S.A.R. Zaidi, Computational methods for network-aware and network-agnostic IoT low power wide area networks (LPWAN), IEEE Internet Things J 6 (2019) 5732−5744.

[79] P. Ruckebusch, S. Giannoulis, I. Moerman, J. Hoebeke, E. De Poorter, Modelling the energy consumption for over-the-air software updates in LPWAN networks: SigFox, LoRa and IEEE 802.15. 4g, Internet Things 3 (2018) 104−119.

11

Energy harvesting—enabled relaying networks

Miroslav Voznak[1], Hoang-Sy Nguyen[1,2], Radek Fujdiak[1,3]

[1]VSB—TECHNICAL UNIVERSITY OF OSTRAVA, OSTRAVA, CZECH REPUBLIC [2]BINH DUONG UNIVERSITY, THU DAU MOT CITY, VIETNAM [3]BRNO UNIVERSITY OF TECHNOLOGY, BRNO, CZECH REPUBLIC

11.1 Introduction

Despite drawing much research attention in recent years because of the growing demand for services requiring wireless connection, limited lifetime is often an enormous challenge for wireless systems, specifically wireless sensor networks (WSNs) or body sensor networks (BSNs). Therefore, energy harvesting (EH) tends to be a prime technology in which energy is harvested directly from the surrounding environment, for example, solar, geothermal, wind, etc., in order to help wireless systems cope with a short lifetime and the frequent replacement of batteries [1,2]. Let us therefore start with an overall picture regarding wireless systems.

RF-EH technology, which can be used to generate energy from RF signals, has recently been seen as a hot research topic [3]. However, since power-constrained wireless networks, that is, WSNs [4], body networks (BNs) [5], and wireless charging systems [6] often face a lack of radio sources that are vital to IT and signal processing, much effort has been made to propose solutions that cope with the limitations of RF-EH. Fortunately, the technique of wireless energy transfer has greatly helped RF-EH by introducing a robust model called simultaneous wireless information and power transfer (SWIPT) [7]. In principle, SWIPT allows energy and information to be carried by RF signals simultaneously.

In Fig. 11–1, an RF-EH system with three primary components is depicted, including information gateways, sources, and nodes. BSs, known as information gateways, are used to illustrate wireless routers or relays. Next, RF sources can be deployed as RF transmitters or ambient RF sources, where nodes represent user equipment (UE) that communicates with the information gateways. A fixed power supply is equipped at each gateway and source, while the energy harvested from RF sources can be used to power the network nodes.

In order to enhance energy efficiency (EE), relaying networks (RNs) have been thoroughly studied and can support short multihop communication without consuming much energy compared to direct communication. Therefore, better coverage and network lifetime can be

LPWAN Technologies for IoT and M2M Applications. DOI: https://doi.org/10.1016/B978-0-12-818880-4.00012-0

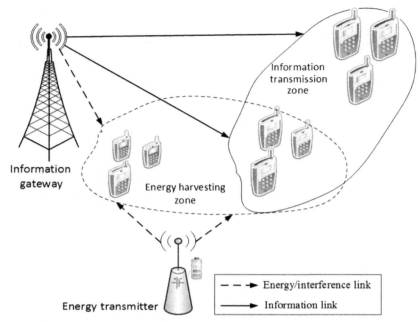

FIGURE 11–1 Infrastructure-based architecture RF-EH.
The devices located in the EH zone (unbroken line) scavenge energy from base station (BS), while components in the information transmission (IT) zone (dashed line) can decode information. Note that each node in the EH and IT zones is equipped with a separate radio frequency (RF) energy harvester and RF transceiver, respectively. Thus, both EH and IT can be done simultaneously.

achieved using RNs. Apart from this, two novel relaying protocols, namely amplify-and-forward (AF) and decode-and-forward (DF), were once deployed in [8]. In particular, the DF scheme lets the R (relay node) receive transmitted signals from the S (source node), which are later decoded and forwarded to the D (destination node). Alternatively, the AF scheme lets R receive, amplify, and forward the source signals to D.

Figs. 11−2 and 11−3 present two receiver architectures based on time switching (TS) or power splitting (PS) architectures, respectively. To begin, TS is technically deployed to coordinate a time for receiving information and RF-EH. Fig. 11−4 depicts a modernized version of TS called the time switching-based relaying (TSR) protocol.

However, the PS architecture with an optimal ratio can split the received RF signals if the circuit power consumption is trivial. Fig. 11−5 illustrates power splitting-based relaying protocol (PSR), which has been deployed widely in different systems, that is, [9,10], and [11].

We also propose a hybrid power time switching-based relay (HPTSR) protocol, which is illustrated in Fig. 11−6. In this protocol [12], T is the time block, where S transmits data to D and αT depicts R harvesting power of the transmitted signal from S $(0 < \alpha < 1)$. The remaining time of $(1 - \alpha)T$ for IT is split into two equal portions of $(1 - \alpha)T/2$ for the S-R link and R-D link, respectively. R consumes all the harvested energy during the energy transfer

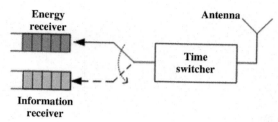

FIGURE 11–2 Time switching architecture.

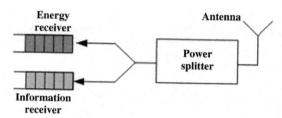

FIGURE 11–3 Power splitting architecture.

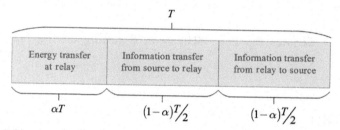

FIGURE 11–4 Time switching-based relay protocol.
The TS receiver model switches between EH and information decoding modes. The received signal denoted by y_R is transmitted to the EH receiver for an amount of time, αT, and then to the information receiver in $(1 - \alpha)T$, where T is the time required to transmit information between the source and destination.

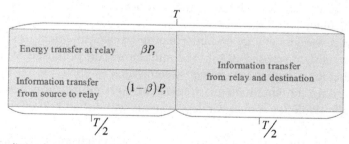

FIGURE 11–5 Power splitting-based relay protocol.
The power of the received signal is split into two parts in PS receiver architecture and is called the PS ratio, denoted by β. A portion, $\sqrt{\beta}y_R$, of the received signal is transmitted to the EH receiver, while the remaining portion of $\sqrt{(1 - \beta)}y_R$ is used for the information receiver.

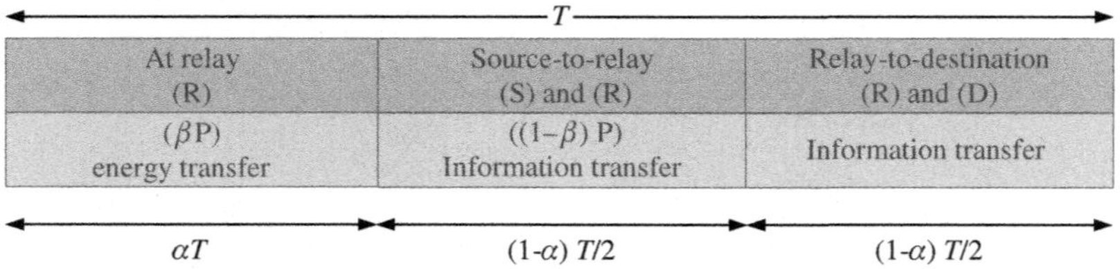

		T
At relay (R)	Source-to-relay (S) and (R)	Relay-to-destination (R) and (D)
(βP) energy transfer	$((1-\beta)\,P)$ Information transfer	Information transfer
αT	$(1-\alpha)\,T/2$	$(1-\alpha)\,T/2$

FIGURE 11–6 The hybrid TSR-PSR protocol.

process to forward signals to D. βP_S is used for EH, where P_S is the transmit power at S while the PS ratio $(1 - \beta)P_S$ is utilized for IT.

The harvested energy at R in TSR, PSR, and HPTSR protocols is therefore defined as [13]

$$E_h^{T\,S\,R} = \eta P_S |h_S|^2 \alpha T, \tag{11.1}$$

$$E_h^{P\,S\,R} = \frac{1}{2}\eta P_S |h_S|^2 \beta T, \tag{11.2}$$

and

$$E_h^{H\,PTS\,R} = \eta P_S |h_S|^2 \alpha \beta T, \tag{11.3}$$

where $0 \leq \eta \leq 1$ denotes the EH efficiency at R relying on the EH circuitry and the rectifier.

11.2 State of the art

In recent years, the power consumption of wireless devices in wireless communication systems has drawn much research interest [14,15]. Particularly, the study in [14] considered a network model of a set of RF charging stations overlaid with an uplink cellular network, while a harvest-then-transmit protocol was proposed for wireless power transfer [15]. Different types of up-to-date beamforming methods have also been implemented to enhance power transfer efficiency [15–17].

In order to assist wireless energy and information transfer, the dual use of RF signals was recently and thoroughly discussed [9,18]. The authors in [19] proposed SWIPT for the purpose of transmitting RF energy in low-power systems such as BSNs. SWIPT therefore has several benefits regarding the simultaneous transfer of both information and energy [18–20] and has proved affordable for robust wireless systems without additional demands on hardware in transmitting devices.

RF signals also have the potential to power low-power electronics such as smart watches, wireless keyboards, and mice because they only consume a small amount of power of roughly microwatts to milliwatts. Therefore, the work in [21] designed an RF circuit to make nonstop charging of mobile electronics possible in crowded areas where more RF signals can be harvested. In fact, the amount of data transmitted is dependent on the RF signals due to entropy fluctuations, while the average squared value of RF signals represents its power. As a consequence, optimizing information and energy simultaneously is not possible and requires innovations in wireless systems.

11.2.1 Scenario 1: Wireless power transfer constraint policies

Unlike conventional wireless communication devices, which often suffer from power constraints, EH transmitters are subject to other EH constraints. In particular, the transmitter in each time slot can only consume at most the amount of stored energy available despite the availability of more energy in upcoming slots. The use of the harvested energy should therefore be taken into consideration.

In fact, a number of studies on the use of energy harvesters as an energy source have been conducted in [22–27]. The study in [22] discussed the technique of dynamic programming. In [24], throughput performance was optimized over an infinite horizon. Adaptive duty cycling was employed for throughput optimization in [25]. In [26], an information-theoretic technique was considered in which energy is harvested at the level of channel use. The work in [27] presented excellent methods for optimizing throughput performance.

In practice, full-duplex (FD) transmission mode allows the transmitter and receiver to operate simultaneously in the same frequency band, assists coverage expansion, and extends network lifetime. This has attracted much research interest in recent years [28,29]. Because of FD transmission mode, spectral efficiency (SE) in future communication systems can be boosted more efficiently than in half-duplex (HD) mode. FD mode is also regarded as an ideal mechanism for improving data rate in order to satisfy the requirements of better transmission rates.

Following that, FD mode tends to bring more benefits than HD mode in terms of coping with self-interference (SI). In [30], the authors evaluated the impact of SI despite several disadvantages discussed in [30]. It is, in fact, intractable and approximations are required because the noise floor increases as a result of the interfering signal. In the SI model discussed in [31], higher transmit power helps obtain better SI cancellation and SI channel estimation. Despite the perfect SI cancellation in order to achieve an upper bound for FD mode performance, it was considered unfeasible [32].

The use of EH transmitters in multihop scenario has also been addressed in many works [33–35]. In [34], EH transmitters must harvest the available energy at any time. In [35], a HD relaying channel using DF protocol was considered, in which two delay constraints were proposed, namely delay-constrained and no-delay-constrained traffic. The study in [36] focused on promising FD methods, such as SI cancellation solutions, which were categorized into passive suppression, analog cancellation, and digital cancellation. Recently, SE loss has been overcome with the help of FD transmission mode by allowing users to exchange information in the same frequency band [37,38].

However, power allocation regarding opportunistic EH-assisted relay schemes has not been studied thoroughly in any previous works; therefore power supply policies will concentrate on DF FD RNs.

11.2.2 Scenario 2: The impact of channel state information using HTPSR protocol

The presence of imperfect channel state information (CSI) in wireless communication systems has attracted much research interest [39–43]. In particular, the authors in [39] studied a transmit power allocation policy for hybrid EH relay networks with channel and energy state uncertainties in order to optimize throughput performance. In [40], the performance of an RN under the impact of imperfect CSI was investigated. Fault-tolerant schemes were also analyzed in the presence of imperfect CSI [41]. In [42], a joint optimization problem was studied in relay selection (RS), subcarrier assignment, and PS ratio.

Although the closed-form expression for OP of two-way FD RNs with a residual SI was obtained [43], EH was not deeply studied in the work. In [44], the impact of CSI at S together with opportunistic regenerative relaying in order to guarantee quality of service (QoS) was studied in a CRN. However, ergodic capacity was not previously discussed concerning PS and TS protocols.

In our work, the end-to-end signal-to-noise ratio (SNR), ergodic capacity, Bit error rate (BER), and the throughput performance in the presence of imperfect CSI was studied. We attempted to optimize TS and PS ratios in HPTSR protocol and compared AF and DF schemes.

11.2.3 Scenario 3: The impact of hardware impairments on cognitive D2D communication

Ideal transceiver hardware is often assumed at all nodes despite the transceiver hardware not always being ideal because of certain issues such as phase noise, I/Q imbalance, and amplifier nonlinearities [45]. The capacity, throughput, and symbol error rate in the presence of hardware impairments (HWIs) were analyzed in [46]. The design for linear precoding and decoding for a two-way RN was studied in [47], examining the impact of both HWIs and CSI.

The impact of HWIs has been addressed in several works. An FD relay in a typical device-to-device (D2D) RN suffering from HWIs was closely studied in [48]. The work in [49] studied the OP of multirelay DF RNs with HWIs, in which RF-EH was fully deployed to support the transceivers.

Previous studies in [9,50–52] mainly focused on the presence of HWIs on particular nodes. In [50] and [51], the impact of HWIs in AF using TSR and PSR protocols was examined, while we worked on DF scheme using TSR protocol in [52]. However, the impact of HWIs in multihop D2D communications has little interest.

To overcome this situation, we derived closed-form expressions for the average EE and SE in order to improve understanding about energy consumption in cases with HWIs at all nodes. We also compared the performance of AF and DF schemes and optimized TS and PS ratios by applying a genetic algorithm (GA). However, the most important focus of this chapter was to optimize the successful transmission probability (STP).

11.3 Performance analysis

11.3.1 Scenario 1: Wireless power transfer constraint policies

In Fig. 11−7, R is deployed to assist S and D in communicating with each other, assuming that S and D are equipped with a single antenna, except R, which has two antennas. The channel coefficients of S-R link, R-D link, and the residual SI are denoted as h, g, and f, respectively. Next, the channel power gains of $|h|^2$, $|g|^2$, and $|f|^2$ are exponentially distributed random variables (RVs) with means $\Omega_h = d_{SR}^{-m}$, $\Omega_g = d_{RD}^{-m}$, and $\Omega_f = d_{LI}^{-m}$, respectively. The distances between S and R and R and D are represented as d_{SR} and d_{RD}, respectively, while the distance between two antennas at R is denoted as d_{LI}. The additive white Gaussian noise (AWGN) is denoted as N_0, with variance σ^2. The path-loss exponent is m.

Because SWIPT is deployed at R, it leads to residual SI, $|f|^2$, despite any applied SI cancellation techniques. To this point, let us consider OP denoted as P_{out}, which is an outage event occurring if the given target rate is higher than the predefined data rate.

11.3.1.1 Separated power mode
Because of an optimal power constraint such as the proposed separated power (SP) mode, we can easily optimize the instantaneous rate and the transmit powers, P_S^* and P_R^* in [53].

Proposition 1 *The analysis of OP for SP can be achieved from the expression*

$$OP^{SP}(\gamma_0) = 1 - \frac{P_S \Omega_h}{\overline{P}_S \Omega_h + \overline{P}_R \Omega_f \gamma_0} \exp\left(- \frac{\sigma^2 \gamma_0}{\overline{P}_R \Omega_g} \right), \tag{11.4}$$

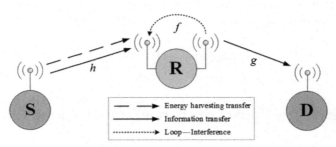

FIGURE 11–7 The system model for EH FD RNs.
Here, TSR protocol is deployed to study EH with *T* being the time block for IT. In particular, IT accounts for αT in the first phase, while the remaining portion, $(1 - \alpha)T$ is for IT in the second phase. Note that P_S and P_R denote the transmit powers at S and R, respectively.

where the SNR threshold is defined as $\gamma_0 = 2^{R_0} - 1$ *with a fixed source transmission rate* R_0 *(bps/Hz), and the average optimal transmit power at R is expressed as* $\overline{P}_R^* = \min\{\overline{P}_S^*, \sqrt{\sigma^2 \Omega_h / \Omega_g \Omega_f}\}$.

Therefore, the expression for asymptotic OP using FD protocol with $(\Omega_f \to 0)$ *is upper-bounded as*

$$OP_{Upper}^{SP}(\gamma_0) \approx 1 - \exp\left(-\frac{\sigma^2 \gamma_0}{\overline{P}_R \Omega_g}\right). \tag{11.5}$$

11.3.1.2 Harvested power-assisted relay

This harvested power (HP) policy follows the mechanism of EH, in which energy is harvested from RF signals because of the lack of a fixed power supply (i.e., batteries).

As a result, HP results in the received signal at R calculated as [54] $P_R = E_h/(1-\alpha)T = \eta\alpha(1-\alpha)^{-1} P_S|h|^2$.

Despite the suitable choice of α_{opt} and knowing that its impact on the instantaneous rate leads to more energy harvested at S and R, the throughput performance is poor at D. We therefore optimize the TS coefficient as

$$\alpha_{opt} = \left[\frac{2\eta\frac{P_S}{\sigma^2}\Omega_h\Omega_f}{\left(\sqrt{1+\frac{P_S}{\sigma^2}\frac{\Omega_h\Omega_f}{\Omega_g}}-1\right)} + 1\right]^{-1} \tag{11.6}$$

Proposition 2 *The expression for OP in HP mode can be given as* [53]

$$OP^{HP}(\gamma_0) = 1 - 2\sigma^2\left(1 - \exp\left(-\frac{(1-\alpha)}{\eta\alpha\gamma_0\Omega_f}\right)\right)\sqrt{\frac{\gamma_0(1-\alpha)}{\eta\alpha P_S\Omega_h\Omega_g}}K_1\left(2\sigma^2\sqrt{\frac{\gamma_0(1-\alpha)}{\eta\alpha P_S\Omega_h\Omega_g}}\right), \tag{11.7}$$

where the modified Bessel function of the second kind with order n is denoted by K_n.

However, the approximation of $xK_1(x) \approx 1$ *with respect to* $(x \to 0)$. *If the high-power regime* $(P_S \to \infty)$ *is used, the asymptotic OP for HP mode is expressed as*

$$OP_{Upper}^{HP}(\gamma_0) \approx \exp\left(-\frac{(1-\alpha)}{\eta\alpha\gamma_0\Omega_f}\right). \tag{11.8}$$

11.3.2 Scenario 2: The impact of channel state information using HTPSR protocol

In Fig. 11–8, $\tilde{h}_1$ and $\tilde{h}_2$ are shown as the dual-hop communication between S, R, and D in a block time denoted as T. Communication between S and R in the first hop lasts αT, which involves EH and IT, while IT accounts for $(1 - \alpha)T$ to conduct IT in the second hop. In our

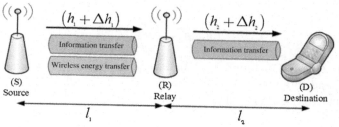

FIGURE 11–8 The system model.
We assume that Rayleigh fading impacts all nodes that are independent and identically distributed (i.i.d) from one block to another. However, the fading channel denoted as the zero mean circularly symmetric complex Gaussian (CSCG) RV is distributed by $\tilde{h}_1 \sim CN(0, \Omega_{\tilde{h}_1})$, $\tilde{h}_2 \sim CN(0, \Omega_{\tilde{h}_2})$. The distances between $S \to R$ and $R \to D$ are l_1 and l_2, respectively.

proposed HTPSR protocol, α and β are the TS and PS ratios in which $\alpha \in (0, 1)$ and $\beta \in (0, 1)$. In particular, βP_S is used in the energy transfer from S to R, while the IT from S to R accounts for $(1 - \beta)P_S$, where P_S is the source transmit power. The received power P_R in the second hop is given as $P_R = \varphi P_S(|h_1|^2 + \Omega_{\Delta h_1})l_1^{-m}$, where $\varphi = \eta \alpha \beta (1 - \alpha)^{-1}$.

In the first hop, the fading channel, $\tilde{h}_1$ is calculated as in [39], $\tilde{h}_1 = h_1 + \Delta h_1$. Similarly, the fading channel $\tilde{h}_2$ in the second hop is $\tilde{h}_2 = h_2 + \Delta h_2$, where h_1, h_2 and Δh_1, Δh_2 are channel errors (CEs) and channel estimation errors (CEEs), respectively. CSCG RVs are denoted by $h_1 \sim CN(0, \Omega_{h_1})$, $h_2 \sim CN(0, \Omega_{h_2})$, and $\Delta h_1 \sim CN(0, \Omega_{\Delta h_1})$, $\Delta h_2 \sim CN(0, \Omega_{\Delta h_2})$, respectively, where $\Omega_{\Delta h_1} = \Omega_{\tilde{h}_1} - \Omega_{h_1}$, and $\Omega_{\Delta h_2} = \Omega_{\tilde{h}_2} - \Omega_{h_2}$.

Next, the instantaneous rate and throughput for HD with imperfect CSI are examined. Both AF and DF schemes for perfect and imperfect CSI are also considered.

11.3.2.1 Calculation of the signal-to-noise ratio
Let us examine the instantaneous rate for the two schemes, beginning with the AF scheme.

11.3.2.1.1 Amplify-and-forward relaying
Here, the end-to-end SNR at D is given by

$$y_{AF} = \frac{|h_1|^2|h_2|^2}{|h_2|^2\mathcal{W}_1 + |h_1|^2\mathcal{W}_2 + \mathcal{W}_3}, \tag{11.9}$$

where $\mathcal{W}_1 = \Omega_{\Delta h_1} + \frac{N_0}{(1-\beta)l_1^{-m}P_S}$, $\mathcal{W}_2 = \Omega_{\Delta h_2}$, and $\mathcal{W}_3 = \Omega_{\Delta h_2}\Omega_{\Delta h_1} + \frac{\Omega_{\Delta h_2}N_0}{(1-\beta)l_1^{-m}P_S} + \frac{l_1^m l_2^m N_0}{\varphi P_S}$.

11.3.2.1.2 Decode-and-forward relaying
Next, the received SNRs at R and D in terms of the DF scheme can be calculated by

$$\gamma_R = \frac{(1 - \beta)P_S|h_1|^2}{(1 - \beta)P_S\Omega_{\Delta h_1} + l_1^m N_0}, \tag{11.10a}$$

$$\gamma_D = \frac{|h_1|^2 |h_2|^2}{(|h_2|^2 \mathcal{Z}_1 + |h_1|^2 \mathcal{Z}_2 + \mathcal{Z}_3)}, \tag{11.10b}$$

where $\mathcal{Z}_1 = \Omega \Delta_{h_1}$, $\mathcal{Z}_2 = \Omega \Delta_{h_2}$, and $\mathcal{Z}_3 = \Omega \Delta_{h_1} \Omega \Delta_{h_2} + \frac{I_1^m I_2^m}{\varphi P_S} N_0$.

Therefore, the end-to-end SNR γ_{DF} can be given as $\gamma_{DF} = \min{(\gamma_R, \gamma_D)}$, where γ_R and γ_D follow from Eqs. (11.10a) and (11.10b).

Remark 1 The appropriate choice of α and β helps the system achieve an instantaneous rate. The deployment of the Bessel functions is useful for obtaining the closed-form expression for ergodic capacity together with the optimal TS and PS ratios. However, this optimization problem can be numerically solved by studying several variables, that is, P_S, noise terms and CEEs. This optimization problem is subsequently examined in the following section.

11.3.2.2 Delay-limited throughput

11.3.2.2.1 Amplify-and-forward relaying

The OP for the AF scheme using HTPSR can be expressed as in [55] by

$$\mathrm{OP}_{AF} \approx 1 - (\mathcal{A}_{\mathrm{AF}})^{-1} \mathcal{B}_{\mathrm{AF}} \times \mathrm{K}_1(\mathcal{B}_{\mathrm{AF}}), \tag{11.11}$$

where $\mathcal{A}_{AF} = \exp\left(\gamma 0 \left(\frac{\mathcal{W}_1}{\Omega_{h_2}} + \frac{\mathcal{W}_2}{\Omega_{h_1}}\right)\right)$ and $\mathcal{B}_{AF} = 2\sqrt{\gamma_0(\mathcal{W}_3 + \gamma_0 \mathcal{W}_1 \mathcal{W}_2)(\Omega_{h_1} \Omega_{h_2})^{-1}}$. Now, we investigate OP in the DF scheme.

11.3.2.2.2 Decode-and-forward relaying

The OP for the DF scheme can be calculated by [55]

$$\mathrm{OP}_{\mathrm{DF}}(\gamma_0) \approx 1 - \exp\left(-\frac{\psi \gamma_0}{\Omega_{h_1}}\right) \times (\mathcal{A}_{DF})^{-1} \mathcal{B}_{DF} \times K_1(\mathcal{B}_{DF}), \tag{11.12}$$

where $\psi = \Omega_{\Delta h_1} + \frac{I_1^m N_0}{(1-\beta) P_S}$, $\mathcal{A}_{DF} = \exp\left(\gamma_0 \left(\frac{\mathcal{Z}_1}{\Omega_{h_2}} + \frac{\mathcal{Z}_2}{\Omega_{h_1}}\right)\right)$, and $\mathcal{B}_{DF} = 2\sqrt{\frac{\gamma_0(\Omega_{h_1} \mathcal{Z}_3 + \gamma_0 \mathcal{Z}_1 \mathcal{Z}_2)}{\Omega_{h_1} \Omega_{h_2}}}$.

11.3.2.2.3 Throughput analysis

Regarding the delay-limited throughput, τ_{dl} is defined as the effective communication time $(1 - \alpha)T$ leading to the given fixed transmission rate R_0. Therefore, knowing OP, the general expression for throughput for both transmission schemes is defined as

$$\tau_i^{dl} \underset{i \in \{AF, DF\}}{=} R_0(1 - OP_i)(1 - \alpha). \tag{11.13}$$

11.3.2.3 Delay-tolerant transmission

In the delay-tolerant transmission mode, the code length is larger than the block time; therefore the code length sees all the possible realizations of the channel during a code-word transmission and average channel conditions. The ergodic capacity can then be derived at an equal rate to the ergodic capacity [56]. The expressions for ergodic capacity for the AF and DF schemes are derived in the following sections.

11.3.2.3.1 Amplify-and-forward relaying

Proposition 3 *The closed-form expression for ergodic capacity is obtained as* [57]

$$C_{AF} \approx \frac{1}{\ln 2} \int_{x=0}^{\infty} \frac{\mathcal{B}_{AF}(x) \times K_1(\mathcal{B}_{AF}(x))}{(1+x)\mathcal{A}_{AF}(x)} dx, \tag{11.14}$$

where $\mathcal{A}_{AF}(x) = \exp\left(x\left(\frac{\mathcal{W}_1}{\Omega_{h_2}} + \frac{\mathcal{W}_2}{\Omega_{h_1}}\right)\right)$ *and* $\mathcal{B}_{AF}(x) = 2\sqrt{\frac{x(\mathcal{W}_3 + x\mathcal{W}_1\mathcal{W}_2)}{\Omega_{h_1}\Omega_{h_2}}}$.

11.3.2.3.2 Decode-and-forward relaying

Proposition 4 *Similarly, the ergodic capacity for DF mode is depicted by* [57]

$$C_{DF} \approx \frac{1}{\ln 2} \int_{x=0}^{\infty} \frac{\exp\left(-\frac{\psi\gamma_0}{\Omega_{h_1}}\right)\mathcal{B}_{DF}(x) \times K_1(\mathcal{B}_{DF}(x))}{(1+x)\mathcal{A}_{DF}(x)} dx, \tag{11.15}$$

where $\mathcal{A}_{DF}(x) = \exp\left(x\left(\frac{\mathcal{Z}_1}{\Omega_{h_2}} + \frac{\mathcal{Z}_2}{\Omega_{h_1}}\right)\right)$ *and* $\mathcal{B}_{DF}(x) = 2\sqrt{\frac{x(\Omega_{h_1}\mathcal{Z}_3 + x\mathcal{Z}_1\mathcal{Z}_2)}{\Omega_{h_1}\Omega_{h_2}}}$.

11.3.2.3.3 Throughput analysis

Knowing the ergodic capacity, the throughput at D is expressed as

$$\tau_i^{dt}\bigg|_{i \in \{AF, DF\}} = \frac{(1-\alpha)T}{T} C_i = (1-\alpha)C_i. \tag{11.16}$$

11.3.2.4 BER consideration

In this section, the BER at D can be given knowing OP in [58] as $BER = E\left[aQ(\sqrt{2b\gamma})\right]$, where the Gaussian Q-Function is denoted as $Q(.)$ defined as $Q(x) = \frac{1}{\sqrt{2\pi}} \int_x^{\infty} e^{-\frac{t^2}{2}} dt$, and the modulation formats are defined for binary-phase shift keying (BPSK) and quadrature phase shift keying (QPSK) as $(a, b) = (1, 2)$ and $(a, b) = (1, 1)$, respectively. As a result, the distribution function of γ is considered before the BER performance can be evaluated. It follows that the expression for BER is given by

$$BER_{i \in \{AF, DF\}} = \frac{a\sqrt{b}}{2\sqrt{\pi}} \int_0^{\infty} \frac{e^{-b\gamma}}{\sqrt{\gamma}} F_{\gamma_i}(\gamma) d\gamma, \tag{11.17}$$

where $F_{\gamma_i}(\gamma) = OP_i(\gamma)$ for the AF or DF protocol.

11.3.2.5 Optimization problems

In this section, the joint optimization problem of both TS and PS ratios for the AF and DF schemes is examined. Let us first look at AF.

11.3.2.5.1 Amplify-and-forward relaying
Here, the OP is given as

$$\max_{\alpha,\beta} R_{i \in \{AF,DF\}}$$
$$\text{subject to} \quad \alpha, \beta \in (0,1). \tag{11.18}$$

Because the above expressions are challenging to solve, it is hard to obtain a closed-form, and the optimal instantaneous rate is a biconvex function of α and β. The complex function above can therefore be numerically evaluated using the NSolve building function of the Mathematica software and will be presented as numerical results.

11.3.2.6 Decode-and-forward relaying
Following from Eqs. (11.10a) and (11.10b), the received SNRs can be rewritten as

$$\gamma_R = \frac{1}{\omega_1 + \frac{\omega_2}{(1-\beta)}}, \tag{11.19a}$$

$$\gamma_D = \frac{1}{\omega_3 + \frac{(1-\alpha)}{\alpha\beta}\omega_4}, \tag{11.19b}$$

where $\omega_1 = \frac{\Omega_{\Delta h_1}}{|h_1|^2}$, $\omega_2 = \frac{I_1^m N_0}{P_S|h_1|^2}$, $\omega_4 = \frac{I_1^m I_2^m N_0}{2\eta P_S|h_1|^2|h_2|^2}$, and $\omega_3 = \frac{\Omega_{\Delta h_1}}{|h_1|^2} + \frac{\Omega_{\Delta h_2}}{|h_2|^2} + \frac{\Omega_{\Delta h_1}\Omega_{\Delta h_2}}{|h_1|^2|h_2|^2}$.

The optimal α_{opt}, β_{opt} can be obtained by solving the following optimization:

$$\max R_{DF} = \arg \max \gamma_{DF}(\alpha, \beta), \tag{11.20}$$

where it is subject to $0 < \alpha < 1, 0 < \beta < 1$.

The optimal values of the above problem are represented as $\gamma_R = \gamma_D$. It follows that

$$\alpha \left[\left(\omega_1 + \frac{\omega_2}{(1-\beta)} - \omega_3 \right) \beta + \omega_4 \right] = \omega_4. \tag{11.21}$$

Now, we set the fixed value for β so that α_{opt} can be obtained as

$$\alpha_{opt} = \frac{\omega_4}{\left[\left(\omega_1 + \frac{\omega_2}{(1-\beta)} - \omega_3 \right) \beta + \omega_4 \right]}. \tag{11.22}$$

Otherwise, as α is fixed, β_{opt} is expressed as

$$\beta_{opt} = \frac{-b + \sqrt{b2 - 4ac}}{2a}, \tag{11.23}$$

where $b = (\alpha\omega_1 + \alpha\omega_2 - \alpha\omega_3 + (1-\alpha)\omega_4)$, $c = (\alpha - 1)\omega_4$, and $a = \alpha(\omega_3 - \omega_1)$.

Eventually, the optimal values for TS and PS are achieved.

11.3.3 Scenario 3: The impact of hardware impairments on cognitive D2D communication

Fig. 11−9 shows a cognitive D2D communication in a typical cellular network with two communication modes. Mode A is a multihop D2D communication, and Mode B is a single-hop peer-to-peer (P2P) communication. Two PU devices (i.e., UE1 and UE2) in Mode A communicate with each other via R. Meanwhile, direct communication between UE2 and UE3 is achieved in Mode B. The HD R can also transfer secondary information to UE3.

The communication model in Mode A comprises two phases. UE1 transmits the information to R in phase 1, in which harvested energy is used to forward the signals to both UE2 and UE3 in phase 2. Subsequently, UE3 can also receive signals to cancel the interference in phase 2. By contrast, UE2 and UE3 exchange information in two time slots in Mode B.

The maximum threshold of the transmit power and circuit power for UEs denoted as E_{Di} and E_{Ci}, where $i = 1, 2, 3$, are assumed to be the same. For simplicity, $E_{Di} = E_D$ and $E_{Ci} = E_C$. Note that the transmit power of R is denoted as E_R, and n_0 is the AWGN with mean power N_0.

The path-loss model denoted as PL_k is defined by $PL_k = 1/\phi r_k^m$, where m is the path-loss exponent, and the path-loss constant is represented by ϕ as in [18]. All channel gains between the two nodes are modeled as Rayleigh fading channels with free-space propagation path loss. In terms of the channel gain coefficient, we have $|h_k|^2 = |k|^2 PL_k$, where k is the complex Gaussian distributed RVs to model fading phenomena with zero mean and variances $\Omega_k \sim CN(0, 1)$.

For this work, we once again deploy HTPSR with T being the time block, where the portion $\alpha_1 T$ is used for EH at R with E_D. Note that the transmitted signals to R from UE1 are divided into two parts. In particular, $\alpha_2 T$, βE_D is used for EH at R, while IT accounts for

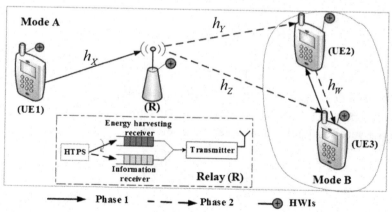

FIGURE 11–9 System model.
The channel gains of the links from UE1 to R and R to UE2, UE3 are represented as h_X, h_Y, h_Z, while the channel gain for peer-to-peer communication is h_W. The distances for the aforementioned links are r_k, where $k \in \{X, Y, Z, W\}$.

$(1 - \beta)E_D$. The remaining time $(T - \alpha_1 T - \alpha_2 T)$ is then used for IT from R to UE2, and UE3 with the power allocation denoted as λ and $(1 - \lambda)$, respectively. As mentioned previously, the PSR protocol is a special case of HTPSR, where $\alpha_1 = 0$ and $\alpha_2 = 0.5$, while the TSR protocol is the same when $\beta = 0$ and $\alpha_2 = 0.5 \times (1 - \alpha_1)$.

We assume that the impact of HWIs is available at all nodes [59], and the practical transceiver impairments at UEs is therefore represented as x in the $a \rightarrow b$ link while the HWIs degrade system performance during the reception phase. HI_a and HI_b are denoted as the aggregate distortions degrading the $a \rightarrow b$ link with zero mean variance $HI_a \sim CN$ $(0, \kappa_a|h|^2 E_D)$ and $HI_b \sim CN (0, \kappa_b|h|^2 E_D)$, respectively, and κ_a and κ_b are the levels of HWIs at HI_a and HI_b, respectively. The received signal in this case can therefore be expressed as $y_b = \sqrt{E_D}h_x + HI_{ab} + n_0$, where the channel gain for the $a \rightarrow b$ link is h, and the aggregate distortion at the receiver b is HI_{ab} with aggregate distortion power $HI_{ab} \sim CN (0, \kappa|h|^2 E_D)$, and $k \triangleq k_a + k_b$ being the aggregate impairment level during the information processing phase. Hence, the distortion noise stemming from HWIs at R is represented by HI_1 with variance $HI_1 \sim CN (0, \kappa E_D PL_X|X|^2)$, while the corresponding distortion noise caused by HWIs at UE2, UE3 are denoted by HI_2 and HI_3 with variance $HI_2 \sim CN (0, \kappa\lambda E_R|Y|^2 PL_Y)$ and $HI_3 \sim CN (0, \kappa(1 - \lambda)E_R|Z|^2 PL_Z)$, respectively.

11.3.3.1 End-to-end signal-to-noise-plus-distortion ratio

In this part, the end-to-end signal-to-noise-plus-distortion ratio (SNDR) is investigated, γ, where $\gamma = E\{|signal|^2\}/E\{|overall\ noise|^2\}$ for the AF and DF schemes. The AF scheme is examined first.

11.3.3.1.1 Amplify-and-forward relaying

The end-to-end SNDR at UE2 is examined when UE2 treats the signal as interference and later decodes the primary information as [60]

$$\gamma_1^{AF} = \frac{\tau_{1,a}|X|^2|Y|^2}{\tau_{1,b}|X|^2|Y|^2 + \tau_{1,c}|Y|^2 + \tau_0}, \tag{11.24}$$

where $\delta_1 = \kappa\lambda + \kappa\lambda (1 - \beta + \kappa) + (1 - \lambda) (1 - \beta + \kappa)$, $\tau_0 = \frac{(1 - \beta + k)(1 - \alpha_1 - a_2)}{\eta(\alpha_1 + a_2\beta)}$, $\tau_{1,a} = \frac{E_D}{N_0}(1 - \beta)\lambda PL_X PL_Y$, $\tau_{1,b} = \frac{E_D}{N_0}\delta_1 PL_X PL_Y$, and $\tau_{1,c} = \lambda PL_Y$.

Similarly, the end-to-end SNDR at UE3 when UE3 treats the signal as interference and later decodes the secondary information can be given as

$$\gamma_2^{AF} = \frac{\tau_{2,a}|X|^2|Z|^2}{\tau_{2,b}|X|^2|Z|^2 + \tau_{2,c}|Z|^2 + \tau_0}, \tag{11.25}$$

where $\delta_2 = \kappa\lambda + \kappa (1 - \lambda) (1 - \beta + \kappa) + \lambda (1 - \beta - \kappa) + \kappa$, τ_0 in the above expression, $\tau_{2,a} = \frac{E_D}{N_0}(1 - \beta)(1 - \lambda)PL_X PL_Z$, $\tau_{2,b} = \frac{E_D}{N_0}\delta_2 L_X PL_Z$, and $\tau_{2,c} = \lambda PL_Z$

Remark 2

- Note that the power allocation λ is principally important because of the need to balance the received information at UE2 and UE3. Therefore, in order to maintain the quality of UE2 at an acceptable level, as λ increases, the amount of signal received at UE3 decreases.
- The end-to-end SNDR at UE2 is obtained to satisfy QoS. In particular, when secondary signals are treated as interference, the desired expression can be easily derived from Eq. (11.24). If $E_D \to \infty$ and all parameters are fixed, then $\lim_{E_D \to \infty} \gamma_1^{AF} = \lambda/(1-\lambda)$. Nevertheless, because of the achieved end-to-end SNDR at UE3 (Eq. 11.25), the primary signals are considered interference.
- If $E_D \to \infty$, then $\lim_{E_D \to \infty} \gamma_2^{AF} = (1-\lambda)/\lambda$. Numerical results are given in order to explain these phenomena.

11.3.3.1.2 Decode-and-forward relaying
Similarly, regarding the DF scheme, the end-to-end SNDR at both UE2 and UE3 can be written as [60] $\gamma_i^{DF} = {}_{i \in \{1,2\}} \min\{\gamma_{i,a}^{DF}, \gamma_{i,b}^{DF}\}$, where the instantaneous SNDR at R, UE2 and UE3 can be calculated, respectively, as follows $\psi_0 = \frac{(1-\alpha_1-\alpha_2)}{\eta(\alpha_1+\alpha_2\beta)}$, $\psi_{1,a} = \frac{E_D}{N_0}\lambda PL_X PL_Y$, $\psi_{2,a} = \frac{E_D}{N_0}(1-\lambda)PL_X PL_Z$, $\psi_{1,b} = \frac{E_D}{N_0}(k\lambda + 1 - \lambda)PL_X PL_Y$, $\psi_{2,b} = \frac{E_D}{N_0}(\kappa(1-\lambda) + \lambda)PL_X PL_Z$, and $\gamma_{1,a}^{DF} = \gamma_{2,a}^{DF} = \frac{(1-\beta)\frac{E_D}{N_0}PL_X|X|^2}{1+\kappa\frac{E_D}{\Omega_0}PL_X|X|^2}$, $\gamma_{1,b}^{DF} = \frac{\psi_{1,a}|X|^2|Y|^2}{\psi_{1,b}|X|^2|Y|^2+\psi_0}$, and $\gamma_{2,b}^{DF} = \frac{\psi_{2,a}|X|^2|Z|^2}{\psi_{2,b}|X|^2|Z|^2+\psi_0}$.

11.3.3.1.3 Peer-to-peer communication
In this section, the SNDR under the impact of HWIs in P2P communication is given as

$$\gamma_3^{PP} = \frac{\frac{E_D}{\Omega_0}PL_W|W|^2}{1+\kappa\frac{E_D}{\Omega_0}PL_W|W|^2}. \tag{11.26}$$

11.3.3.2 Successful transmission probability
In principle, the STP is considered the probability of the receiver being able to receive packets in a time slot successfully. In terms of the STP in P2P communication, packets are successfully received if the SNDR is higher than its threshold Γ_D, $\Pr(\gamma \geq \Gamma_D)$. Therefore, UE2 and UE3 will receive negative feedback, and the packets will be allocated first in the queue for retransmission. The STP in this communication is given as

$$\Pr(\gamma_3^{PP} \geq \Gamma_D) = \Pr\left(\frac{\frac{E_D}{\Omega_0}PL_W|W|^2}{1+\kappa\frac{E_D}{\Omega_0}PL_W|W|^2} \geq \Gamma_D\right)$$

$$= e^{-\frac{\Omega_0\Gamma_D}{\Omega_W E_D PL_W(1-\kappa\Gamma_D)}} \tag{11.27}$$

The STP in D2D communication with large-scale path loss and small-scale Rayleigh fading at UE2 and UE3 in the AF and DF schemes will be examined in the next proposition.

Proposition 5 *The STP at UE2 and UE3 in the AF scheme* [60]
* *when* $\Gamma_D \geq \tau_{i,a}/\tau_{i,b}$ *can be calculated as*

$$\Pr\left(\gamma_i^{AF} \geq \Gamma_D\right)_{i \in \{1,2\}} = 1, \tag{11.28}$$

* *or when* $\Gamma_D < \tau_{i,a}/\tau_{i,b}$ *as*

$$\Pr\left(\gamma_i^{AF} \geq \Gamma_D\right)_{i \in \{1,2\}} = 2e^{-\omega_i^{AF}}\sqrt{\vartheta_i^{AF}}K_1\left(2\sqrt{\vartheta_i^{AF}}\right), \tag{11.29}$$

where $\omega_i^{AF} = \frac{\Gamma_D\tau_{i,c}}{\Omega_X(\tau_{i,a} - \Gamma_D\tau_{i,b})}$ *and* $\vartheta_i^{AF} = \frac{\Gamma_D\tau_0}{\Omega_X\Omega_Y(\tau_{i,a} - \Gamma_D\tau_{i,b})}$.

Proposition 6 *Similarly, the STP at UE2 and UE3 DF scheme* [60] *can be calculated as*

$$\Pr\left(\gamma_i^{DF} \geq \Gamma_D\right)_{i \in \{1,2\}} = 2e^{-\omega_i^{DF}}\sqrt{\vartheta_i^{DF}}K_1\left(2\sqrt{\vartheta_i^{DF}}\right), \tag{11.30}$$

where $\omega_i^{DF} = \frac{\Gamma_D\Omega_0}{\Omega_X E_D PL_X((1 - \beta) - \Gamma_{D^\kappa})}$ *and* $\vartheta_i^{DF} = \frac{\Gamma_D\psi_0}{\Omega_X\Omega_Z(\psi_{i,a} - \Gamma_D\psi_{i,b})}$.

Remark 3 The joint optimization of TS and PS can help us achieve the optimal STP, which is difficult to evaluate because of the Bessel function for maximum transmission power, distances, power allocation, and HWIs level. Being nonconvex, the above optimization problem can be solved with a GA, which is discussed in next section.

11.3.3.3 Average energy efficiency and average spectral efficiency

In this section, the expressions for the average EE and SE are provided, in which EE is defined as the average transmission rate per unit energy consumption. In order to achieve an energy efficient system, both the transmission power and circuit power should be carefully examined [61].

Proposition 7 *The expressions for the average EE and average SE in the AF scheme are derived as* [60]

$$ee_i^{AF} = \frac{B}{2E_{sum}}\int_{x=0}^{\tau_{i,a}/\tau_{i,b}}(M_i^{AF} + N_i^{AF})\log_2(1 + x)dx, \tag{11.31a}$$

and

$$se_i^{AF} = \frac{1}{2}\int_{x=0}^{\tau_{i,a}/\tau_{i,b}}(M_i^{AF} + N_i^{AF})\log_2(1 + x)dx, \tag{11.31b}$$

where $M_i^{AF} = \frac{2\tau_{i,a}e^{-\omega_i^{AF}}\vartheta_i^{AF}K_0\left(2\sqrt{\vartheta_i^{AF}}\right)}{x(\tau_{i,a}-\mathcal{X}\tau_{i,b})}$, $N_i^{AF} = \frac{2\tau_{i,a}\omega_i^{AF}e^{-\omega_i^{AF}}\sqrt{\vartheta_i^{AF}}K_1\left(2\sqrt{\vartheta_i^{AF}}\right)}{x(\tau_{i,a}-\mathcal{X}\tau_{i,b})}$, *and the total power consumption of Mode A is defined as* $E_{sum} = 2E_D + 2E_C + E_R$.

The closed-form average EE and average SE in Mode A using the DF scheme are given as

$$ee_i^{DF} = \frac{B}{2E_{sum}}\int_{x=0}^{\psi_{i,a}/\psi_{i,b}}(M_i^{DF}+N_i^{DF})\log_2(1+x)dx, \tag{11.32a}$$

and

$$se_i^{DF} = \frac{1}{2}\int_{x=0}^{\psi_{i,a}/\psi_{i,b}}(M_i^{DF}+N_i^{DF})\log_2(1+x)dx, \tag{11.32b}$$

where $M_i^{DF} = \frac{2\psi_{i,a}e^{-\omega_i^{AF}}\vartheta_i^{AF}K_0\left(2\sqrt{\vartheta_i^{DF}}\right)}{x(\psi_{i,a}-\mathcal{X}\psi_{i,b})}$ and $N_i^{DF} = \frac{2(1-\beta)\omega_i^{DF}e^{-\omega_i^{DF}}\sqrt{\vartheta_i^{DF}}K_1\left(2\sqrt{\vartheta_i^{DF}}\right)}{x(\psi_{i,a}-\mathcal{X}\psi_{i,b})}$.

When $\Gamma_D < 1/\kappa$ and the sum of the transmit power and the circuit power is defined as $P_{sum} = 2E_D + 2E_C$, the expressions for both average EE and average SE in P2P communication are

$$ee_3^{PP} = \frac{B}{P_{sum}\ln2}\int_{x=0}^{1/\kappa}e^{-\frac{\Omega_0 x}{\Omega_W E_D PL_W(1-\kappa x)}}(1+x)^{-1}dx, \tag{11.33a}$$

and

$$se_3^{PP} = \frac{1}{\ln2}\int_{x=0}^{1/\kappa}e^{-\frac{\Omega_0 x}{\Omega_W E_D PL_W(1-\kappa x)}}(1+x)^{-1}dx. \tag{11.33b}$$

11.3.3.4 Optimization problem

The joint optimization of TS and PS is solved by optimizing the STP, which is first generally expressed as

$$\max_{\alpha_1,\alpha_2,\beta}\left\{2e^{-\omega_i^j}\sqrt{\vartheta_i^j}K_1\left(2\sqrt{\vartheta_i^j}\right)\right\}, \tag{11.34}$$

which is subject to $\alpha_1, \alpha_2, \beta \in (0, 1]$, and $\omega_i^j, \vartheta_i^j$ are already defined.

As mentioned above, the expression (11.34) is a nonconvex function, and a GA is applied to obtain the optimal values of TS and PS, explained in detail as follows:

Definition 1 The generation of a random population is defined as a set of chromosomes comprising a group of genes, and it is assumed to contain the optimal values for the considered variables [62]. Since the chromosome is against an objective function, the fitness of each one is determined. Only the best chromosomes can exchange information (via crossover or mutation) to produce offspring chromosomes so that simulations for the natural survival of the fittest process can be provided. If the offspring solutions are more feasible than weak population members, they are investigated and used for population evolution. The process continues for many generations in order to find a best-fit (near-optimum) solution. The parameters such as the number of generations, population size, crossover rate, and mutation rate affect the performance of GAs [63,64].

TS and PS denoted as α and β are considered genes. A chromosome is created by combining α and β. To obtain each chromosome's fitness, an objective function in Eq. (11.34) is used. H_{max} denotes the optimal solution of the tth generation, and the predefined precision with the GA constraint tolerance is denoted by ϵ. As a result, the use of GA optimization helps us achieve the joint optimal TS and PS ratios in order to guarantee the best STP [60].

11.4 Numerical results

11.4.1 Scenario 1: Wireless power transfer constraint policies

In scenario 1, simulations are provided to prove the correctness of the numerical results. For simplicity, all the primary parameters and default values are listed in Table 11–1.

Figs. 11–10 and 11–11 show the throughput in the first hop with $\overline{\gamma}_{S\,R} = \overline{\gamma}_{RD}$. We can see that $\alpha = 0.5$ is dedicated to SP mode, while the optimal TS ratio α_{opt} is used in HP mode.

Figs. 11–12 and 11–13 show the optimal throughput performance in both delay-limited and delay-tolerant transmission modes obtained with the optimal TS ratio in HP mode under different impacts of SI $\overline{\gamma}_{LI} = 5\ dB$ or $\overline{\gamma}_{LI} = 10$ dB and α_{opt}.

In the considered protocol, throughput increases as α climbs from 0 to the optimal value $\alpha = 0.1$, resulting in the better system performance than in SP mode with $\alpha = 0.5$. When α is less than the optimal value of α_{opt}, more time is consumed for EH. We can clearly see that SI compromises the throughput performance at any time. However, in both transmission modes, a suitable choice of α for HP can lead to better throughput performance regardless of the impact of SI.

Table 11–1   Main simulation parameters.

Parameters	Values
Transmission fixed rate, R_0	2 (bps/Hz)
Transmit power at S, P_S	1 (Joules/s)
EH efficiency, η	0.5
Bandwidth, B	1
Noise variances for all nodes, σ^2	1/2
SNRs S-R	$\overline{\gamma}_{S\,R} = \Omega_h$
SNRs R-D	$\overline{\gamma}_{S\,R} = \Omega_g$
SNRs SI	$\overline{\gamma}_{LI} = \Omega_f$

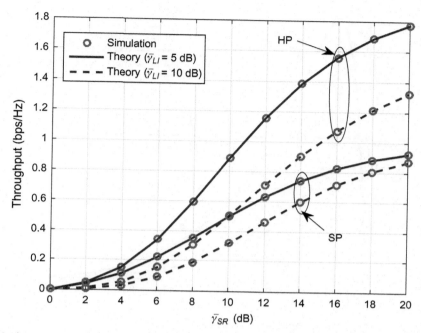

FIGURE 11–10 Theory and simulation results of throughput in delay-limited mode. Throughput versus $\bar{\gamma}_{SR}$ under different impacts of SI $\bar{\gamma}_{LI} = 5$ dB or $\bar{\gamma}_{LI} = 10$ dB and α_{opt}.

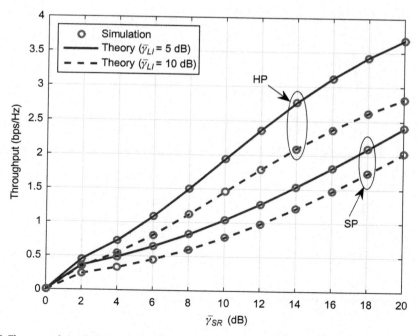

FIGURE 11–11 Theory and simulation results of throughput in delay-tolerant mode.

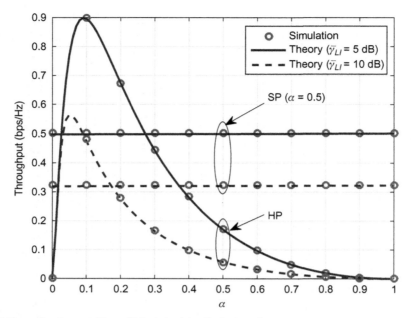

FIGURE 11–12 Throughput versus TS coefficients in delay-limited mode.

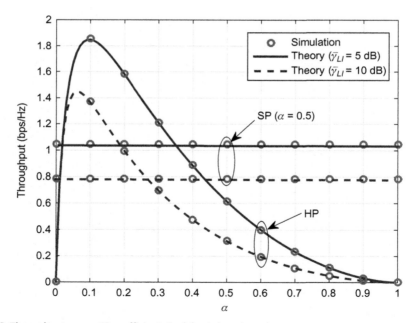

FIGURE 11–13 Throughput versus TS coefficients in delay-tolerant mode.

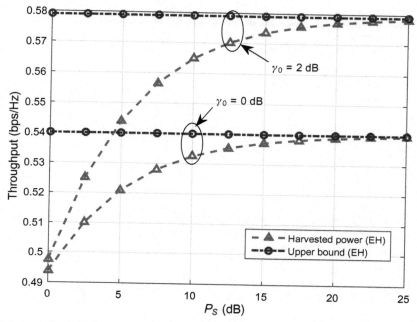

FIGURE 11–14 Asymptotic throughput performance using transmit power in the delay-limited mode for different transmit powers $\overline{\gamma}_{S\,R} = \overline{\gamma}_{RD} = 10$ dB and $\overline{\gamma}_{LI} = 5$ dB.

The delay-limited throughput of HP mode in Fig. 11–14 reaches its optimal value, which then drops slightly. The highest throughput performance can be achieved regardless of the increase in the transmit power at S and the tight upper bound. However, we see fluctuations in throughput performance with different SNR thresholds. Therefore, with a suitable SNR level, R can satisfy proper applications for wireless communications.

Fig. 11–15 shows the throughput performance in the presence of SI. Throughput falls as SI rises when the upper bound of SP approaches 0. Because of both predefined SNR thresholds, that is, 0 and 2 dB, throughput performance is degraded.

Fig. 11–16 shows the transmission rate in terms of the optimal TS. It is clear that only a small amount of energy is harvested to assist communication in SP mode so that the transmission rate remains stable. By contrast, a higher transmission rate is seen in HP mode as a result of the EH capacity and energy storage deployed at R.

11.4.2 Scenario 2: The impact of channel state information using HTPSR protocol

For this scenario, all parameters and default values for the simulation results are summarized in Table 11–2.

Figs. 11–17 and 11–18 show the OP with perfect CSI and imperfect CSI in the AF scheme, which outperforms the DF scheme. More precisely, as α and β vary between 0 and

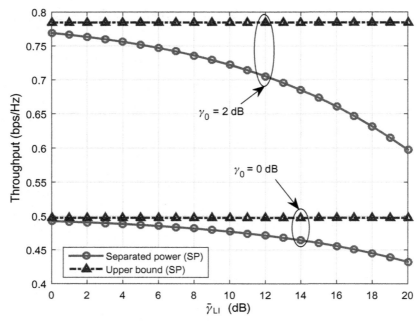

FIGURE 11–15 Throughput performance under the impact of self-interference with the transmit power $\overline{\gamma}_{SR} = \overline{\gamma}_{RD} = 20$ dB.

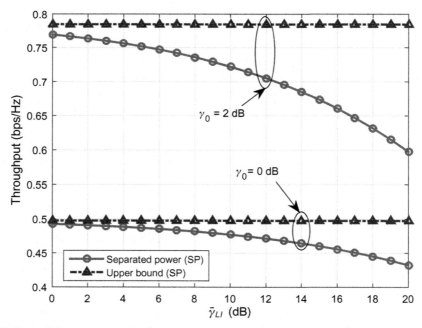

FIGURE 11–16 Transmission rate versus P_S other parameters: $\overline{\gamma}_{LI} = 5$ *dB*, $\overline{\gamma}_{SR} = \overline{\gamma}_{RD} = 20$ *dB*, $\alpha = 0.1$, and α_{opt}.

Table 11–2 Main simulation parameters.

Parameters	Values
Transmission fixed rate, R_0	3 (bps/Hz)
Transmit power at S, P_S	1 (Joules/s)
EH efficiency, η	1
Noise variances for all nodes, $\Omega_{\Delta h_1} = \Omega_{\Delta h_2}$	0.03
PS ratios, β	0.3
TS ratios, α	0.3
Path-loss exponent, m	2.7

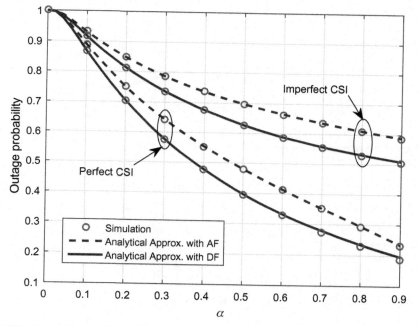

FIGURE 11–17 OP of the perfect and imperfect CSI for AF and DF RNs for α.

0.9, dramatic decreases in the OP of AF and DF can be seen, especially when α is around 0.9. Unlike TS, OP declines as β varies between 0 and 0.7, as shown in Fig. 11–18. Because more energy is harvested at R, the system enjoys better outage performance. However, worse OP arises from an increase of PS because of less power for information processing, and the performance gap between imperfect CSI and perfect CSI is clear at approximately $\alpha = 0.9$ and $\beta = 0.7$ because of CEEs. The instantaneous rate of imperfect CSI and perfect CSI is depicted in Fig. 11–19 for the two schemes with different values of SNR. Here, only imperfect CSI is discussed, and a comparison between PSR, TSR, and HTPSR is given to help us discover that TSR outperforms the other two. This performance depends on the

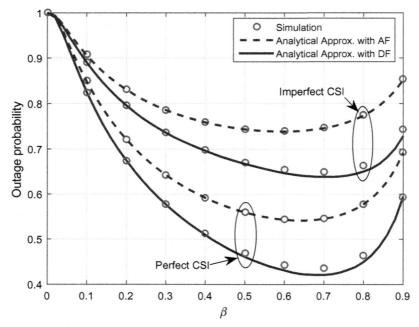

FIGURE 11–18 OP of the perfect and imperfect CSI for AF and DF RNs for β.

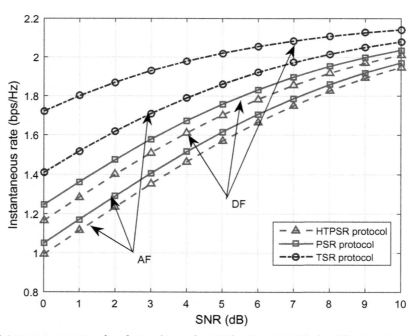

FIGURE 11–19 Instantaneous rate of perfect and imperfect CSI for AF and DF RNs for different values of SNR (dB).

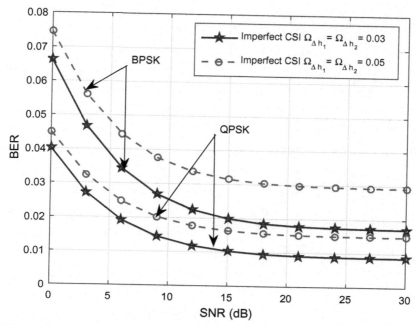

FIGURE 11–20 BER performance of AF RNs for various values of P_S.

instantaneous values of the channel because of the transmit power at S, and SNR intends to supply the EH circuit at the relay node in the TSR protocol, while a small fraction of this power is used for HTPSR. Furthermore, as SNR increases, the system throughput with imperfect CSI rises.

Figs. 11–20 and 11–21 show the BER of AF and DF RNs. As P_S rises from 0 Δh to 30, QPSK is better than BPSK in both AF and DF. However, $\sigma^2_{\Delta h_2}$ increases as the BER performance in the presence of imperfect CSI falls. Both the AF and DF schemes see the same tendency.

Fig. 11–22 illustrates the instantaneous rate as a function of the transmit power at S. The instantaneous rate is better with optimal α and β. It is clear that as P_S climbs from 0 to 0.6, the instantaneous rate increases considerably. It then sees only a gradual increase from 0.6 to 1.

Fig. 11–23 shows the OP in the presence of imperfect CSI, which is compared in three cases, that is, AF, DF, and the optimal power allocation (OPA) proposed in [65]. Here, the CCEs are set to 0.001, $\eta = 0.9$, $\alpha - 0.3$, $\beta = 0.5$, and the number of relays is set to $K = 1$. It is evident that outage performance of HTPSR is better than that in the OPA scheme.

11.4.3 Scenario 3: The impact of hardware impairments on cognitive D2D communication

For this scenario, all the primary parameters and default values are listed in Table 11–3, which are used when UE1, R, and UE2 are placed at (0, 0), (0.5, 0), and (1, 0) on the XY plane, respectively, while UE3 is located at (1, 0.5).

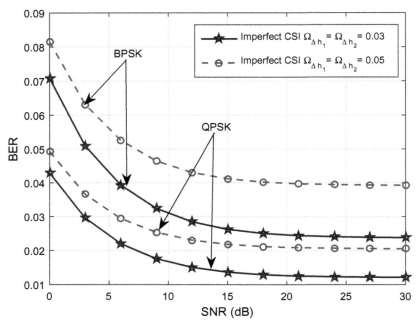

FIGURE 11–21 BER performance of DF RNs for various values of P_S.

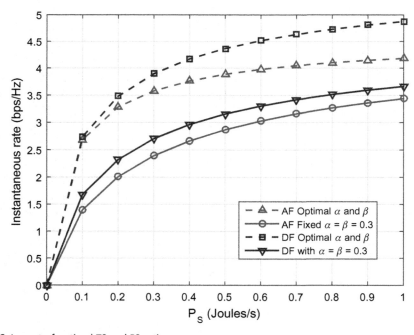

FIGURE 11–22 Impact of optimal TS and PS ratios.

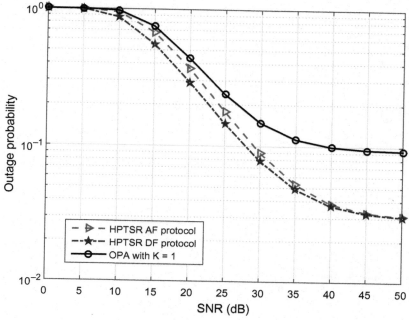

FIGURE 11–23 Comparison between our model and recent work.

Table 11–3 Main simulation parameters.

Parameters	Values
Circuit power at UEi, E_C	100 (W)
Channel bandwidth, B	10 (MHz)
Path loss for all links in the system, PL_k, $k \in [X, Y, Z, W]$ in Ref. [66]	$(148 + 40\log_{10}(r_k))^{-1}$ (dB)
Thermal noise density, N_0	-174 (dBm/Hz)
Energy conversion efficiency, η	1
TS and PS ratios, $\alpha_1 = \alpha_2 = \beta$	0.1
Power allocation, λ	0.7
Hardware impairments level, κ	0.15
SNDR threshold at UE2, Γ_D	3 (dB)
SNDR threshold at UE3, Γ_D	1 (dB)

Figs. 11−24 and 11−25 show STP as a function of the maximum transmission E_D at UE2 and UE3, respectively. In this situation, we compare AF and DF using HTPSR and TSR, which indicates that more E_D contributes to better STP and that the DF scheme outperforms the AF scheme, in which HTPSR is better than TSR in terms of STP. UE2 without spectrum sharing also performs better than UE3, for example, $E_D = 10$ dB.

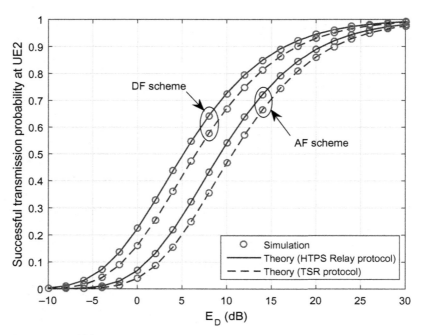

FIGURE 11–24 STP versus E_D (dB) at UE2.

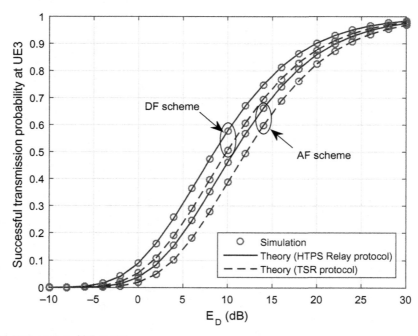

FIGURE 11–25 STP versus E_D (dB) at UE3.

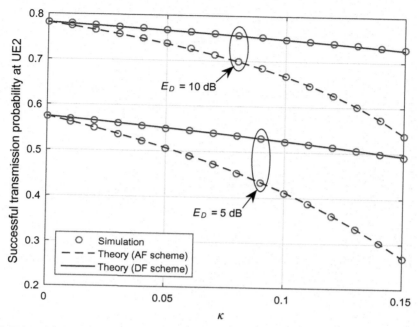

FIGURE 11–26 STP versus HWIs level, κ with $E_D = 10$ dB and $E_D = 5$ dB at UE2.

Figs. 11–26 and 11–27 show STP for AF and DF and indicate that when the HWIs level κ rises, STP is compromised. In this case, STP in the AF scheme drops considerably when $E_D = 5$ dB.

Figs. 11–28 and 11–29 chart the average EE and average SE at UE2 as a function of E_D (W). Here, the average EE rises as E_D increases before dropping as E_D approaches approximately 0.2. Since E_D rises, E_{sum} increases, while the average SE is linear. In Fig. 11–28, when two values of HWIs are considered, that is, $\kappa = 0.15$ and $\kappa = 0.3$, the average EE is ruined. However, the placement of R in Case 1 at (0.5, 0) and Case 2 at (0.3, 0), as shown in Fig. 11–29, indicates that Case 2 enjoys better SE as a result of the close distance between UE2 and R. Similarly, the average EE at UE3 falls as E_D rises, as shown in Figs. 11–30 and 11–31, as the average SE rises. Note that the performance gap between Case 1 and Case 2 is clear when R and UE3 are placed closely together.

Fig. 11–32 shows the average EE and average SE as functions of E_D in P2P communication when the distance between UE2 and UE3 is set to 5 km. Fig. 11–33 compares two cases of the distance between two nodes, Case 1 being at 5 km and Case 2 at 10 km, respectively. It is clear that the EE first increases along with SE.

After achieving the highest level ($\kappa = 0.15$ and $\kappa = 0.3$), it drops when the average SE increases. When $\kappa = 0.15$, SE is not affected much by HWIs and drastically increases compared to when $\kappa = 0.3$.

Figs. 11–34 and 11–35 show different TS and PS ratios for UE2 and UE3 under the impact of HWIs. The STP in the case of HTPSR achieves its optimal values when the joint

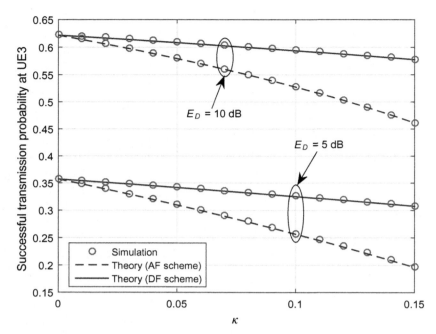

FIGURE 11–27 STP versus HWIs level, κ with $E_D = 10$ dB or $E_D = 5$ dB at UE3.

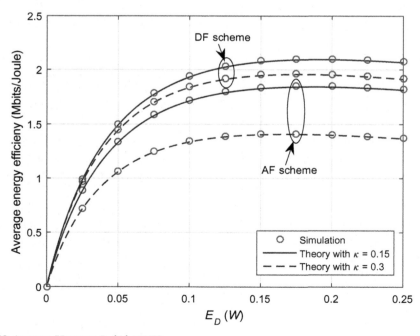

FIGURE 11–28 Average EE versus E_D (W) at UE2.

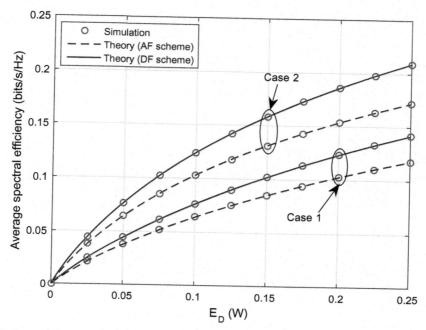

FIGURE 11–29 Average SE versus E_D(W) in Case 1 and Case 2 at UE2.

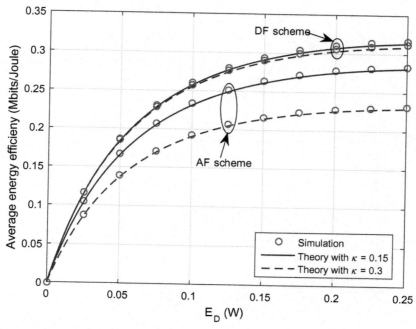

FIGURE 11–30 Average EE versus E_D(W) at UE3.

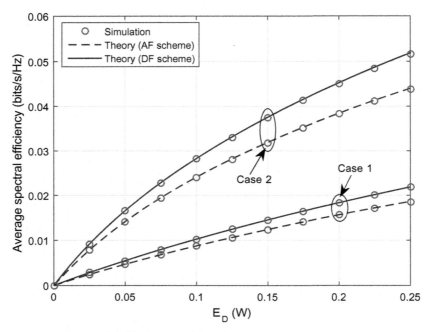

FIGURE 11–31 Average SE versus E_D(W) in Case 1 and Case 2 at UE3.

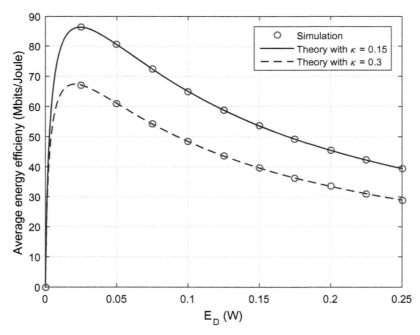

FIGURE 11–32 Average EE versus E_D(W) in P2P communication under the impact of HWIs.

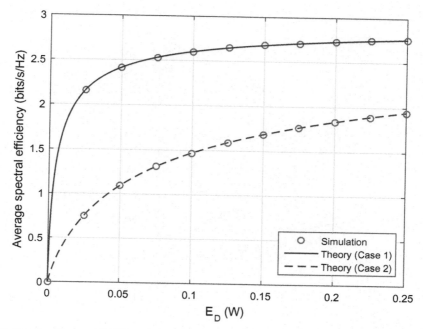

FIGURE 11–33 Average SE versus $E_D(W)$ in P2P communication under the impact of HWIs.

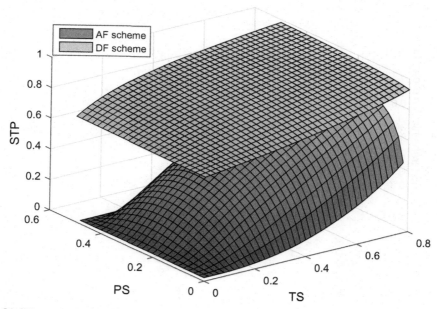

FIGURE 11–34 STP versus TS and PS at UE2.

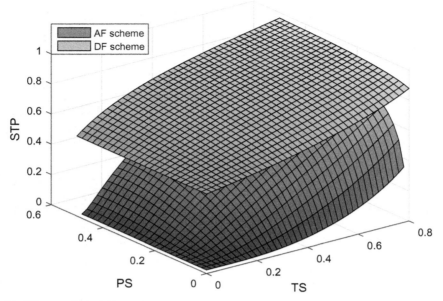

FIGURE 11–35 STP versus TS and PS at UE3.

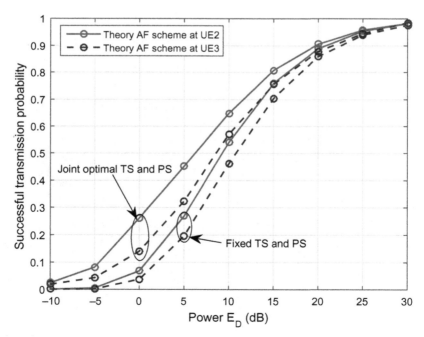

FIGURE 11–36 STP versus E_D(dB) in the AF scheme.

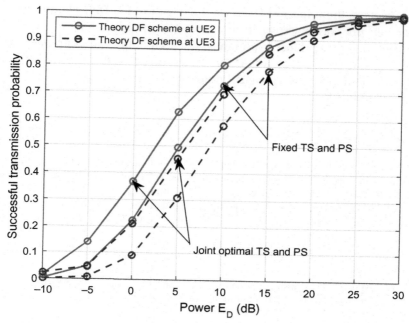

FIGURE 11–37 STP versus E_D(dB) in the DF scheme.

optimal values of TS and PS vary between 0 and 0.7 and 0 and 0.4, respectively, with $E_D = 10$ dB and $\kappa = 0.15$.

In Section 11.3.3.4, the joint optimization of TS and PS was already discussed, so the number of generations, constraint tolerance, and mutation probability were set as $N = 100$, $\epsilon = 10^{-5}$ and $p_m = 0.05$, respectively, in order to simulate the results shown in Figs. 11–36 and 11–37. Comparing the fixed values of TS and PS ratios with $\alpha_1 = \alpha_2 = \beta = 0.1$ at UE2 and UE3 in the AF and DF schemes, upward trends for all curves as E_D increases can be observed as a result of the higher E_D. The performance gain of the jointly optimized TS and PS also achieves maximum value when E_D approaches approximately 25 dB.

11.5 Summary

Regarding power supply policies, the results show that appropriate optimal TS in HP mode can help R consume less energy than in SP mode. Acceptable outage performance can also be achieved with both HP and SP. Most importantly, the impact of SI, TS ratio and transmit power on system throughput was examined thoroughly.

For the impact of CSI using HTPSR protocol, the effect of imperfect CSI on a system of AF and DF RNs was investigated, in which analytical expressions of achievable throughput, ergodic capacity, and EE for both schemes were achieved. The numerical analysis showed that a DF system in the presence of various parameters outperforms an AF system.

In the final section, the impact of HWIs on the cognitive EH D2D communication underlaying cellular network was investigated and closed-form expressions for STP, average EE, and average SE were derived. Most importantly, the joint optimization problem of PS and TS ratios was successfully solved with a GA. The simulation results compared the AF and DF transmission scheme and showed that DF outperforms AF in every aspect. Despite HWIs, the STP is still guaranteed.

Acknowledgments

This work was supported by the Ministry of Education, Youth and Sports from the Large Infrastructures for Research, Experimental Development and Innovations project "IT4Innovations National Supercomputing Center—LM2015070" and partially received financial support under grant No. SGS SP2019/41 conducted at VSB—Technical University of Ostrava, Czech Republic and under grant LO1401 (The National Sustainability Program). For the research, the infrastructure of the SIX Center was used.

References

[1] De Rango, M. Tropea, Energy saving and load balancing in wireless ad hoc networks through ant-based routing, in: Proceedings of the 2009 International Symposium on Performance Evaluation of Computer and Telecommunication Systems, SPECTS 2009, 2009, pp. 117–124.

[2] De Rango, M. Tropea, Swarm intelligence based energy saving and load balancing in wireless ad hoc networks, in: Proceedings of the 2009 Workshop on Bio-inspired Algorithms for Distributed Systems, BADS '09, 2009, pp. 77–83.

[3] H.J. Visser, R.J.M. Vullers, RF energy harvesting and transport for wireless sensor network applications: principles and requirements, Proc. IEEE 101 (6) (2013) 1410–1423. Available from: https://doi.org/10.1109/JPROC.2013.2250891.

[4] H. Nishimoto, Y. Kawahara, T. Asami, Prototype implementation of ambient RF energy harvesting wireless sensor networks, Proceedings of IEEE Sensors, IEEE, Kona, 2010, pp. 1282–1287. Available from: https://doi.org/10.1109/ICSENS.2010.5690588.

[5] X. Zhang, H. Jiang, L. Zhang, C. Zhang, Z. Wang, X. Chen, An energy-efficient ASIC for wireless body sensor networks in medical applications, IEEE Trans. Biomed. Circuits Syst. 4 (1) (2010) 11–18. Available from: https://doi.org/10.1109/TBCAS.2009.2031627.

[6] X. Lu, D. Niyato, P. Wang, D.I. Kim, Z. Han, Wireless charger networking for mobile devices: fundamentals, standards, and applications, IEEE Wirel. Commun. 22 (2) (2015) 126–135. Available from: https://doi.org/10.1109/MWC.2015.7096295.

[7] D. Mishra, S. De, S. Jana, S. Basagni, K. Chowdhury, W. Heinzelman, Smart RF energy harvesting communications: challenges and opportunities, IEEE Commun. Mag. 53 (4) (2015) 70–78. Available from: https://doi.org/10.1109/MCOM.2015.7081078.

[8] K. Ishibashi, H. Ochiai, Analysis of instantaneous power distributions for non-regenerative and regenerative relaying signals, IEEE Trans. Wirel. Commun. 11 (1) (2012) 258–265.

[9] R. Zhang, C.K. Ho, MIMO broadcasting for simultaneous wireless information and power transfer, IEEE Trans. Wirel. Commun. 12 (5) (2013) 1989–2001.

[10] L. Liu, R. Zhang, K.C. Chua, Wireless information transfer with opportunistic energy harvesting, IEEE Trans. Wirel. Commun. 12 (1) (2013) 288–300.

[11] A.A. Nasir, X. Zhou, S. Durrani, R.A. Kennedy, Relaying protocols for wireless energy harvesting and information processing, IEEE Trans. Wirel. Commun. 12 (7) (2013) 3622–3636.

[12] H.S. Nguyen, T. Nguyen, V.T. Vo, M. Voznak, Hybrid full-duplex/half-duplex relay selection scheme with optimal power under individual power constraints and energy harvesting, Comput. Commun. 124 (2018) 31−44. Available from: https://doi.org/10.1016/j.comcom.2018.04.014.

[13] H.S. Nguyen, T. Nguyen, M. Voznak, Relay selection for swipt: performance analysis of optimization problems and the trade-off between ergodic capacity and energy harvesting, AEU Int. J. Electron. Commun. 85 (2018) 59−67. Available from: https://doi.org/10.1016/j.aeue.2017.12.012.

[14] L. Liu, R. Zhang, K.C. Chua, Multi-antenna wireless powered communication with energy beamforming, IEEE Trans. Wirel. Commun. 62 (12) (2014) 4349−4361.

[15] G. Yang, C.K. Ho, Y.L. Guan, Dynamic resource allocation for multiple-antenna wireless power transfer, IEEE Trans. Signal Process. 62 (14) (2014) 3565−3577.

[16] Varshney, Transporting information and energy simultaneously, in: Proceedings of International Symposium on Information Theory, July 2008, pp. 1612−1616.

[17] X. Chen, X. Wang, X. Chen, Energy-efficient optimization for wireless information and power transfer in large-scale mimo systems employing energy beamforming, IEEE Wirel. Comm. Lett. 2 (6) (2013) 667−670.

[18] X. Lu, P. Wang, D. Niyato, Z. Han, Resource allocation in wireless networks with RF energy harvesting and transfer, IEEE Netw. 29 (6) (2015) 68−75. Available from: https://doi.org/10.1109/MNET.2015.7340427.

[19] Grover, A. Sahai, Shannon meets tesla: wireless information and power transfer, in: Proceedings of International Symposium on Information Theory, June 2010, pp. 2363−2367.

[20] C.A. Balanis, Antenna Theory: Analysis and Design, John Wiley & Sons, 2012.

[21] R. Jabbar, Y.S. Song, T.T. Jeong, RF energy harvesting system and circuits for charging of mobile devices, IEEE Trans. Consum. Electron. 56 (1) (2010) 247−253.

[22] D.P. Bertsekas, Dynamic Programming and Optimal Control, vol. 1, Athena Scientific, 1995. ISBN 1-886529-26-4.

[23] A. Fu, E. Modiano, J. Tsitsiklis, Optimal energy allocation and admission control for communications satellites, IEEE/ACM Trans. Netw. 11 (2003) 488−500. Available from: https://doi.org/10.1109/TNET.2003.813041.

[24] V. Sharma, U. Mukherji, V. Joseph, S. Gupta, Optimal energy management policies for energy harvesting sensor nodes, IEEE Trans. Wirel. Commun. 9 (2010) 1326−1336. Available from: https://doi.org/10.1109/TWC.2010.04.080749.

[25] A. Kansal, J. Hsu, S. Zahedi, M.B. Srivastava, Power management in energy harvesting sensor networks, ACM Trans. Embed. Comput. Syst. 6 (2007) :32. Available from: https://doi.org/10.1145/1274858.1274870.

[26] Ozel, S. Ulukus, Information-theoretic analysis of an energy harvesting communication system, in: International Workshop on Green Wireless (W-GREEN) at IEEE Personal, Indoor and Mobile Radio Communications, 2010. < https://doi.org/10.1109/PIMRCW.2010.5670389 >.

[27] O. Ozel, K. Tutuncuoglu, J. Yang, S. Ulukus, A. Yener, Transmission with energy harvesting nodes in fading wireless channels: optimal policies, IEEE J. Sel. Areas Commun. 29 (2011) 1732−1743. Available from: https://doi.org/10.1109/JSAC.2011.110921.

[28] K. Yamamoto, K. Haneda, H. Murata, S. Yoshida, Optimal transmission scheduling for hybrid of full-duplex and half-duplex relaying, IEEE Commun. Lett. 15 (3) (2011) 305−307. Available from: https://doi.org/10.1109/LCOMM.2011.011811.101925.

[29] T. Riihonen, S. Werner, R. Wichman, Mitigation of loopback self-interference in full-duplex mimo relays, IEEE Trans. Signal Process. 59 (12) (2011) 5983−5993. Available from: https://doi.org/10.1109/TSP.2011.2164910.

[30] B. Day, A. Margetts, D. Bliss, P. Schniter, Full-duplex mimo relaying: achievable rates under limited dynamic range, IEEE J. Sel. Areas Commun. 30 (8) (2012) 1541−1553. Available from: https://doi.org/10.1109/JSAC.2012.120921.

[31] Aggarwal, M. Duarte, A. Sabharwal, N.K. Shankaranarayanan, Full-duplex or half-duplex? A capacity analysis with bounded radio resources, in: IEEE Information Theory Workshop, 2012, pp. 207−211. < https://doi.org/10.1109/ITW.2012.6404659 >.

[32] Barghi, A. Khojastepour, K. Sundaresan, Characterizing the throughput gain of single cell MIMO wireless systems with full-duplex radios, in: Proceedings of 10th International Symposium on Modeling and Optimization in Mobile, Ad Hoc and Wireless Networks, 2012, pp. 6874. < https://doi.org/10.1016/0003-4916(63)90068-X >.

[33] Aryafar, M.A. Khojastepour, K. Sundaresan, S. Rangarajan, M. Chiang, Midu: enabling MIMO full-duplex, in: Proceedings of 18th Annual International Conference on Mobile Computing and Networking, 2012, pp. 257−268.

[34] Khafagy, A.E. Shafie, A. Sultan, Throughput maximization for buffer-aided hybrid half-/full-duplex relaying with self-interference. In: Proceedings of 2015 IEEE International Conference on Communications (ICC), 2015, pp. 1926−1931 < https://doi.org/10.1109/ICC.2015.7248607 >.

[35] K. Tutuncuoglu, A. Yener, Sum-rate optimal power policies for energy harvesting transmitters in an interference channel, J. Commun. Net. 14 (2) (2012) 151−161. Available from: https://doi.org/10.1109/JCN.2012.6253063.

[36] Z. Zhang, K. Long, A.V. Vasilakos, L. Hanzo, Full-duplex wireless communications: challenges, solutions, and future research directions, Proc. IEEE 104 (7) (2016) 1369−1409. Available from: https://doi.org/10.1109/JPROC.2015.2497203.

[37] O. Orhan, E. Erkip, Optimal transmission policies for energy harvesting two-hop networks, in: Proceedings of Annual Conference on Information Sciences and Systems (CISS), 2012, pp. 16. < https://doi.org/10.1109/CISS.2012.6310831.19 >.

[38] C. Huang, R. Zhang, S. Cui, Throughput maximization for the Gaussian relay channel with energy harvesting constraints, IEEE J. Sel. Areas Commun. 31 (8) (2013) 1469−1479. Available from: https://doi.org/10.1109/JSAC.2013.130811.

[39] A. Imtiaz, I. Aissa, W.K.N. Derrick, S. Robert, Power allocation for a hybrid energy harvesting relay system with imperfect channel and energy state information, in: Proceedings of IEEE Wireless Communications and Networking Conference (WCNC), 2014, pp. 990−995. < https://doi.org/10.1109/WCNC.2014.6952243 >.

[40] P. Binod, D.R. Sanjay, K. Sumit, Secondary throughput in underlay cognitive radio network with imperfect CSI and energy harvesting relay, in: Proceedings of IEEE International Conference on Advanced Networks and Telecommunications Systems (ANTS), 2015, pp. 1−6. < https://doi.org/10.1109/ANTS.2015.7413619 >.

[41] T. Kamel, A.Q. Khalid, A. Mohamed-Slim, Outage analysis for underlay cognitive networks using incremental regenerative relaying, IEEE Trans. Veh. Technol. 62 (2) (2012) 721−734. Available from: https://doi.org/10.1109/TVT.2012.2222947.

[42] W. Fei, Z. Xi, Resource allocation for multiuser cooperative overlay cognitive radio networks with RF energy harvesting capability, Proceedings of IEEE Global Communications Conference (GLOBECOM), 2016, pp. 1−6. Available from: https://doi.org/10.1109/GLOCOM.2016.7842221.

[43] D. Li, C. Shen, Z. Qiu, Sum rate maximization and energy harvesting for two way AF relay systems with imperfect CSI, in: Proceedings of International Conference on Acoustics, Speech and Signal Processing, May 2013, pp. 4958−4962.

[44] Z. Xiang, M. Tao, Robust beamforming for wireless information and power transmission, IEEE Commun. Lett. 1 (4) (2012) 372−375.

[45] T. Schenk, RF Imperfections in High-Rate Wireless Systems: Impact and Digital Compensation., Springer, 2008.

[46] E. Bjornson, M. Matthaiou, M. Debbah, A new look at dual-hop relaying: performance limits with hardware impairments, IEEE Trans. Commun. 61 (11) (2013) 4512−4525.

[47] O. Taghizadeh, V. Radhakrishnan, A.C. Cirik, R. Mathar, L. Lampe, Hardware impairments aware transceiver design for bidirectional full-duplex MIMO OFDM systems, IEEE Trans. Veh. Technol. 67 (8) (2018) 7450—7464. Available from: https://doi.org/10.1109/TVT.2018.2839661.

[48] X. Xia, D. Zhang, K. Xu, W. Ma, Y. Xu, Hardware impairments aware transceiver for full-duplex massive MIMO relaying, IEEE Trans. Signal Process. 63 (24) (2015) 6565—6580. Available from: https://doi.org/10.1109/TSP.2015.2469635.

[49] N.T. Do, D.B. da Costa, B. An, Performance analysis of multirelay RF energy harvesting cooperative networks with hardware impairments, IET Commun. 10 (18) (2016) 2551—2558. Available from: https://doi.org/10.1049/iet-com.2016.0392.

[50] T.P. Huynh, H.S. Nguyen, D.T. Do, M. Voznak, Impact of hardware impairments in AF relaying network for wipt: TSR and performance analysis, in: Proceedings of International Conference on Electronics, Information, and Communications (ICEIC), 2016, pp. 1—4. <https://doi.org/10.1109/ELINFOCOM.2016.7562967>.

[51] H.S. Nguyen, H. Nguyen, S.D. Sau, M. Voznak, T. Huynh, Impact of hardware impairments for power splitting relay with wireless information and EH, in: Proceedings of International Conference on Electronics, Information, and Communications (ICEIC), 2016, pp. 1—4. <https://doi.org/10.1109/ELINFOCOM.2016.7562941>.

[52] T.N. Nguyen, P.T. Tran, H.G. Hoang, H.S. Nguyen, M. Voznak, On the performance of decode-and-forward half-duplex relaying with time switching based energy harvesting in the condition of hardware impairment, in: Proceedings of International Conference on Advances in Information and Communication Technology (ICACT), 2017, pp. 421—430. <https://doi.org/10.1007/978-3-319-49073-1/46>.

[53] H.S. Nguyen, M. Voznak, M.T. Nguyen, L. Sevcik, Performance analysis with wireless power transfer constraint policies in full-duplex relaying networks, Elektron. Elektrotech. 23 (4) (2017) 70—76. Available from: https://doi.org/10.5755/j01.eie.23.4.18725.

[54] H. Ju, R. Zhang, User cooperation in wireless powered communication networks, Proceedings of IEEE GLOBECOM, 2014, pp. 1430—1435. Available from: https://doi.org/10.1109/GLOCOM.2014.7037009.

[55] H.S. Nguyen, A.H. Bui, D.T. Do, M. Voznak, Imperfect channel state information of AF and DF energy harvesting cooperative networks, China Commun. 13 (10) (2016) 11—19. Available from: https://doi.org/10.1109/CC.2016.77320058.

[56] D.N. Van, D.V. Son, S. Oh-Soon, Opportunistic relaying with wireless energy harvesting in a cognitive radio system, Proceedings of IEEE on Wireless Communications and Networking Conference (WCNC), 2015, pp. 87—92. Available from: https://doi.org/10.1109/WCNC.2015.7127450.

[57] H.S. Nguyen, D.T. Do, M. Voznak, Exploiting hybrid time switching-based and power splitting-based relaying protocol in wireless powered communication networks with outdated channel state information, Automatika 58 (1) (2017) 111—118. Available from: https://doi.org/10.1080/00051144.2017.1372124.

[58] A.J. Goldsmith, Wireless Communications, Cambridge University Press, Cambridge, UK, 2005.

[59] M. Matthaiou, A. Papadogiannis, E. Bjornson, M. Debbah, Two-way relaying under the presence of relay transceiver hardware impairments, IEEE Commun. Lett. 17 (6) (2013) 1136—1139.

[60] H.S. Nguyen, T. Nguyen, M. Voznak, Successful transmission probability of cognitive device-to-device communications underlaying cellular networks in the presence of hardware impairments, EURASIP J. Wirel. Commun. 2017: (208) (2017). Available from: https://doi.org/10.1186/s13638-017-0994-0.

[61] Z. Wang, Z. Chen, B. Xia, L. Luo, J. Zhou, Cognitive relay networks with energy harvesting and information transfer: design, analysis and optimization, IEEE Trans. Wirel. Commun. 15 (4) (2016) 2562—2576. Available from: https://doi.org/10.1109/TWC.2015.2504581.

[62] D.E. Goldberg, Genetic Algorithms in Search, Optimization and Machine Learning, Addison-Wesley Longman Publishing Co., Inc., Boston, MA, 1989.

[63] K. Man, K. Tang, S. Kwong, Genetic algorithms: concepts and applications, IEEE Trans. Ind. Electron. 43 (5) (1996) 519–534. Available from: https://doi.org/10.1109/41.538609.

[64] X.K. GDu, Y. Zhang, Z. Qiu, Outage analysis and optimization for time switching-based two-way relaying with energy harvesting relay node, KSII Trans. Intern. Inf. Syst. 9 (2) (2014) 1–19. Available from: https://doi.org/10.3837/tiis.2015.02.004.

[65] Z. Yangyang, G. Jianhua, M. Jinjin, O. Fengchen, Z. Chensi, Joint relay selection and power allocation in energy harvesting AF relay systems with ICSI, IET Microw. Antennas Propag. 10 (15) (2016) 1656–1661. Available from: https://doi.org/10.1049/iet-map.2016.0028.

[66] Selection procedures for the choice of radio transmission technologies of the UMTS. 3GPP TR 30.03U, version 3.2.0, 1998.

12

Energy-efficient paging in cellular Internet of things networks

O. Vikhrova[1], S. Pizzi[1], A. Iera[1], A. Molinaro[1], K. Samuylov[2], G. Araniti[1]

[1]DIIES DEPARTMENT, MEDITERRANEAN UNIVERSITY OF REGGIO CALABRIA, REGGIO CALABRIA, ITALY [2]PEOPLES' FRIENDSHIP UNIVERSITY OF RUSSIA (RUDN UNIVERSITY), MOSCOW, RUSSIA

12.1 Introduction

The future landscape envisioned by the Internet of things (IoT) paradigm foresees a collection of heterogeneous smart devices that interact on a collaborative basis to fulfill a common goal. IoT-based massive machine-type-communications (mMTC) market category mainly incorporates *smart wearables* and *sensor networks* industrial vertical features. The number of devices to be used in different service areas such as security, tracking and tracing, payment service, health care, automatic control or remote control, metering, and consumer electronics is exponentially increasing. Most of the sensor-like IoT devices are deployed with no access to a main power source and entirely rely on battery power and must be able to operate autonomously for at least 10 years in the most extreme coverage situations assuming a 5-watt-hour battery [1].

Wireless mobile networks have been employing *discontinuous reception* (DRX) to conserve the power of their end points since early releases. DRX allows an *idle* device to power down its radio interface for a predefined period called *DRX cycle* instead of continuously listening to the radio channel. However, MTC requires more efficient power saving mechanisms to facilitate operation of low-cost and battery-constrained devices for years. Moreover, IoT applications, for example, smart meters, require very sparse periods of device activity (once a day/week/month).

The Third Generation Partnership Project (3GPP) study on MTC [2] introduced two important solutions to optimize the device power consumption, namely *power saving mode* (PSM) and *enhanced discontinuous reception* (eDRX). The later allows to configure much longer inactivity period than that of a conventional DRX, while PSM is a new operation mode.

From the network point of view, MTC traffic is either *mobile-terminated (MT)* or *mobile-originated (MO)*. Generally, IoT devices generate a sporadic light payload in the uplink (UL)

LPWAN Technologies for IoT and M2M Applications. DOI: https://doi.org/10.1016/B978-0-12-818880-4.00013-2

direction and occasionally receive data in downlink (DL). Most of the time devices are inactive; thus they may power down or even switch off their circuits in order to prevent battery drain while no data communication is expected. PSM or *deep sleep mode* was specified both for GSM and LTE and let a device turn off most of its circuits to conserve the battery and reduce energy consumption level to a bare minimum. In the power saving state, the device is unreachable for MT services. Unlike the idle state where a device still consumes some energy to perform neighbor cell measurements and discontinuously listen for paging messages, energy consumption in PSM is hardly above 0 watt-hour. An important difference between PSM and complete power-off state is that in PSM the device stays registered in the network and maintains its connection configurations, so it does not need to attach to the network after wake-up. This feature reduces the signaling overhead and optimizes the device power consumption.

In IoT applications, mainly with MO traffic, network specifies time when devices should wake up for data transmission and go back to sleep (idle), based on the application requirements for data reporting and network capabilities to support devices in sleep mode. The longer devices remain in sleep mode, the higher the energy saving. However, PSM may introduce a long delay for MT traffic since data cannot be immediately delivered to devices in sleep mode. Thus it does not well suit critical or latency-sensitive applications in which end points should be informed in a timely manner by the radio access network (RAN) or core network (CN) about upcoming data through the paging procedure. The right balance between wake-up and sleep periods allows to improve device energy consumption without a significant data latency degradation.

12.2 Power saving solutions for cellular Internet of Things

Any two-way communications between device and network is possible only when the device keeps its UL timing synchronized with the network and has valid radio resource allocation for signaling or data transmission, that is, when it is in the *connected mode*. A device establishes a connection with the BS through the exchange of radio resource control (RRC) messages. While in connected mode, devices can be individually addressed by their *cell radio network temporary identifier* (C-RNTI) assigned by the BS, are available for signaling, and can transmit or receive data. If all data packets have been successfully uploaded and the device does not expect any further communication, it sends the *RRC_connection Release* or *RRC_connection Suspend* message to inform the network about its intention to go to sleep. Alternatively, if no signaling data arrives in DL for a certain time, the device reports a failure to decode control information and is considered out of sync. If a device remains in out of sync condition for 200 ms, defined by N310 timer, it should get back UL synchronization within a time interval specified by T310 timer. Otherwise, the device has to switch to the sleep mode.

As mentioned before, from 3GPP release 12, IoT devices can choose between two energy saving options: DRX or eDRX in idle mode and PSM. All battery-constrained IoT devices must support eDRX functionality, while PSM implementation is optional.

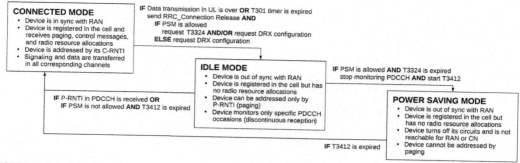

FIGURE 12–1 Transition diagram between IoT-enabled states. *StateTransitions.png.*

Devices in idle mode are unreachable most of the time except for the periodic paging occasions specified by RAN or CN. They discontinuously monitor physical downlink control channel (PDCCH) for a specific *paging RNTI* (P-RNTI) in d*ownlink control information* (DCI). If received P-RNTI does not address the device, it will remain in idle mode. All devices that have been addressed by P-RNTI and successfully decoded the paging message have to switch to connected mode.

Devices in PSM periodically wake up to deliver MO traffic or perform t*racking area update* (TAU) procedure to confirm its registration in the cell. PSM cycle could take as long as the maximum value of the TAU timer *T3412*. If an IoT device supports PSM and requires its activation, it should request *active time* for T3324 timer during which it can be reached for paging (i.e., the time in idle mode). Upon the expiration of active time, the device goes to deep sleep. The description of each mode and transition triggers are summarized in Fig. 12−1.

When a device indicates the end of any data transmission, *inactivity timer* starts during which the device keeps UL synchronization and monitors PDCCH for indication of upcoming data schedule in DL direction or for system change information. Long-term evolution for machines (LTE-M) and narrowband-Internet of things (NB-IoT) devices receive paging messages in direct indication information on LTE-M PDCCH (MPDCCH) and NB-IoT PDCCH (NPDCCH), respectively. CN specifies the first subframe of MPDCCH or NPDCCH repetitions for devices in coverage enhancement. If inactivity timer expires before any message arrives, the device has to enter idle mode. Otherwise, after the reception of paging message, the timer will be extended. This ensures that device will not go to sleep if the BS has a valid schedule for an upcoming transmission. Frequent transitions between states negatively impact battery life and increase the overhead; thus network could either let a device go to sleep in case of delay-tolerant applications or keep it in connected mode periodically sending paging messages.

12.2.1 Discontinuous reception

DRX mechanism defines how many PDCCH subframes device should monitor before the DRX cycle. Paging occasion (PO) can be explicitly defined by a paging frame (PF) number in

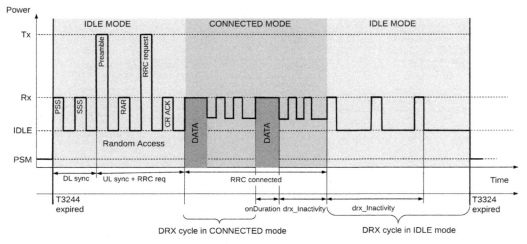

FIGURE 12–2 Device energy consumption in connected and idle modes. *PowerDiagram.png.*

terms of system frame number (SFN) and paging subframe within the frame. Up to 10 paging subframes in one frame can be configured to support massive IoT communications [3].

After a device informs the network that it supports DRX mechanism, the later replays whether DRX is configured and communicates its parameters. DRX can be used in two different modes: connected and idle also referred to as paging (see Fig. 12–2).

12.2.2 Discontinuous reception in connected mode

This mechanism addresses two main issues related to the device battery drain during the connected stage. When the data report uploading is over, the device sends the *RRC_connection Release* or *RRC_connection suspend* message to stop UL synchronization and signaling in RAN and then goes to idle state. However, if there is any upcoming data in the DL direction, the device has to obtain RRC_connection again in order to receive it. Frequent transitions between idle and connected states should be avoided since they increase signaling overhead and energy consumption. Another issue that DRX is aimed to deal with is a long time required for reporting *out_of_sync* status after the last data transmission in connected mode. DRX defines *inactivity timer* to inform devices for how long they should continuously monitor PDCCH opportunities after the subframe where the P-RNTI in DCI has been successfully received and decoded. If no control message arrives while the inactivity timer is running, devices go idle upon its expiration. The *onDuration timer* is used to indicate the time while devices should monitor all PDCCH opportunities within a single DRX cycle, that is, the time interval between two consequent onDuration periods. If no data in PDCCH is detected during the onDuration time, devices stop listening to the DL and power down their circuits until the end of the ongoing DRX cycle. Otherwise, the inactivity timer starts. If a control message arrives with a new DRX parameters allocation while onDuration or inactivity is running, both timers have to be set to zero and the next DRX cycle will apply new parameters. Therefore, the onDuration timer allows faster transition to the idle mode if a device is inactive, while the inactivity timer prevents oscillations between connected and idle states. DRX parameters should be requested by devices in RRC-related messages, for example, RRC_Reconfiguration and RRC_Connection requests. Devices may request individual values by adding them into the request. The BS agrees on individual DRX parameters if they are not in conflict with the higher layer requirements. Otherwise, default values are sent.

12.2.3 Discontinuous reception in idle mode

The mechanism is used as a power saving solution for devices in idle state. It is applied in case when devices generally do not intend to send any data and would like to remain in sleep mode as long as possible. However, RAN or CN might schedule data session while devices are unreachable. In order to minimize the number of unsuccessful attempts to reach devices and not to keep them awake for a long time, the network should configure DRX cycle taking into consideration network capabilities and application requirements.

There are two important parameters for the paging configuration that define the exact time instant when the network can reach a device: *PF* and *PO*. The first one refers to a specific radio frame (RF) containing one or several POs. PO itself defines the subframe number within the PF when a device must listen for the paging message. For NB-IoT devices, the concept of paging narrowband is used, instead of PO, to indicate the narrowband on which the device performs the paging message reception.

The paging message is the same for both RAN-initiated paging and CN-initiated paging. The UE initiates RRC connection resume procedure upon receiving RAN paging. If the UE receives a CN-initiated paging in RRC_INACTIVE state, the UE moves to RRC_IDLE and informs NAS.

If both eNodeB (eNB) and the device agreed on the paging cycle time T, they may independently calculate the SFN that refers to PF when the condition $SFN \bmod T = (T/N) \cdot (UE\_ID \bmod N)$ is met, where $UE\_ID = IMSI \bmod 1024$ if a device has International Mobile Subscriber Identity (IMSI) (otherwise, it is set to 0), and N stands for the frequency of PF appearing in a system frame and takes value of 1, 1/2, 1/4, 1/8, 1/16, and 1/32.

Note: if P-RNTI is monitored on NPDCCH, then UE_ID = IMSI mod 4096, and if device supports paging on a nonanchor carrier and P-RNTI is monitored on MPDCCH or NPDCCH, then UE_ID = IMSI mod 16384.

For NB-IoT devices, N is defined as the minimum between T and possible nB value of 4T, 2T, T, T/2, T/4, T/8, T/16, T/32 included in system information message. The subframe numbers where POs can be scheduled within each PF are defined in Ref. [3], based on the possible number of POs within a single PF. For BL IoT devices, only subframe #1 and subframe #6 are available for POs. Thus a paging message might be sent once, twice, or maximum of four times in each PF at the specified POs.

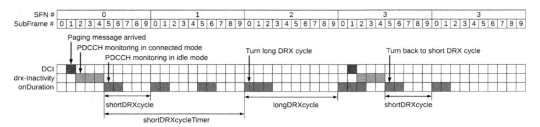

FIGURE 12–3 DRX in idle mode. *DRX.png.*

Beside onDuration timer and inactivity timer allocation, *long DRX cycle* must be configured as a default DRX cycle if the *short DRX cycle* is not specified. It is recommended to set a short DRX cycle for application with critical or delay-sensitive data communication. The example of long and short DRX parameters is shown in Fig. 12−3. The short DRX cycle is the first cycle a device follows after the successful reception of paging message in PDCCH. The number of consecutive short DRX cycles before using long DRX cycle is defined by *short DRX cycle timer*. The subframe number where the long DRX cycle should start is indicated in the s*tart offset* value.

12.2.4 Extended discontinuous reception

SFN and subframe number counters are used for time synchronization between the network and devices underpinning DRX mechanism. The longest time interval available for the synchronization is limited by the maximum number of SFNs, and it is equal to 1024. It means that devices cannot remain in sleep mode for a time period longer of than 10.24 seconds. The DRX cycle can be as long as 8, 16, 32, 64, 128, 256, and 512 ms for eDRX cycle configuration. Machine-type applications may require less frequent communications than every 10.24 seconds. eDRX is a necessary step forward to address a wide set of IoT use cases. In order to allow a longer eDRX cycle, a new SFN counter, namely hyper SFN (H-SFN), was introduced. One H-SFN contains 1024 SFNs and builds an interval of 10,485,760 ms (almost 3 hours), while the system time is sufficiently extended to the 1024 H-SFNs. With the new H-SFN feature, the maximum eDRX cycle value is extended to 43.69 minutes.

Since only one paging message is expected to be sent in a single DRX cycle, any timing inaccuracy could cause the loss of paging message. Therefore, every eDRX cycle can be configured with a *paging transmission window* to increase the number of paging opportunities for improving the paging reliability.

12.2.5 Power saving mode

PSM stands for a deep sleep state similar to the hibernation mode while almost all device circuits are switched off except for the critical ones. PSM targets extremely delay-tolerant

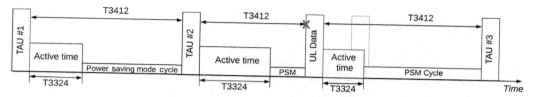

FIGURE 12-4 Power saving mode. *PSM.png.*

applications or can be applied only for MO or MT applications with relaxed latency requirements. Anytime when a device wants to report its status change or send data to the network, it can leave PSM.

PSM is based on two timers: T3324 (Active Time), and T3412 (extended TAU time), which together frame PSM cycle (see Fig. 12−4). Note that any time when a device leaves PSM for MO transmission, timer T3412 is restarted. Active time refers to the time interval when a device is reachable for paging. The minimum recommended value for T3324 is 2 DRX cycles plus 10 seconds [4]. Since the DRX cycle value may vary, the relation between T3324 and T3412 should meet the condition (T3412 − T3324)/T3412 > 0.9. T3412 is expected to be not less than 4 hours but cannot exceed the maximum length of 413 days, as defined in Ref. [5].

A device that wants to use the PSM feature has to request Active Time during each wake-up period in any RRC-related message. The network replies with an Active Time value indicating that the PSM is enabled. Upon T3324 expiration, a device enters PSM and becomes unreachable while T3412 is running or until it requires to send any data to the network. Active Time value can be either allocated by RAN or CN or requested by a device. After receiving the request for T3324 configuration, the network will choose an appropriate value taking into account, for instance, whether the DRX or eDRX is currently enabled. If no MO data are expected within T3412, an IoT device can remain in deep sleep for as long as the upper limit determined by the maximum value of the TAU timer.

12.2.6 Wake up signal

To reduce the device energy consumption during the idle mode even further, a *wake-up signal* (WUS) was introduced in 3GPP release 15 [6]. The WUS sequence design is based on one Zadoff−Chu sequence and a scrambling code. A WUS is an indication sent to a device or a group of devices, whether to listen for the very next PO following after the subframe where the signal is received ignoring current DRX/eDRX configuration.

If device supports WUS, it first shall get WUS configuration from the system information to know exactly when and where to monitor WUS. When the device detects WUS, it monitors a single PDCCH opportunity. When eDRX is used, a device monitors a batch of following POs. It may skip POs monitoring after paging message is received. If the device does not detect WUS or detects WUS for the other device, it will listen to the PO according to its DRX/eDRX cycle.

12.3 Paging strategies

Paging is also used as a pull-based technique for RAN overload control in mMTC scenarios. Signaling overload in ultra-dense IoT environments cannot be avoided due to the huge number of simultaneous access attempts in the shared radio channel and high control message payload required to keep massive connections. Radio shared channel capacity is limited by its nature and can be evaluated in terms of *random access opportunities (RAOs)*. Each RAO is a combination of a unique radio-frequency sequence called a *preamble*, and the subframe position within an RF when the randomly generated preamble can be sent by a device. The network informs all devices in the tracking area when they can send a preamble and specifies all the preambles it can decode. The number of unique sequences is relatively small; thus more than one device can choose the same preamble. If two or more devices simultaneously transmit the same preamble sequence, a collision may occur, and the BS will not be able to discern collided devices and properly address them with the resource allocation.

Differently from push-based techniques which may lead to a high number of arrivals to an RAO, the paging mechanism represents a proactive approach for handling access requests in a controlled manner. Thanks to DRX or PSM, arrivals are scattered over the time and, moreover, the network is aware of their activity patterns.

MT services use paging to inform devices about new schedules for data delivery or system changes. Devices are addressed by their unique identity or by a *group identifier* (GID). One paging message transmitted in PDSCH contains a list of device identities. The number of device IDs also knows as paging records carried in a paging message depends on the size of each device identity. The size of paging record usually varies between 25 and 61 bits, corresponding to three and eight device IDs [7].

Provided that the capacity of an individual paging is very much limited, the use of GID introduces a good scalability and flexibility in terms of the number of simultaneously paged devices. However, the bigger the number of paged devices, the higher the collision probability during the access stage following the paging. Another issue is that IoT devices have different POs and only devices with the same DRX cycle may be acknowledged by one paging message.

Different paging approaches may fulfill the heterogeneity of IoT application requirements in terms of latency, reliability, and scalability.

In the following, we will describe three different paging approaches and, for each of them, we discuss their strong points, potential issues, and applicability for IoT.

12.3.1 Standard paging

In the legacy 3GPP paging, up to 16 devices can be addressed by a single paging message. In NB-IoT due to the limited bandwidth, the number of paging records is twice less. In order to page all relevant devices, the network has to consequently send a number of messages.

Paging delay, that is, the time to page these devices, can be roughly estimated as product of paging message inter-arrival time and number of sent messages plus average access delay.

Since the paging delay linearly increases with the number of IoT devices, the SP approach is valid until the service latency requirement is less than estimated paging delay.

12.3.2 Group paging

When the issue of connecting millions of IoT devices arose, the legacy paging was no longer acceptable due to the extremely long paging delay. The concept of GID was designed to overcome this limitation. During camping procedure, a group of devices can be assigned with the same IMSI or a GID. Thus all devices in the group will listen to the same PO in the PF calculated from their IMSI or GID. Only one paging message is needed to page all members of the same group.

The more devices are grouped together, the higher the collision probability at the access stage. Devices have a number of attempts to successfully transmit preambles. If devices transmit their preambles for the first time, a portion of collided devices will try to send preambles in the next attempt. Collision probability depends on the number of available preambles and the number of contending devices. This number will gradually decrease. All devices in a group are expected to successfully connect to the network only if they are able to complete RA within the predefined number of attempts.

12.3.3 Enhanced group paging

Due to the high collision rate experienced in group paging (GP) and increasing paging delay of SP under the increasing number of devices to be paged, a balanced paging approach for mMTC is required. Enhanced GP (eGP) is based on the idea of group-by-group paging. The concept assumes to page more devices each PO than in SP using the GID and, at the same time, to improve the collision probability at the expense of increasing interval between two consequent paging POs. The interval should be long enough to ensure that the previous group of paged devices can complete RA. The failure of access attempt could be caused not only by the preamble collision but also by the resource shortage in DL for the RA response and RRC-related message.

The size of paging group and paging periodicity of the eGP concept should be properly adjusted to meet the use case requirements or the network capabilities.

12.4 Applications for paging in cellular Internet of things

Device manufacturers need to perform device update in order to keep IoT infrastructure secure and up-to-date. In most of MT applications, IoT devices are receivers of the same content (e.g., firmware/software update, configuration file, schedule, routine task, etc.). Such applications may greatly benefit from point-to-multipoint (i.e., multicast), rather than unicast, communications since multiple receivers can be fed by a single data transmission.

More general, the use cases where MTC devices might benefit from the group communications are [8]:

- Planned data delivery,
- Initially unplanned noncritical data delivery, and
- Initially unplanned critical data delivery.

Initially unplanned noncritical data delivery:

When an update file is available for download, the network shall inform MTC devices about the new schedule when they are reachable, that is, at the very next paging opportunity or right after the waking up for the periodic TAU procedure.

The time interval between the subframe when multicast session schedule was announced and subframe when the announced session starts must be bigger than the longest PSM cycle in the multicast group to ensure that all group members are informed about the forthcoming data delivery session.

Initially unplanned critical data delivery:

A critical software/firmware update, when available, must be delivered as soon as possible. However, the eNB can inform devices about a new multicast session schedule only when they are awake and listen to the PDCCH. Data can be repeated in several transmissions until all devices receive the content. The time between two successive transmissions is assumed to be fixed and is called a *critical interval*. The time between when a schedule for the critical file delivery is announced and when the first group-oriented transmission starts shall be less than the shortest PSM cycle of all group members. eNB can repeat data delivery only to devices that have not received the schedule in the previous sessions.

12.4.1 Group communications

Cellular IoT supports multimedia broadcast and multicast services (MBMS) in the form of single-cell point-to-multipoint (SC-PTM) communications. In the SC-PTM framework, a new single-cell multimedia radio bearer and two channels, namely single-cell multicast control channel (SC-MCCH) and single-cell multicast transport channel (SC-MTCH), were introduced for group-oriented data delivery. The bearer service (and multicast session) can be identified by the group radio network temporary identifier (G-RNTI). Newly designed channels are scheduled and carried by PDCCH and PDSCH, respectively [9]. Multicast transmissions related to the MBMS specific procedures, such as *service announcement, session start, data transfer,* and *session stop* [10], are scheduled with a periodicity specified by *SC-PTM DRX cycle* and delivered to the devices in idle mode. A new broadcasted System Information Block-20 carries scheduling information for one SC-MCCH per cell, while SC-MCCH contains scheduling information for SC-MTCH per multicast service. This information contains SC-MTCH scheduling cycle, *SC-MTCH onDuration time* and *SC-MTCH inactivity timer.*

The periodic monitoring of the SC-MCCH required for service announcement is an extremely energy and resource-consuming approach for NB-IoT. In fact, devices have to listen to the channel even if there is currently no available service for them. *On-demand paging* for the service announcement might be an efficient solution to avoid continuous listening to the SC-MCCH and improve the paging reliability.

12.4.2 Solutions for improving battery lifetime in Internet of things group communications

In this subsection, we present two group-based delivery strategies for critical and noncritical firmware/software update applications. The first strategy can be generally applied to any noncritical delay application with a low-periodic (planned or unplanned) MT traffic. When the new software/firmware file is available for download, the RAN informs a group of devices specified by device owner or device manufacturer through the paging procedure and initiates data delivery after the last device of the group receives MBMS configuration and session scheduling information. The second strategy deals with the class of delay-sensitive IoT applications characterized by a sporadic critical MT traffic. In order to communicate update file with a reasonable delay, RAN initiates a multicast session at the beginning of each critical interval only for those devices that has been successfully paged in the last critical interval.

Any of the paging strategies (SP, GP, or eGP) can be utilized by RAN to inform IoT devices about upcoming data delivery session. Note that each member of the same paging group must follow the same DRX cycle in order to simultaneously listen to the PDCCH opportunity.

12.5 Paging enhancement in 5G

12.5.1 Secure paging

The latest 5G standard includes privacy safeguards against IMSI catchers to ensure the privacy of paging message distribution. 5G exploits a privacy enhancement in uplink communication by using a concealed identifier *SUCI* (Subscription Concealed Identifier) instead of an IMSI analog called *SUPI* (Subscription Permanent Identifier). It is generated every time when a device wants to transmit data by using asymmetric cryptography. As for the downlink protection in 5G, paging protocols were enhanced by using new temporary identifiers 5G-S-TMSI and I-RNTI (to identify device context in a new *RRC_INACTIVE* state) instead of long-term IMSI and SAE Temporary Mobile Subscriber Identity (S-TMSI identifiers) [11].

5G-S-TMSI is used as an IMSI (long-term identifier) analog in legacy paging. In 4G paging, timing was determined based on IMSI, while in 5G both PF and PO are based on a 5G-S-TMSI. This novelty makes it difficult for an over-the-air attacker to deduce information about a device's IMSI by monitoring the air interface and detecting which paging occasions the device is monitoring.

In legacy paging, IMSI and S-TMSI are used as a paging identifier, to indicate paged devices in a paging message. In early generations, IMSI was used for a paging identifier in order to restore the connection in case of lost or corruption of device context information (such as temporary identifier) in the CN. With native virtualization and cloud-based RAN support, only temporary identifier (5G-S-TMSI or I-RNTI) can be used as a paging identifier in 5G networks. If an attacker somehow obtains the device's long-term identifier, in 5G

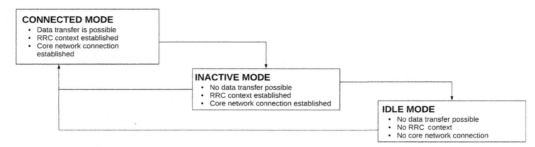

FIGURE 12–5 RRC states in 5G. *RRC_5G.png*.

network it still cannot attack the device because there is no long-term identifier-based paging to start with. Moreover, a new paging identifier I-RNTI is used as one-time paging identifier and must be refreshed after each paging.

Unlike the S-TMSI that could be refreshed optionally, it is compulsory to refresh 5G-S-TMSI in 5G as well. By refreshing the 5G-S-TMSI, PO and PF are also changed, making it more difficult for an attacker to track a device during the paging [12].

12.5.2 Random access network paging in 5G

In 5G systems [13], the mobility states are introduced based on current LTE mobility states but add some modifications. Unlike the two RRC states in LTE—*RRC idle* and *RRC Connected*—5G has designed a new state called *RRC inactive* to indicate whether the UE exchanges data packets with its serving RAN.

RRC idle and *RRC Connected* state transitions are the source of a significant signaling overhead, especially when a huge number of UEs wake up to transmit or receive data simultaneously. Moreover, most of RRC connected states usually send data of less than 1 kilobyte and then switch back to RRC idle state. In order to keep signaling overhead at a reasonable level, a new RRC inactive state has been introduced.

Fig. 12−5 depicts RRC state transitions in 5G, including the transition from and to the *RRC inactive* state. The new state keeps the UE connected from the 5G CN perspective, that is, the UE access stratum context is stored in both the UE and serving RAN. By introducing RRC inactive stage the paging procedure initiation has been changed, providing several advantages for TAU and paging procedures as compared with LTE. In 5G, this procedure can be triggered by RAN, which meaningfully reduces the paging latency to meet the 5G requirements and signaling overhead [14].

12.6 Open issues and Third Generation Partnership Project study in Release 16

3GPP has established new working groups for mMTC in 5G namely "Further NB-IoT enhancement" and "Even further enhanced MTC for LTE" [15,16]. One objective of these two new research directions is power consumption reduction for physical channels for NB-IoT and eMTC. Since 3GPP has started discussions and work on the design of the wake-up and go-to-sleep signals/channels, it has been agreed on the need for Group WUS and early termination for NPDCCH or WUS. Also, a combination of WUS and *discontinuous transmission* is under the discussion.

For NB-IoT, one wake-up signal can be applied to all the devices associated to a PO. It is under the discussion whether to further divide devices associated to a PO into several subgroups [17]. The intention of subgrouping is for reducing the false alarm probability of paging. It is beneficial for device power consumption but with the expense of increased network overhead.

Group WUS sequence should consider the fallback to legacy device behavior, inter-cell interference randomization, group ID for different subgroups for WUS, and effect of sequence detection on device complexity.

In Ref. [18], authors use stochastic geometry to analyze the performance of an energy-efficient joint downlink and uplink radio-frequency wake-up solution for IoT devices over cellular networks. Instead of waking up from sleep state on a timer basis and wasting power if the device has no data to transmit, the IoT devices wake-up after receiving dedicated RF signals from their serving BSs. By adding a simple wake-up receiver at the front-end of the IoT device, the power of a specific wake-up signal transmitted from the serving BS along with the ambient interference can be captured and used to activate the device and initiate uplink communication. This solution is well suited to the infrequent transmissions in various IoT use cases where a periodic wake-up is not needed. It enhances the energy efficiency of IoT over cellular but at the expense of increased false wake-up rate.

Authors in Refs. [19,20] utilized *microsleep* and DRX to improve battery lifetime of IoT devices in 5G networks by allowing a device to process only some small predefined messages during Active Times and consequently remove empty subframe buffering. The simulation results show that such a scheme can reduce power consumption of IoT device by up to 70%.

In Ref. [21], authors proposed an approach, according to which the mobile device monitors WUS at the specific time instants and subcarriers, in order to decide whether to process the actual upcoming PDCCH or not. A low-complexity wake-up receiver concept was introduced to decode the WUS and to acquire the time and frequency synchronization.

Paging has an important application for device grouping problem in cellular networks. For example, in Ref. [22], grouping is based on similar device quality of service, or on device priority in the RA process [23]. Paging is also used as a pull-based approach to tackle overload problem when a large number of IoT devices try to access RAN [24].

In Refs. [25,26], devices are grouped according to their location, and a device is selected as a head, which coordinates the paging of the other devices, while in Ref. [27], the head of the group is responsible for the data aggregation and its transmission to the network. However, the communication between the head and other devices in a group should be organizes via Device-to-Device (D2D) links or other short-distance communication technologies, which may require extra procedures that further increase the energy consumption.

References

[1] Third Generation Partnership Project, Technical Report 45. 820v15.0. 0, Cellular system support for ultralow complexity and low throughput Internet of things, 2018.

[2] Third Generation Partnership Project, Technical specification 22. 368v15.0. 0, service requirements for machine-type communications (MTC); Stage 1, 2016.

[3] Third Generation Partnership Project, Technical Specification 36. 304v15.0. 0, Evolved universal terrestrial radio access (E-UTRA); User equipment (UE) procedures in idle mode, 2018.

[4] LTE-M deployment guide to basic feature set requirements. GSMA White Paper, Version 2.0, April 2018.

[5] Third Generation Partnership Project, Technical Specification 24. 008v15.0. 0, Mobile radio interface layer 3 specification; Core network protocols; Stage3, 2018.

[6] Third Generation Partnership Project, Technical Specification 36. 211v15.0. 0, Evolved universal terrestrial radio access (E-UTRA); Physical channels and modulation, 2018.

[7] Third Generation Partnership Project, Technical Specification 36. 213v15.0. 0, Evolved universal terrestrial radio access (E-UTRA); Physical layer procedures, 2018.

[8] Third Generation Partnership Project, Technical Specification 26.850 v.15.0.0, Technical Specification Group services and system aspects MBMS for IoT, 2018.

[9] Third Generation Partnership Project, Technical Specification 23.246 v.15.0.0, Multimedia Broadcast/Multicast Service (MBMS); Architecture and functional description, 2018.

[10] Third Generation Partnership Project, Technical Specification 25.346 v.15.0.0, Introduction of the multimedia broadcast/multicast service (MBMS) in the radio access network (RAN); Stage 2, 2017.

[11] Third Generation Partnership Project, Technical Specification 33.501 v.15.5.0, 5G; Security architecture and procedures for 5G System, 2019.

[12] Third Generation Partnership Project, Technical Specification 23.501 v.16.0.0, System architecture for the 5G System (5GS), 2019.

[13] Third Generation Partnership Project, Technical Specification 38.304 v.15.4.0, New radio; User equipment (UE) procedures in idle mode and in RRC inactive state, 2019.

[14] Third Generation Partnership Project, Technical Specification 23.502 v.16.0.0, Procedures for the 5G system (5GS), 2019.

[15] RP-172063, "Further NB-IoT enhancements," Huawei, HiSilicon, RAN#77, Saporo, Japan, September 11−14, 2017.

[16] RP-171427, "Even further enhanced MTC for LTE," Ericsson, Qualcomm, RAN#76, West Palm Beach, USA, June 5−8, 2017.

[17] RP-182070, "Additional enhancements for NB-IoT," Huawei, RAN#81, Gold Coast, Australia, September 10−13, 2018.

[18] N. Kouzayha, Z. Dawy, J.G. Andrews, H. ElSawy, Joint downlink/uplink RF wake-up solution for IoT over cellular networks, IEEE Trans. Wirel. Commun. 17 (3) (2018) 1574−1588.

[19] M. Lauridsen, Studies on Mobile Terminal Energy Consumption for LTE and Future 5G (PhD dissertation), Aalborg University, 2015.

[20] S. Rostami, K. Heiska, O. Puchko, K. Leppanen, M. Valkama, Robust pre-grant signaling for energy-efficient 5G and beyond mobile devices, in: Proc. IEEE ICC'18 GCSN, Kansas City, USA, May 2018.

[21] S. Rostami, K. Heiska, O. Puchko, J. Talvitie, K. Leppanen, M. Valkama, Novel wake-up signaling for enhanced energy-efficiency of 5G and beyond mobile devices, in: Proceedings of IEEE Globecom 2018, December 2018, pp. 1−7.

[22] P. Si, J. Yang, S. Chen, H. Xi, Adaptive massive access management for QoS guarantees in M2M communications, IEEE Trans. Veh. Technol. 64 (7) (July 2015) 3152−3166.

[23] M. Tavana, V. Shah-Mansouri, V.W.S. Wong, Congestion control for bursty M2M traffic in LTE networks, in: 2015 IEEE International Conference on Communications (ICC), June 2015, pp. 5815–5820.

[24] O. Arouk, A. Ksentini, T. Taleb, Group Paging-Based Energy Saving for Massive MTC Accesses in LTE and Beyond Networks, IEEE J. Sel. Areas Commun. 34 (5) (2016) 1086–1102.

[25] S. Xu, Y. Liu, W. Zhang, Grouping based discontinuous reception for massive narrowband internet of things systems, IEEE Internet Things J. PP (99) (2018), pp. 1–1.

[26] K. Lee, J. Shin, Y. Cho, K. S. Ko, D. K. Sung, H. Shin, A group based communication scheme based on the location information of MTC devices in cellular networks, in: 2012 IEEE International Conference on Communications (ICC), June 2012, pp. 4899–4903.

[27] L. Karim, A. Anpalagan, N. Nasser, J. N. Almhana, I. Woungang, An energy efficient, fault tolerant and secure clustering scheme for M2M communication networks, in: 2013 IEEE Globecom Workshops, December 2013, pp. 677–682.

13

Guidelines and criteria for selecting the optimal low-power wide-area network technology

Guillermo del Campo, Igor Gomez, Guillermo Cañada, Luca Piovano, Asuncion Santamaria

TECHNICAL UNIVERSITY OF MADRID, UNIVERSIDAD POLITECNICA DE MADRID, MADRID, SPAIN

13.1 Introduction

As mentioned in previous chapters, there are different families of communication technologies that are part of the Internet of Things (IoT) ecosystem such as low-power wireless personal area network (LoWPAN), wireless local area network (WLAN), Cellular networks or low-power wide-area network (LPWAN). LoWPAN technologies (e.g., 6LoWPAN, ZigBee, Z-Wave, Bluetooth) are designed for short-range, low data rate, and battery-powered applications [1]. Coverage can be extended using mesh networking, thus increasing deployment costs due to the necessary number of gateways and devices. WLAN (e.g., Wi-Fi) provides higher data rates at the expense of large power consumption and shorter ranges [2]. On the other hand, traditional cellular communications [e.g., long-term evolution (LTE), global system for mobile communications (GSM)] provide wider area coverages, but while optimized for voice and high data rate services, rise cost and power consumption [3]. LPWAN differs from these communication technologies as it aims at applications that simultaneously demand low-power and low-cost connectivity for a large number of devices over an extended range [4]. LPWAN applications range from smart city services [5] to infrastructure monitoring [6], including wildlife supervision [7] or logistics [8].

However, once identified LPWAN as the optimal communication type for our application, it surges another dilemma: which specific LPWAN technology do we choose? There is an extended variety of LPWAN technologies, including SigFox, long-range wide-area network (LoRaWAN), Weightless, narrowband-Internet of things (NB-IoT), LTECat-M1, or Ingenu-RPMA (random phase multiple access). Every technology has its advantages and weaknesses. The final selection should attend to the specific requirements of the IoT application, mainly determined by the factors that affect aforementioned LPWAN objectives: low power, low cost, high number of devices, and medium-long range [9].

LPWAN Technologies for IoT and M2M Applications. DOI: https://doi.org/10.1016/B978-0-12-818880-4.00014-4

The low-power objective is established by factors such as network topology, duty cycling, media access control (MAC) mechanism, bidirectionality, packet size, latency, or energy consumption. Similarly, the cost relies on different factors: for example, hardware (HW) complexity and commercial availability, type of infrastructure, proprietary/free band, MAC mechanism, data plans, and development feasibility (tools, documentation, communities). Meanwhile, the range is defined by characteristics such as the frequency band, the modulation technique, or the link budget. On the other hand, features such as diversity techniques, densification, packet size, or latency determine the amount of devices, that is, scalability. Finally, there are some factors that affect all the LPWAN objectives transversally: for example, security, data rate, existence of standards and alliances, or location functionalities.

This chapter introduces the most relevant factors that affect the LPWAN technology selection process, classified according to their category. Section 13.2 focuses in technical factors, divided by their network communication layer: physical layer (e.g., modulation method or frequency band), link layer (e.g., MAC mechanism), and network layer (e.g., latency or network topology). Security mechanisms are also part of the technical factors. Implementation factors are depicted in Section 13.3 and are categorized according to their influence with respect to the cost (e.g., communication infrastructure), development complexity (e.g., tool kits availability), and status (e.g., coverage or commercial devices). Section 13.4 describes functional factors such as energy consumption, integration of location services, or IP support.

In order to facilitate the technology selection, Section 13.5 presents a comparative analysis of technologies attending to the category of the factors and a global cross-factor analysis. Different visualizations are presented to help the identification of the LPWAN technology that best matches the IoT application requirements.

Finally, Section 13.6 illustrates the LPWAN technology selection process with examples of application in different IoT domains such as smart city or logistics.

For every factor, besides its explanation, the related concrete properties of each LPWAN technology are presented. For this end, apart from the most known (and already introduced in previous chapters) LPWAN technologies (LoRa, SigFox, NB-IoT, LTE-CatM, and sensor network over white spaces-SNOW), answering to references, existing deployments, and future trends, the following have also been considered: Weightless, Ingenu-RPMA, Telensa, GSM-IoT, Wi-SUN, DASH7, IQRF, and MIOTY.

13.1.1 Weightless

Weightless is an open protocol created in 2011 by a special interest group (SIG) formed by different companies such as Neul or ARM. Formerly, there were three communications standards: Weightless-P (sub-GHz ISM (industrial, scientific and medical) band), Weightless-N (Uplink only), and Weightless-W (TV white space). Based on the market traction, Weightless-P has prevailed. It offers c. 100 kbps, uses frequency division multiple access (FDMA), and time division multiple access (TDMA) techniques, and requires the user to implement its own network [10].

13.1.2 Ingenu-RPMA

Ingenu-RPMA, previously known as On-Ramp Wireless, was created in 2008. Unlike other LPWAN technologies, it uses the 2.4 GHz ISM band, benefiting from the relaxed regulation of the band: increased transmission power and not duty-cycling limitations. It requires private deployments and based on spread spectrum modulation techniques, offers data rates of 80 kbps for up to 15 km ranges [11].

13.1.3 Telensa

Telensa is a UK-based company mainly focused in smart street lighting and therefore in smart cities. Their solution, called Telensa PLANet, was created in 2005 and uses ultra-narrowband (UNB) technology to offer 500 bps data rates in both directions, thus working also for control applications. It is deployed in 1.7 million streetlights and other city services, such as waste analytics or measurements of the air quality [12].

13.1.4 GSM-IoT

EC-GSM-IoT, also known as GSM-IoT or extended coverage-GSM (EC-GSM), was ratified in 2016 in the Third Generation Partnership Project (3GPP) specification. GSM-IoT is an enhancement of 2G/3G/4G networks, being implemented with just a software update in current GSM networks. It provides data rates between 70 and 240 kbps for a range of up to 10 km. As it is an operator-based technology, it requires the subscription of data plan for each end-device [13].

13.1.5 Wi-SUN

Created in 2011, Wi-SUN, also known as IEEE 802.15.4g, it creates a new PHY (physical layer) to handle huge networks with a small infrastructure. Wi-SUN relies on a mesh topology that reduces the black spots, increasing the communication range up to 10 km. It involves private deployments and the data rate varies between 5 and 800 kbps [14].

13.1.6 DASH7

DASH7 is an open protocol initially based on the ISO/IEC 18000-7 specification, which describes the physical layer for communication of active radio-frequency identification (RFID) devices. Since its creation in 2007, DASH7 has been modified until the current version (v1.1, January 2017), offering communications up to 2 km, mesh topology, low latency, mobility services, data rates up to 167 kbps, and support for AES128 encryption [15].

13.1.7 IQRF

IQRF is a private protocol developed by the Company IQRF Technology (year 2004), which offers both hardware components (end-devices, transceivers, and gateways) and software modules, allowing the communication with devices using high-level commands. Direct communication range is tenths and hundreds of meters in closed and open environments respectively, with data rates up to 20 kbps. Coverage can be extended, forming a mesh network of up to 240 devices [16].

13.1.8 MIOTY

MIOTY is a new LPWAN technology developed in the year 2016 by the Fraunhofer Institute from Germany and commercialized by the Company BehrTech from Canada. It offers a communication range of up to 5 km in urban areas and up to 15 km in open spaces, with a data rate of 512 bps. It is an open protocol, recently standardized by the ETSI (European Telecommunications Standards Institute), and provides long battery life of end-devices [17].

13.2 Technical factors

This section describes the communication techniques. LPWAN technologies employ to achieve long range with low-power consumption and low cost. These techniques are classified by their belonging to the hardware layers of the OSI (Open Systems Interconnection) reference model: physical layer, link layer, and network layer.

13.2.1 Physical layer

13.2.1.1 Frequency band
The majority of the LPWAN technologies use the sub-GHz band, which offers longer ranges and lower power consumption. Compared to the 2.4 GHz band (crowded by Wi-Fi and LoWPAN technologies), the sub-GHz band is less congested and suffers less attenuation and multipath fading caused by obstacles and walls.

On the other hand, most of the technologies make use of the unlicensed bands (both in sub-GHz and 2.4 GHz bands), as they require fewer time to set up, are easy to maintain and more cost-effective. On the other hand, the licensed bands offer higher bandwidth and more reliability as interference is minimized.

Logically, NB-IoT, LTE-CatM, and GSM-IoT, which make the most of the cellular network, use the licensed (L) Sub-GHz band (900 MHz). The rest of the LPWAN technologies (SigFox, LoRa, Dash7, IQRF, MIOT, Telensa) operate in the most common unlicensed bands (433 and 868 MHz in Europe, 915 MHz in the United States). Exceptions are Weightless (also in 138, 470, 780, and 923 MHz bands), Wi-SUN (also in the 2.4 GHz band), Ingenu-RPMA (works just in the 2.4 GHz band) and SNOW (470−790 MHz band).

13.2.1.2 Modulation method
Achieving long range requires lowering of the modulation rate to put more energy in each transmitted bit (or symbol), easing decoding duties of receivers. Accordingly, LPWAN modulation techniques can be divided into two families: narrowband (NB) and spread spectrum (SS) techniques.

Narrowband techniques concentrate the signal energy within a very narrowband (below 25 kHz), minimizing the noise level, reducing the transceiver complexity, and allowing to share the spectrum between multiple links. Some LPWAN technologies reduce the width of the carriers to ultra narrow band-UNB (down to 100 Hz), lowering the noise and incrementing the number of supported devices. At the same time, UNB reduces the data rate and requires higher transmission times.

On the other hand, spread spectrum techniques expand the same signal energy over a wider frequency band. Received signals are usually below the noise floor, preventing interferences and eavesdropping, but requiring transceivers that are more complex.

The majority of the LPWAN technologies use narrowband with different modulation techniques: NB-IoT (QPSK—quadrature phase-shift keying), LTE-CatM (16QAM—quadrature amplitude modulation), Weightless (GMSK—Gaussian minimum-shift keying), SNOW (BPSK —binary phase-shift keying), GSM-IoT (GMSK, 8PSK—phase-shift keying), IQRF (GFSK— Gaussian frequency-shift keying), DASH7 (GFSK). SigFox (DBPSK—differential BPSK, GFSK), MIOTY (GMSK), and Telensa (2-FSK) adopt UNB modulation techniques. Different spread spectrum techniques are used by LoRA (CSS—chirp spread spectrum), and Ingenu-RPMA (DSSS—direct sequence spread spectrum), while Wi-SUN uses both (DSSS, FSK).

13.2.1.3 Data rate

Transmission data rate is determined by several factors, ranging from technical ones such as the frequency band and the modulation technique, to implementation features such as the complexity of the transceiver or the scalability.

On the other hand, in every communication system, there is a direct relation between data rate, link budget, and achievable range. Therefore, and considering the distance dispersion of LPWAN nodes, some technologies offer an adaptive data rate (ADR) that changes depending on the link constraints.

Starting with data rates below 1 kbps, SigFox offers 100 bps and MIOTY offers 512 bps. In the same range is Telensa, providing 500 bps for downlink (DL) and 62.5 bps for uplink (UL). LoRa and Weightless implement ADR, varying, respectively, from 300 and 625 bps to 50 and 100 kbps. Then, we can find various technologies that offer tenths of kbps: IQRF (20 kbps), SNOW (50 kbps) or Ingenu-RPMA (80 kbps), and others, providing modes with different data rates: Wi-SUN (4.8—800 kbps), DASH7 (9.6—166 kbps), and GSM-IoT (70—240 kbps). Finally, it ranks the cellular technologies, which present the higher data rates: NB-IoT (250 kbps) and LTE-CatM (up to 1 Mbps).

13.2.1.4 Range

Similarly, the communication range is defined by pure communication characteristics, but also by other factors such as the emitted power or the sensibility of the receivers. Therefore, there is not a direct relation between data rate and communication range.

Ranking from higher to lower communication ranges, we can find SigFox (40 km), LoRa (20 km), NB-IoT/Weightless/LTE-CatM/Ingenu-RPMA/MIOTY (15 km), GSM-IoT/Wi-SUN (10 km), and Telensa/DASH7/IQRF (5 km). It is important to notice that these are maximum range, that is, with clear line of sight and low data rate modes when applicable.

Table 13−1 summarizes the technical factors belonging to physical layer.

Table 13–1 Technical factors belonging to physical layer.

LPWAN technology	Frequency band (MHz)	Modulation method	Data rate (kbps)	Range (km)
LoRa	433, 868 (EU), 915 (United States)	SS (CSS)	50	20
SigFox	433, 868 (EU), 915 (United States)	UNB (DBPSK, GFSK)	0,1	40
NB-IoT	900 MHz (L)	NB (QPSK)	250	15
LTE-CatM	900 (L)	NB (16QAM)	1000	15
SNOW	470–790 MHz	NB (BPSK)	50	5
Weightless	138, 433, 470, 780, 868 (EU), 915 (United States), 923	NB (GMSK)	100	15
Ingenu-RPMA	2.4 GHz	SS (DSSS)	80	15
Telensa	433, 868 (EU), 915 (United States)	UNB (2-FSK)	0,5	5
GSM-IoT	900 (L)	NB (GMSK, 8PSK)	240	10
Wi-SUN	433, 868 (EU), 915 (United States), 2.4 GHz	SS (DSSS), NB (FSK)	800	10
DASH7	433, 868 (EU), 915 (United States)	NB (GFSK)	166	5
IQRF	433, 868 (EU), 915 (United States)	NB (GFSK)	20	5
MIOTY	433, 868 (EU), 915 (United States)	UNB (GMSK)	0,512	15

13.2.2 Link layer

13.2.2.1 MAC protocol

The use of traditional cellular medium access control (MAC) protocols that employ time and frequency diversity, requires precise synchronization, and are incompatible with the low-cost LPWAN end-devices. One of the most adopted MAC protocols for LPWAN is the carrier-sense multiple access with collision avoidance (CSMA/CA), widely implemented in WLAN and LoWPAN technologies. However, CSMA/CA becomes less effective when the number of nodes increases (an intrinsic characteristic of LPWAN). The use of virtual carrier sensing solves this problem, but it does not behave well with massive deployments. Alternatives to CSMA/CA are the use of ALOHA, a carrier-sensing-less random access protocol for simple and low-cost transceivers; or TDMA/orthogonal FDMA (OFDMA)-based protocols, resulting in more complex and expensive end-devices.

LoRa (TDMA), SigFox (FDMA), Telensa (FDMA), and Wi-SUN (PCA—pure collective ALOHA) use the ALOHA protocol. CSMA/CA is implemented by SNOW (also DOFDM), DASH7, and Wi-SUN. NB-IoT and LTE-CatM employ OFDMA for the downlink communication and single-carrier FDMA (SC-FDMA) for the uplink. GSM-IoT and Weightless use FDMA and TDMA. Finally, Ingenu-RPMA utilizes code division multiple access (CDMA) and MIOTY employs telegram splitting multiple access (TSMA)

13.2.2.2 Bidirectionality

Depending on the final application, the system may demand different types of bidirectionality. For iinstace, it may range from an almost inexistent downlink (just for maintenance duties) in monitoring services, to a total symmetric channel for control applications.

Table 13–2 Technical factors belonging to link layer.

LPWAN technology	MAC protocol	Bidirectionality	Packet size (bytes)
LoRa	ALOHA (TDMA)	YES	51
SigFox	ALOHA (FDMA)	Limited	12
NB-IoT	OFDMA, SC-FDMA	YES	125
LTE-CatM	OFDMA, SC-FDMA	YES	1000
SNOW	DOFDM, CSMA/CA	YES	28
Weightless	FDMA, TDMA	YES	48
Ingenu-RPMA	CDMA	YES	64
Telensa	ALOHA (FDMA)	YES	65
GSM-IoT	TDMA, FDMA	YES	65
Wi-SUN	CSMA/CA, ALOHA (PCA)	YES	2047
DASH7	CSMA/CA	YES	256
IQRF	IQMESH	YES	128
MIOTY	TSMA	YES	192

Bidirectionality is supported by all the technologies, though there are some exceptions. For SigFox, the downlink is very limited: it depends on the subscription level and can only occur after uplink messages. Likewise, for some LoRa and MIOTY class end-devices, the downlink communication only can happen immediately after an uplink transmission.

13.2.2.3 Packet size

The packet or payload size limits the utilization of LPWAN technologies on the applications that need to send large data sizes. In some cases, messages can be fragmented and reconstructed at destination, but it increases the complexity of the systems and the number of transmission packets needed for delivering a certain message or data, resulting in an increase in the power consumption of the nodes, increase in latency, and so on

Ranking LPWAN technologies by packet size, it begins with the lowest being SigFox (12 bytes for uplink and 8 bytes for downlink), SNOW (28 bytes), Weightless (varying from 10 to 48 bytes) and LoRa (51 bytes for UL and 14 for DL). Then, there are Ingenu-RPMA, Telensa, and GSM-IoT, which offer 64 bytes per packet. NB-IoT and IQRF provide packet sizes of 128 bytes, while MIOTY varies from 10 to 192 bytes. Higher packet sizes are offered by DASH7 (256 bytes), LTE-CatM (1000 bytes), and Wi-SUN (2047 bytes).

Table 13−2 summarizes the technical factors belonging to link layer.

13.2.3 Network layer

13.2.3.1 Network topology

Most of the LPWAN technologies use the star topology, connecting the nodes directly to the base stations. In comparison with mesh topologies (employed by LoWPAN technologies), whose nodes actuate as repeaters, the star topology simplifies and reduces the cost and power consumption of the end-devices, while resulting in higher infrastructure costs (gateways, routers, etc.).

Exceptions to this rule are Telensa, Ingenu-RPMA, and DASH7, which apart from star, also support tree topology; Wi-SUN, which also supports mesh; and IQRF, which just implements mesh topology.

13.2.3.2 Duty cycling

The most power-consuming component of LPWAN end-devices is the transceiver. Thus for achieving low-power consumption, it is essential to turn it off as much as possible. This behavior is determined by the duty-cycling mechanisms, which are adapted based on bidirectionality, type of power source, applications, or hardware and firmware design.

The ISM frequency bands implement regulations about the channel maximum use time for every IoT node. The 868 MHz band in Europe limits the duty cycling to 1%, while the 915 MHz band in United States limits the air time to maximum 400 ms. The 2.4 GHz band and the licensed bands present no restrictions in terms of duty cycling.

On the other hand, some LPWAN technologies, such as LoRa and SigFox, implement mechanisms to reduce the duty cycling and increase the life of batteries.

13.2.3.3 Scalability

The LPWAN technologies that use the ISM bands and employ simple MAC protocols (ALOHA or CSMA/CA) have lower throughput and device capacity due to the collision and interference probability. To support the massive and ever-increasing number of end-devices, there are different techniques that exploit diversity of space, time, and radio channels. Taking into account that LPWAN nodes are by definition low-cost and low-power devices, the diversity management has to be done at network infrastructure elements (base stations and back ends). Examples of these techniques are dense base station deployments, ADR, and channel selection or parallel transmission. However, not every technology supports these techniques, resulting in reduced scalability in real deployments.

LoRa implements both adaptive channel and ADR techniques. A LoRa device transmits randomly in one of the eight available channels and can be received simultaneously by different gateways. Besides, the data rate is adapted depending on the signal level received. Similarly, SigFox end-devices transmit in a random channel and repeat the transmission in three different frequencies, being received in various base stations at the same time. MIOTY employs TSMA, a random MAC in which the transmission of a message is divided into several short packets that are transmitted randomly distributed in different channels and times.

Weightless, GSM-IoT, NB-IoT, and LTE-CatM uses both FDMA and TDMA, increasing the network capacity.

13.2.3.4 Latency

Latency is determined by how long it takes the message from the node to the final application. Thus it is direct function of the data rate and therefore power consumption. In most of the LPWAN technologies, the gateways (or base stations) actuate as a transparent bridge, sending received data to the cloud. Some fog computing techniques may reduce latency.

Ranking from lower to higher latency times (average), we start with LTE-CatM (100 ms), NB-IoT (200 ms), and SNOW/Wi-SUN (400 ms). It follows GSM-IoT (1 second), DASH7/LoRa

Table 13–3   Technical factors belonging to network layer.

LPWAN technology	Network topology	Duty cycling score (1 = worst, 5 = best)	Scalability score (1 = worst, 5 = best)	Latency (s)
LoRa	Star	5	3	5
SigFox	Star	3	3	10
NB-IoT	Star	2	5	0,2
LTE-CatM	Star	1	5	0.1
SNOW	Star	4	2	0.4
Weightless	Star	4	3	6
Ingenu-RPMA	Star, tree	3	4	10
Telensa	Star, tree	4	1	10
GSM-IoT	Star	1	4	1
Wi-SUN	Star, mesh	3	2	0.4
DASH7	Star, tree	4	2	5
IQRF	Mesh	4	2	14
MIOTY	Star	5	4	10

(5 seconds), and Weightless (6 seconds). SigFox, Ingenu-RPMA, Telensa, and MIOTY provide average latency of 10 seconds. The slowest LPWAN technology is IQRF, with an average latency of 14 seconds.

Table 13–3 summarizes the technical factors belonging to the network layer. Duty cycling and scalability are presented using a quality score ("1" = worst, "5" = best), which consider the use of different techniques to respectively reduce the duty cycling and improve scalability.

13.2.4 Security

In order to comply with the low-power, low-cost, and low-complexity requirements of the end-devices, the LPWAN technologies are forced to implement simple security techniques. Most of them use symmetric key cryptography (e.g., AES128), defining secret keys previously shared by end-devices and network servers. The cellular technologies (NB-IoT, LTE-CatM, and GSM-IoT) use techniques implemented by the 3GPP, such as the integration of the eSIM technology. LoRa uses AES128 at both network and application layers. SigFox only uses a secret key in the end-device registration process. Weightless employs AES128/256 for encryption and authentication of both the end-device and the network. Wi-SUN uses AES128 for encryption and IEEE802.1x/EAP-TLS for network authentication. For IQRF, network encryption is done with AES128 while end-device uses a key specified by the user. MIOTY implements AES128 for network encryption.

13.3 Implementation factors

When selecting an LPWAN technology, apart from the technical characteristics, there are other extremely important factors, especially during the system implementation phase, such as the cost and the development complexity on the associations/standards support.

13.3.1 Cost

Regarding the cost of implementing an IoT application based on LPWAN technologies, different cost aspects need to be considered: nodes and devices price, infrastructure cost, data plans, etc.

13.3.1.1 Nodes and devices cost

As explained above, compared to other communication technologies, LPWAN transceivers need to manage less complex waveforms, minimizing hardware complexity and cost.

For applications that deploy a huge number of end-devices, it is worthwhile to develop ad-hoc nodes based on communication chipset. There is a broad availability for the following technologies: LoRa (€2.5), SigFox (€1), and NB-IoT/LTE-CatM (€6−10) (commercial prices for more than 1000 units).

In those applications with reduced scales, it is preferable to use communication modules: in the range of €10 for SigFox and LoRa; €15 for IQRF; €30 for Wi-SUN and Ingenu-RPMA; around €70 for NB-IoT and LTE-CatM (commercial prices for a single unit).

13.3.1.2 Communication infrastructure cost

The communication infrastructure cost is a relevant factor for those technologies that demand the deployment of private network elements (e.g., LoRa or Weightless). The infrastructure cost depends directly on the necessary network elements. For the LPWAN technologies that use star topology, a single base station/gateway provides coverage to several km^2, reducing the cost to this unique element.

The gateway price varies from around €100 (IQRF) and €200 (LoRa, Wi-SUN, and DASH7), to up to €1000 (Weightless) (commercial prices for a single unit).

13.3.1.3 Data plans

For those LPWAN technologies that provide the network infrastructure (e.g., SigFox or NB-IoT), communication service is provided with data plans, which may vary depending on the number of messages and data rates ranges.

Considering country regulations and market competitiveness dependency of data plans price, the average values are presented. SigFox yearly subscription fee per end-device varies between €1 and €14, depending on the number of devices. NB-IoT and LTE-CatM end-device fees start, respectively, in 6 and 36 €/year and can be as high as 50 and 100 €/year, depending of the monthly data to be transmitted. Finally, in those locations where there are LoRa networks deployed by telecommunication operators, the fees range from 5 to 20 €/year.

Table 13−4 summarizes the deployment cost factors.

13.3.2 Development

The design and development phases of any IoT application require many tasks that can be optimized with tool kits and the support from developer's communities.

Table 13–4 Deployment cost factors.

LPWAN technology	Communication module (€)	Gateway (€)	Data plans min (€)	Data plans max (€)
LoRa	10	200	5	20
SigFox	10	n/a	1	14
NB-IoT	70	n/a	6	50
LTE-CatM	70	n/a	36	100
SNOW	–	–	n/a	n/a
Weightless	–	1000	n/a	n/a
Ingenu-RPMA	30	n/a	–	–
Telensa	–	–	n/a	n/a
GSM-IoT	–	n/a	–	–
Wi-SUN	30	200	n/a	n/a
DASH7	–	200	n/a	n/a
IQRF	15	100	n/a	n/a
MIOTY	–	–	n/a	n/a

n/a, Not applicable.

13.3.2.1 HW and SW tool kits

For most of the LPWAN technologies, there are tool kits that can be used to get familiar with the technologies and explore their working principles.

For both LoRa and SigFox, there is a wide offer of development kits, starting from €30. Similarly, there are some kits for NB-IoT and LTE-CatM, starting from €45. Weightless offers a development kit that includes the base station and several nodes (€1200). There are also some kits for Wi-SUN, Ingenu-RPMA, DASH7, IQRF, and MIOTY. On the other hand, there are no development kits for SNOW and Telensa.

13.3.2.2 Documentation availability

Together with the existence of development tool kits, the access to the technology documentation is essential. In this sense, the landscape is heterogonous, ranging from total availability of the specification at the website (LoRa, SigFox, Ingenu-RPMA, Weightless, Wi-SUN, DASH7, and IQRF), passing through partial access to technical documentation (NB-IoT, LTE-CatM, GSM-IoT, and MIOTY), to no documentation at all (Telensa, SNOW).

13.3.2.3 Users and developers community

Other important footholds are the user and developer communities, in which information, doubts, and best practices can be interchanged. The LPWAN technologies that are opener providing information usually present wider communities. The most established communities are the Things Network (LoRa) and the City User Group (SigFox). The mobile IoT innovators community supports NB-IoT and LTE-CatM, while there are developers' forums for Ingenu-RPMA, IQRF, and DASH7.

Table 13–5 summarizes the deployment factors that affect the development process.

Table 13–5 Deployment process factors.

LPWAN technology	Tool kits availability	Documentation availability	User communities
LoRa	Very high	Very high	Very high
SigFox	Very high	Very high	Very high
NB-IoT	High	Moderate	Moderate
LTE-CatM	High	Moderate	Moderate
SNOW	Very low	Very low	Very low
Weightless	Moderate	High	High
Ingenu-RPMA	Moderate	High	High
Telensa	Very low	Very low	Very low
GSM-IoT	Moderate	Low	Low
Wi-SUN	Moderate	High	Low
DASH7	Moderate	High	High
IQRF	Moderate	High	High
MIOTY	Low	Low	Low

13.3.3 Status

Finally, there are other factors that are related to the current situation of every technology (and therefore may vary with time) and are relevant when opting for a specific LPWAN technology.

13.3.3.1 Coverage/availability

Depending on the type of infrastructure implementation (operator-based or private) and the LPWAN technologies themselves, the coverage of each technology differs. Operator-based technologies provide higher coverages, but at the same time cannot be extended.

SigFox offers coverage for most of European and American countries, plus Australia and some areas in Asia and Africa. NB-IoT and LTE-CatM provide service in Europe, North America, Brazil, Argentina, Russia, China, and Oceania. GSM-IoT coverage matches GSM networks availability. Ingenu-RPMA is mainly deployed in the United States, and is being expanded to China, Italy, South Africa, and some areas of Oceania and Asia.

Although LoRa is a private infrastructure technology, in several countries of Europe, America, Asia, and Oceania and some in Africa, telecommunication operators have deployed their own infrastructure to provide communication services.

13.3.3.2 Standards and alliances

Although none of the LPWAN technologies is a real standard, most of them either rely on them or are strongly supported by powerful alliances.

Ingenu-RPMA and Wi-SUN are based on the standard IEEE 802.15.4k (Wi-SUN also on the specification g for nonurban areas). SigFox, LoRa, and Telensa are in conversations with ETSI, trying to be included in the upcoming LPWAN standard named low-throughput network (LTN).

On the other hand, the 3GPP defines the specifications for NB-IoT, LTE-CatM, and GSM-IoT. The LoRa Alliance is an association that develops and fosters the LoRaWAN

Table 13–6   Implementation factors.

LPWAN technology	Coverage	Alliances and standards	Commercial devices availability
LoRa	Private/some countries of Europe, America, Asia, Oceania, and Africa	LoRa Alliance ETSI LTN	Very high
SigFox	Europe, America, Australia, and areas of Africa and Asia	ETSI LTN	Very high
NB-IoT	Europe, North America, Brazil, Argentina, Russia, China, and Oceania	3GPP	High
LTE-CatM	Europe, North America, Brazil, Argentina, Russia, China, and Oceania	3GPP	High
SNOW	—	—	Very low
Weightless	Private	Weightless Special Interest group	Low
Ingenu-RPMA	United States, China, South Africa, and Italy	Wi-SUN Alliance IEEE 802.15.4k,g	Very low
Telensa	—	ETSI LTN	Low
GSM-IoT	Global (GSM coverage)	EC-GSM-IoT Group 3GPP	Low
Wi-SUN	Private	IEEE 802.15.4k	Low
DASH7	Private	DASH7 Alliance	Moderate
IQRF	Private	IQRF Alliance	Low
MIOTY	Private	—	Low

specification. Similarly, the Weightless Special Interest Group is an organization that develops and boosts Weightless technology. DASH7 and IQRF are promoted by respective alliances.

13.3.3.3 Commercial devices

Many IoT solutions are focused on the upper layers of the systems (e.g., visualization applications, data analytics) and demand the availability of commercial IoT nodes and devices, that is, already mounted sensors or actuators. Not all the LPWAN technologies offer the same level of commercial availability. For LoRa and SigFox, there is a wide offer of devices, including chipsets, communication modules, and sensor nodes. Similarly, there is relevant catalog for NB-IoT and LTE-CatM. With some exception such as DASH7, for the rest of the technologies, there is no commercial availability.

Table 13–6 summarizes the implementation factors that define the status of each LPWAN technology.

13.4 Functional factors

Finally, there are some factors that affect everyday working of IoT applications, ranging from the autonomy of the devices to IP connectivity.

13.4.1 Energy consumption

Energy consumption, and therefore battery life, is directly related to the time the end-devices are turned on. LPWAN technologies with higher data rates and more complex MAC protocols will drain batteries earlier.

All the LPWAN technologies claim that their devices can last for at least 10 years without replacing the battery, though it is in the best-case scenario, with sporadic transmissions.

13.4.2 Remote firmware updating

In order to facilitate maintenance, security, and update tasks, it is essential that the LPWAN technologies support remote firmware updating. It is fully supported by LTE-CatM, Weightless, Wi-SUN, Ingenu-RPMA, DASH7, and IQRF. For LoRa, NB-IoT, and MIOTY, it is feasible though implies some adaptations. For SigFox, considering its downlink constraints, it implies higher complexity.

13.4.3 Location services

For some applications, for example, logistics or waste management services, real-time location is one of the main assets. Some LPWAN technologies implement their own location algorithms, making use of the network infrastructure, while others rely on the addition of a GPS chip.

SigFox geolocation is based on machine learning algorithms that use the received signal strength indicator (RSSI) from the different base stations and it is available for customers paying a monthly fee. Similarly, LoRa and DASH7 geolocation mechanisms work when the message received by an end-device is received by at least three gateways (triangulation method). As with any cellular network, LTE-M, NB-IoT, and GSM-IoT can make free use of the chipset mobile location, while increasing precision (with an extra cost) through observed time difference of arrival. Weightless, Wi-SUN, SNOW, and IQRF offer the option to use RSSI levels to configure location services.

13.4.4 IP support

IP connectivity facilitates many of the upper layer tasks and eases the interaction of devices from different applications. Considering the LPWAN constraints (data size, periodicity, asymmetry, etc.), it is difficult to adapt IPv6 to these technologies. IP connectivity is supported by LTE-CatM and Wi-SUN, while there are ongoing researches for LoRa, SigFox, NB-IoT, and Weightless.

13.4.5 Network interoperability

Network interoperability among different regions, that is, roaming, enables the end-devices to automatically send and receive messages when moving from the home network to a visited network.

Table 13–7 Functional factors.

LPWAN technology	Remote firmware updating score (1 = worst, 5 = best)	Location services score (1 = worst, 5 = best)	IP support	Roaming support
LoRa	2	3	Under research	Under research
SigFox	3	5	Under research	Yes
NB-IoT	3	3	Under research	Under research
LTE-CatM	5	3	Yes	Under research
SNOW	1	2	No	—
Weightless	5	2	No	Yes
Ingenu-RPMA	4	2	Yes	No
Telensa	1	1	No	—
GSM-IoT	1	3	No	Under research
Wi-SUN	5	3	No	Yes
DASH7	5	3	No	Yes
IQRF	5	2	No	Yes
MIOTY	3	1	No	—

It is already implemented by SigFox (extra cost), Weightless, Ingenu-RPMA, DASH7, and IQRF. For LoRa and the cellular network technologies (NB-IoT, LTE-CatM, and GSM-IoT), it is under development.

Table 13–7 summarizes the other factors related to relevant functionalities of long-range IoT applications. Remote firmware updating and location services are presented using a quality score ("1" = worst, "5" = best), which consider the use of feasibility of using both functionalities for each technology.

13.5 Comparative analysis

In order to ease the LPWAN technology selection process, this section introduces a comparative analysis among the technologies. The format of the factors is very heterogeneous, varying from text to numerical values and passing to frequency bands or rating scales. For the quantifiable factors or those that can be ranked, we have normalized them using a Likert-style rating system. Additionally, information is depicted using different charts to facilitate comprehension.

13.5.1 Technical analysis

Fig. 13–1 represents the frequency bands used by each LPWAN technology. It can be observed that, with the exception of Ingenu-RPMA and Wi-SUN, all the technologies use the sub-GHz band. The 2.4 GHz band offer less communication ranges, but provides global compatibility. The cellular technologies (NB-IoT, LTE-CatM, and GSM-IoT) employ the licensed

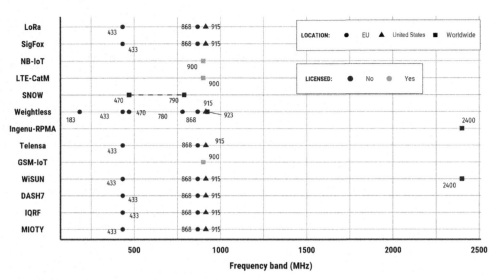

FIGURE 13–1 Frequency bands of LPWAN technologies.

	Data rate max (kbps)	Range (km)	Packet size (bytes)	Latency (s)
LoRa	50 (score: 2)	20 (score: 4)	51 (score: 2)	5 (score: 2)
SigFox	0.1 (score: 1)	40 (score: 5)	12 (score: 1)	10 (score: 1)
NB-IoT	250 (score: 4)	15 (score: 3)	125 (score: 3)	0.2 (score: 5)
LTE-CatM	1000 (score: 5)	15 (score: 3)	1000 (score: 4)	0.1 (score: 5)
SNOW	50 (score: 2)	5 (score: 1)	28 (score: 1)	0.4 (score: 4)
Weightless	100 (score: 3)	15 (score: 3)	48 (score: 2)	6 (score: 2)
Ingenu-RPMA	80 (score: 3)	15 (score: 3)	64 (score: 2)	10 (score: 1)
Telensa	0.5 (score: 1)	5 (score: 1)	65 (score: 2)	10 (score: 1)
GSM-IoT	240 (score: 4)	10 (score: 2)	65 (score: 2)	1 (score: 3)
WiSUN	800 (score: 1)	10 (score: 2)	2047 (score: 5)	0.4 (score: 4)
DASH7	166 (score: 3)	5 (score: 1)	256 (score: 3)	5 (score: 2)
IQRF	20 (score: 2)	5 (score: 1)	128 (score: 3)	14 (score: 1)
MIOTY	512 (score: 1)	15 (score: 3)	192 (score: 3)	10 (score: 1)

FIGURE 13–2 Values and scores of technical factors.

900 MHz band, while the other technologies take advantage of the unlicensed ISM bands, with higher interferences and transmission time constraints.

Fig. 13−2 presents a bubble chart of the technical factors with numerical values: data rate, range, packet size, and latency. Together with the actual value (e.g., 50 kbps for LoRa

	Data rate max	Range	Packet size	Latency	Duty cycling	Scalability	Technical score
LoRa	2	4	2	2	5	3	3
SigFox	1	5	1	1	3	3	2.33
NB-IoT	4	3	3	5	2	5	3.67
LTE-CatM	5	3	4	5	1	5	3.83
SNOW	2	1	1	4	4	2	2.33
Weightless	3	3	2	2	4	3	2.83
Ingenu-RPMA	3	3	2	1	3	4	2.67
Telensa	1	1	2	1	4	1	1.67
GSM-IoT	4	2	2	3	1	4	2.67
WiSUN	1	2	5	4	3	2	2.83
DASH7	3	1	3	2	4	2	2.5
IQRF	2	1	3	1	4	2	2.17
MIOTY	1	3	3	1	5	4	2.83

SCORE: 1 2 3 4 5

FIGURE 13–3 Heat map of technical factors.

max. data rate), we have introduced the quality score ("2"). This factor is calculated, ranking and ranging the numerical values. We have used a five-level Likert-style rating system, whereas the "5" corresponds to the best value and the "1" to the worst. Therefore, for high data rates, the score will be "5," while for high latency times the score will be "1."

In order to include in the comparison with respect to the other quantifiable technical factors, that is, duty cycling and scalability, we present the information using a heat map chart (Fig. 13–3). This graphic represents the quality scores of all the technical factors that can be quantified. The last column introduces the average score within these factors. As it can be appreciated, the best LPWAN technologies in terms of technical factors are LTE-CatM and NB-IoT, with LoRa leading the rest of the field.

13.5.2 Implementation analysis

Fig. 13–4 presents a bubble chart of the implementation factors related to deployment costs: nodes and devices, gateway, and data plans. It can be observed how there are some missing values, represented by "n/a" (not applicable) and "−" (not found/not existing). The operator-based technologies do not apply for gateway cost. Similarly, the technologies that require private deployments do not require the subscription to data plans.

In order to compare the different LPWAN technologies according to the implementation aspects, we have defined quality scores for the following factors: average cost, availability of tool or development kits, documentation availability, existence of user communities and forums, and availability of commercial devices. Fig. 13–5 represents these scores in a heat

	Nodes and devices (€)		Gateway (€)		Data plans min (€/year)		Data plans max (€/year)	
LoRa	○	10	○	200	○	5	○	20
SigFox	○	10		n/a	○	1	○	14
NB-IoT	●	70		n/a	○	6	●	50
LTE-CatM	●	70		n/a	●	36	●	100
SNOW		—		—		n/a		n/a
Weightless		—	●	1000		n/a		n/a
Ingenu-RPMA	●	30		n/a		—		—
Telensa		—		—		n/a		n/a
GSM-IoT		—		n/a		—		—
WiSUN	●	30	○	200		n/a		n/a
DASH7		—	○	200		n/a		n/a
IQRF	●	15	○	100		n/a		n/a
MIOTY		—		—		n/a		n/a

FIGURE 13–4 Values and scores of implementation factors.

	Average cost	Tool-kits	Documentation	Communities	Commercial devices	Implementation score
LoRa	5	5	5	5	5	5
SigFox	5	5	5	5	5	5
NB-IoT	3	4	3	3	4	3.4
LTE-CatM	2	4	3	3	4	3.2
SNOW	1	1	1	1	1	1
Weightless	2	3	4	4	2	3
Ingenu-RPMA	3	3	4	4	1	3
Telensa	1	1	1	1	2	1.2
GSM-IoT	1	3	2	2	2	2
WiSUN	4	3	4	2	2	3
DASH7	3	3	4	4	3	3.4
IQRF	4	3	4	4	2	3.4
MIOTY	1	2	2	2	2	1.8

SCORE: ▨ 1 ▨ 2 ▨ 3 ▨ 4 ■ 5

FIGURE 13–5 Heat map of implementation factors.

map, where the last column corresponds to the average score of all the implementation factors. It can be observed how LoRa and SigFox are the best-ranked technologies, facilitating the implementation process.

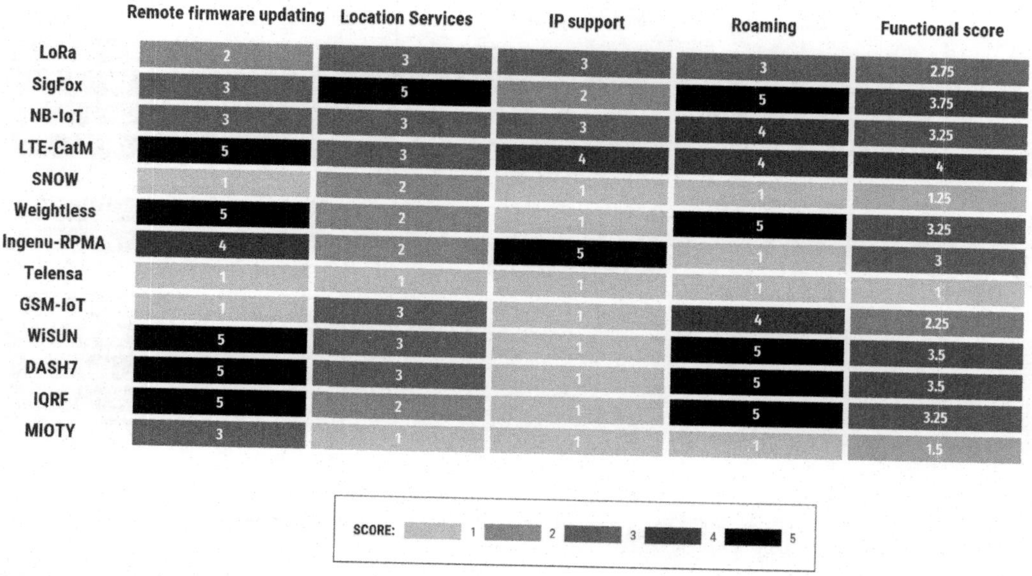

	Remote firmware updating	Location Services	IP support	Roaming	Functional score
LoRa	2	3	3	3	2.75
SigFox	3	5	2	5	3.75
NB-IoT	3	3	3	4	3.25
LTE-CatM	5	3	4	4	4
SNOW	1	2	1	1	1.25
Weightless	5	2	1	5	3.25
Ingenu-RPMA	4	2	5	1	3
Telensa	1	1	1	1	1
GSM-IoT	1	3	1	4	2.25
WiSUN	5	3	1	5	3.5
DASH7	5	3	1	5	3.5
IQRF	5	2	1	5	3.25
MIOTY	3	1	1	1	1.5

SCORE: 1 2 3 4 5

FIGURE 13–6 Heat map of functional factors.

13.5.3 Functional analysis

Fig. 13–6 presents a heat map of the functional factors: remote firmware updating, location services, IP support, and roaming. Battery life has not been included, as all the technologies claim similar duration of over 10 years, and it strongly depends on the constraints of the final application. At the last column, which represents the average score for functional factors, it can be appreciated how the best-rated technologies are SigFox, LTE-CatM, Wi-SUN, and DASH7.

13.5.4 Global analysis

Finally, Fig. 13–7 represents the global score, which is the mean of the technical, implementation, and functional scores. It can be observed that there is not a single LPWAN technology above the "4" score, meaning that there is no perfect technology and its selection should depend on the final application requirements. Nevertheless, SigFox, LoRa, NB-IoT, and LTE-CatM are the best-rated.

13.6 Use-case examples

As introduced in the previous sections, the different LPWAN technologies have their advantages and their limitations and therefore cannot comply with all the requirements of every IoT application. In this section, we introduce several use-cases that demand diverse communication and other inherent characteristics.

	Technical score	Implementation score	Functional score	Global score
LoRa	3	5	2.75	3.58
SigFox	2.33	5	3.75	3.69
NB-IoT	3.67	3.4	3.25	3.44
LTE-CatM	3.83	3.2	4	3.68
SNOW	2.33	1	1.25	1.53
Weightless	2.83	3	3.25	3.03
Ingenu-RPMA	2.67	3	3	2.89
Telensa	1.67	1.2	1	1.29
GSM-IoT	2.67	2	2.25	2.31
WiSUN	2.83	3	3.5	3.11
DASH7	2.5	3.4	3.5	3.13
IQRF	2.17	3.4	3.25	2.94
MIOTY	2.83	1.8	1.5	2.04

SCORE: 1 2 3 4 5

FIGURE 13–7 Global scores.

13.6.1 Agroindustry and forestry

Smart agriculture and forestry monitoring applications use low-cost sensors to measure different parameters such as humidity, temperature, wind speed and direction, rain gauge, soil moisture, or CO_2 emissions. These sensors will send small-sized messages with a low periodicity. Both farms and forest cover large areas, demanding long-range and low-power consumption (i.e., long battery life). Fig. 13−8 shows the heat map for the factors that have more weight in this type of applications: range, duty cycling (power consumption), and cost. It can be observed that SigFox and LoRa are the most suitable candidates, as they provide a higher average score. Examples of real-world implementations are the monitoring of different ambient parameters in cow farms in New Zealand rural areas using LoRa [18] or the monitoring of weather parameters for improving farming in Ireland employing SigFox [19].

13.6.2 Transport and logistics

These applications are characterized by mobility of nodes (including roaming), location services and large coverages. Transport applications that depend on information exchange among vehicles or on management centers to organize traffic, demand reliable high data rate, and real-time transmissions. Fig. 13−9 represents the heat map of the factors that affect transport applications: data rate, latency, location services, and roaming. It can be observed how LTE-CatM and NB-IoT are the LPWAN technologies that comply best with these requirements (higher average score). Ericsson is implementing an advanced traffic management system in Dallas (United States) using NB-IoT [20]. On the other hand, logistics

	Range	Duty cycling	Average cost	Average score
LoRa	4	5	5	4.67
SigFox	5	3	5	4.33
Ingenu-RPMA	3	3	3	3
IQRF	1	4	4	3
MIOTY	3	5	1	3
Weightless	3	4	2	3
WiSUN	2	3	4	3
DASH7	1	4	3	2.67
NB-IoT	3	2	3	2.67
LTE-CatM	3	1	2	2
SNOW	1	4	1	2
Telensa	1	4	1	2
GSM-IoT	2	1	1	1.33

SCORE: ▢ 1 ▢ 2 ▢ 3 ▢ 4 ▢ 5

FIGURE 13–8 Ranking of LPWAN technologies for smart agro applications.

	Data rate	Latency	Location services	Roaming	Average score
LTE-CatM	5	5	3	4	4.25
NB-IoT	4	5	3	4	4
GSM-IoT	4	3	3	4	3.5
DASH7	3	2	3	5	3.25
WiSUN	1	4	3	5	3.25
SigFox	1	1	5	5	3
Weightless	3	2	2	5	3
IQRF	2	1	2	5	2.5
LoRa	2	2	3	3	2.5
SNOW	2	4	2	1	2.25
Ingenu-RPMA	3	1	2	1	1.75
MIOTY	1	1	1	1	1
Telensa	1	1	1	1	1

SCORE: ▢ 1 ▢ 2 ▢ 3 ▢ 4 ▢ 5

FIGURE 13–9 Ranking of LPWAN technologies for transport applications.

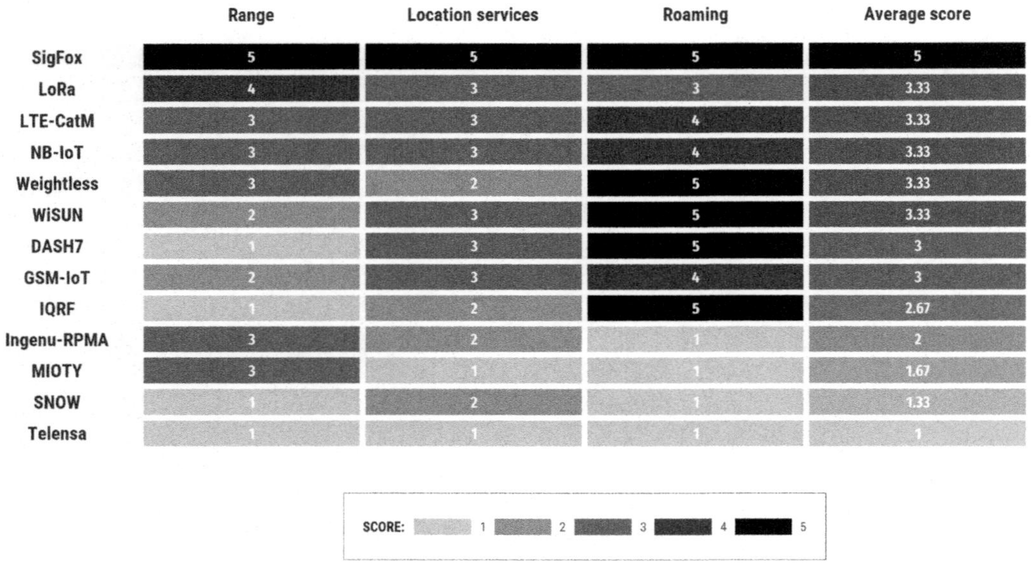

	Range	Location services	Roaming	Average score
SigFox	5	5	5	5
LoRa	4	3	3	3.33
LTE-CatM	3	3	4	3.33
NB-IoT	3	3	4	3.33
Weightless	3	2	5	3.33
WiSUN	2	3	5	3.33
DASH7	1	3	5	3
GSM-IoT	2	3	4	3
IQRF	1	2	5	2.67
Ingenu-RPMA	3	2	1	2
MIOTY	3	1	1	1.67
SNOW	1	2	1	1.33
Telensa	1	1	1	1

SCORE: 1   2   3   4   5

FIGURE 13–10 Ranking of LPWAN technologies for logistics applications.

applications are not so restrictive in terms of throughput and latency. Therefore, range, location, and roaming are the defining factors. Fig. 13−10 shows the heat map corresponding to logistics factors whereas SigFox seems the best option. One example is the use of SigFox technology for geolocation of sharing bikes in Singapore [21].

13.6.3 Smart city

There is a wide spectrum of smart city applications, ranging from air quality monitoring to street lighting controlling or traffic management. Air quality monitoring demands the sending of small-sized messages from nodes distributed along the city. Street lighting management requires bidirectionality and the sending of many messages at the same time, though it allows some delay. Fig. 13−11 represents the heat map of the two most deciding factors for smart city systems: cost and scalability. It can be observed how the LPWAN technologies that achieve the best average scores are LoRa, NB-IoT, and SigFox, although the last one does not comply with the bidirectionality requirement. Many cities are opting to deploy LPWAN networks to offer connectivity as a resource for councils, business, or schools, such as in Tasmania (Australia) [22].

13.6.4 Infrastructure management

Electricity, gas, or water supply networks are in need of continuous monitoring to prevent breakdowns and ease maintenance duties. For example, considering the power distribution grids, various electrical parameters of different network elements have to be measured: for

	Average cost	Scalability	Average score
LoRa	5	3	4
NB-IoT	3	5	4
SigFox	5	3	4
Ingenu-RPMA	3	4	3.5
LTE-CatM	2	5	3.5
IQRF	4	2	3
WiSUN	4	2	3
DASH7	3	2	2.5
GSM-IoT	1	4	2.5
MIOTY	1	4	2.5
Weightless	2	3	2.5
SNOW	1	2	1.5
Telensa	1	1	1

SCORE: 1 2 3 4 5

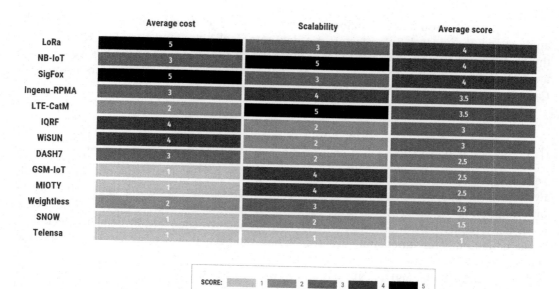

FIGURE 13–11 Ranking of LPWAN technologies for smart city applications.

	Data rate	Scalability	Average cost	Average score
LTE-CatM	5	5	2	4
NB-IoT	4	5	3	4
Ingenu-RPMA	3	4	3	3.33
LoRa	2	3	5	3.33
GSM-IoT	4	4	1	3
SigFox	1	3	5	3
DASH7	3	2	3	2.67
IQRF	2	2	4	2.67
Weightless	3	3	2	2.67
WiSUN	1	2	4	2.33
MIOTY	1	4	1	2
SNOW	2	2	1	1.67
Telensa	1	1	1	1

SCORE: 1 2 3 4 5

FIGURE 13–12 Ranking of LPWAN technologies for infrastructure management applications.

example, optical sensors at substations or electric discharge at high-power towers. For these applications, the most defining factors are data rate, scalability, and cost. Fig. 13−12 shows the heat map of these factors, where it can be depicted that the most suitable technologies are LTE-CatM and NB-IoT. One real-world example is the use of water maters using LTE-CatM in Texas City (United States) [23].

13.7 Conclusions

Selecting the optimal LPWAN technology for a specific IoT application is not an easy task. Every technology has advantages and weak points. In addition, every IoT application has its own requirements, which can be depicted in different factors (e.g., range, data rate, cost, etc.). This chapter has introduced the most relevant factors, classified by category: technical, implementation, and functional. Besides, it has established a scoring system, which has served to rank the LPWAN technologies, both by category and global. The higher-scoring technologies are LoRa, SigFox, NB-IoT, and LTE-CatM. On the other hand, when observing a concrete final IoT application (e.g., farm monitoring or smart parking) these rankings should be adapted to the specific requirements of the IoT application. Attending to these rankings, which help to facilitate the technology selection process, the best ranked are again LoRa, SigFox, NB-IoT, and LTE-CatM. However, there are some applications, such as power distribution networks management, where the diversity in location characteristics (power and Internet connections) and sensor requirement (data rate, bidirectionality), make impossible the use of a unique LPWAN technology. In those cases, two technologies with different properties, such as LoRa and NB-IoT, can be combined, implementing a hybrid communication architecture [24].

References

[1] S. Al-Sarawi, M. Anbar, K. Alieyan, M. Alzubaidi, Internet of Things (IoT) communication protocols: review, in: 8th International Conference on Information Technology (ICIT), Amman, 2017, pp. 685−690. <https://doi.org.10.1109/ICITECH.2017.8079928>.

[2] S. Aust, R.V. Prasad, I.G.M.M. Niemegeers, Outdoor long-range WLANs: a lesson for IEEE 802.11ah, IEEE Commun. Surv. Tutor. 17 (3) (2015) 1761−1775. Available from: https://doi.org/10.1109/COMST.2015.2429311.

[3] R. Tadayoni, A. Henten, M. Falch, Internet of Things—the battle of standards, in: 2017 Internet of Things Business Models, Users, and Networks, Copenhagen, 2017, pp. 1−7. <https://doi.org.10.1109/CTTE.2017.8260927>.

[4] J. Bardyn, T. Melly, O. Seller, N. Sornin, IoT: the era of LPWAN is starting now, in: 42nd European Solid-State Circuits Conference, Lausanne, 2016, pp. 25−30. <https://doi.org.10.1109/ESSCIRC.2016.7598235>.

[5] J. Santos, P. Leroux, T. Wauters, B. Volckaert, F. De Turck, Anomaly detection for smart city applications over 5G low power wide area networks, in: IEEE/IFIP Network Operations and Management Symposium, Taipei, 2018, pp. 1−9. <https://doi.org.10.1109/NOMS.2018.8406257>.

[6] M. Saravanan, A. Das, V. Iyer, Smart water grid management using LPWAN IoT technology, in: Global Internet of Things Summit (GIoTS), Geneva, 2017, pp. 1–6. <https://doi.org.10.1109/GIOTS.2017.8016224>.

[7] E.D. Ayele, K. Das, N. Meratnia, P.J.M. Havinga, Leveraging BLE and LoRa in IoT network for wildlife monitoring system (WMS), in: IEEE 4th World Forum on Internet of Things (WF-IoT), Singapore, 2018, pp. 342–348. <https://doi.org.10.1109/WF-IoT.2018.8355223>.

[8] K. Mekki, E. Bajic, F. Chaxel, F. Meyer, A comparative study of LPWAN technologies for large-scale IoT deployment, ICT Express 5 (1) (2019) 1–7. Available from: https://doi.org/10.1016/j.icte.2017.12.005. ISSN 2405-9595.

[9] U. Raza, P. Kulkarni, M. Sooriyabandara, Low power wide area networks: an overview, IEEE Commun. Surv. Tutor. 19 (2) (2017) 855–873. Available from: https://doi.org/10.1109/COMST.2017.265232.

[10] Weightless SIG. <http://www.weightless.org/keyfeatures/weightless-specification> (accessed 18.07.19).

[11] Ingenu Inc. <https://www.ingenu.com/technology/rpma/> (accessed 12.04.19).

[12] Telensa. <https://www.telensa.com/technology#top> (accessed 15.04.19).

[13] Extended coverage—GSM—Internet of Things (EC-GSM-IoT). <https://www.gsma.com/iot/mobile-iot-technology-ec-gsm-iot/> (accessed 07.04.19).

[14] IETF Wi-SUN FAN Overview. <https://tools.ietf.org/id/draft-heile-lpwan-wisun-overview-00.html> (accessed 17.04.19).

[15] DASH7 specification. <https://dash7-alliance.org/product/dash7-alliance-protocol-specification-v1-2/> (accessed 19.03.19).

[16] IQRF technology. <https://www.iqrfalliance.org/technology> (accessed 26.03.19).

[17] MIOTY. <https://www.iis.fraunhofer.de/en/ff/lv/net/tech/telemetrie.html> (accessed 22.04.19).

[18] Actility success story spark & sensys. <https://www.actility.com/wp-content/uploads/Downloads/Success_stories/Customer-stories_SenSys_01_DIGITAL.pdf> (accessed 17.07.19).

[19] Sencrop Weather Station for Irish farmers. <https://www.farmersjournal.ie/irish-farmers-journal-weather-with-sencrop-445771> (accessed 17.07.19).

[20] NB-IoT traffic management in Dallas. <https://www.zdnet.com/article/ericsson-to-power-dallas-smart-city-traffic-solutions/> (accessed 17.07.19).

[21] Sharing bike geolocation. <https://www.sigfox.com/en/news/obike-partners-unabiz-adopting-geolocation-one-million-bikes-sigfox-global-network> (accessed 17.07.19).

[22] Launceston LoRa network launch. <https://www.criticalcomms.com.au/content/radio-systems/news/launceston-s-lora-iot-network-goes-live-370662674#axzz5u8jTkbnX> (accessed 18.07.09).

[23] Water metering using LTE-CatM. <https://www.rcrwireless.com/20170308/internet-of-things/AT%26T-Captone-Metering-tag4-tag99> (accessed 18.07.09).

[24] G. del Campo, I. Gomez, G. Canada, A. Santamaria, Hybrid LPWAN communication architecture for real-time monitoring in power distribution grids, in: IEEE 5th World Forum on Internet of Things (WF-IoT), Limerick, 2019.

14

Internet of wearable low-power wide-area network devices for health self-monitoring

Raluca Maria Aileni[1], George Suciu[1], Carlos Alberto Valderrama Sukuyama[2], Sever Pasca[1], Rajagopal Maheswar[3]

[1]POLITEHNICA UNIVERSITY OF BUCHAREST, FACULTY OF ELECTRONICS, TELECOMMUNICATIONS AND INFORMATION TECHNOLOGY, BUCHAREST, ROMANIA
[2]DEPARTMENT OF ELECTRONICS AND MICROELECTRONICS, FACULTY OF ENGINEERING, UNIVERSITY OF MONS, MONS, BELGIUM [3]SCHOOL OF ELECTRICAL & ELECTRONICS ENGINEERING (SEEE), VIT BHOPAL UNIVERSITY, BHOPAL

14.1 Self-monitoring solutions, strategies, and risks

The popularity of the low-power wide-area network (LPWAN) concept is continuously increasing in the IoT world mainly due to its low operating and development cost, its low power consumption, and its long transmission range. For those key advantages to be obtained, this concept should trades the low bit transmission rates. The applications, where the data volume to be transmitted is not large, or the data transmission time is not relevant, can benefit from LPWAN technology. In addition, the big data application, for example, real-time health monitoring (e.g., fall detection), can be implemented, deployed, and can benefit from LPWAN technology by applying Fog/Edge computing together with LPWAN, those real-time health monitoring applications (e.g., fall detection) [1].

Over the last decade, significant improvements in the wireless sensor network (WSN) field were achieved. New wireless communication protocols such as long range (LoRa) and SigFox were designed, implemented, and continuously improved.

One of the most popular LPWAN protocols, with numerous applications implemented worldwide, is LoRa/LoRa for wide-area networks (LoRaWAN). Developed in France from 2010 onward, it is based on a physical layer with a proprietary modulation scheme owned by Semtech, a chip manufacturer. The frequency bands it uses are less than 1 GHz—more specifically 868 MHz in Europe and 915 MHz in the United States. In open wide spaces, the protocol's range is up to 10 km. Frequency shift key modulation is one of the possible modulations to use with LoRa, but the defining LoRa modulation is Chirp spread spectrum

LPWAN Technologies for IoT and M2M Applications. DOI: https://doi.org/10.1016/B978-0-12-818880-4.00015-6

(CSS) [2]. LoRa is using coding gain, spread spectrum modulation techniques, to improve the receiver's sensitivity, and uses the full channel bandwidth to transmit a signal. In this way, the channel becomes robust to noise and insensitive to the frequency compensations caused by the use of low-cost crystals [3].

LoRa defines only the lower physical layer of the network. LoRa Alliance is a society specially created to support LoRaWAN. LoRaWAN was developed to define the upper layers of the network, and it is a media access control (MAC) layer protocol. However, its role is close to a routing protocol, acting as a network layer protocol for managing communication between LPWAN gateways and end-node devices. The LoRa Alliance maintains these devices. Version 1.0 of the LoRaWAN specification was released in June 2015 [4]. One of the essential advantages of LoRaWAN is its scalability. Depending on the message size, the number of LoRa channels, and the number of modulation channels used, LoRaWAN has the potential of connecting millions of devices.

Similar to LoRa, the Sigfox technology was developed in France and is similar narrow bands 863–870 MHz in European Telecommunications Standards Institute (ETSI) and Association of Radio Industries and Businesses (ARIB) regions and at 902–928 MHz in federal communications commission region [5].

Sigfox protocol is described as "ultra-narrowband (UNB)" and consists of three layers: frame, medium access control (MAC), and physical. The main advantage of the UNB concept Sigfox uses is deficient noise levels, leading to minimize power consumption, high receiver sensitivity, and low-cost antenna design [6]. The one-hop star, topology is used and also a mobile operator is needed to carry the generated traffic [7].

The signal can quickly cover large areas and can reach underground objects [8]. Differential binary phase-shift keying modulation is used to uplink while Gaussian frequency shift keying modulation is used to downlink. Only the uplink transmission was used at first and later the bidirectional communication was developed.

Another narrowband LPWAN protocol that is becoming popular is the narrowband-Internet of things (NB-IoT) protocol. It is designed for indoor use and high connection density. Its main advantages are low cost and low energy consumption. Another feature of NB-IoT is the integration in long-term evolution (LTE) or GSM (under licensed frequency bands). For example, in LTE, a narrow band of 200 kHz is used. In terms of modulation, orthogonal frequency-division multiplexing (OFDM) is used for downlink and single-carrier frequency-division multiple access (SC-FDMA) is used for uplink [9].

NB-Fi is an open protocol that has a similar range in urban areas to LoRa, and it operates under an unlicensed radio band (also similar to LoRa). NB-Fi protocol was developed by WAVIoT [10]. The company developed a transceiver for the protocol defining its physical layer. The main advantages of the transceiver are the low cost, very low-power consumption and high availability, being manufactured with widespread electronic components. The topology used is the one-hop star.

Security is one of the critical issues in the IoT fields. Both low-range protocols and LPWAN technologies are still vulnerable to cyberattacks. For example, the most massive distributed denial-of-service attack ever recorded was launched through an IoT botnet [11].

In addition, several IoT devices (cardiac devices, baby heart monitors) presented huge vulnerabilities that could allow third-party entities to take control of the devices [12].

In terms of security, LoRa uses AES-128 algorithm for message encryption. The network and application key secure the packets of data. However, a key issue (the length of the message being the same before and after the encryption) is making LoRa vulnerable to jamming, wormhole, and replay attacks.

NB-IoT consists of three layers: perceptron, transmission, and application, inheriting LTE's authentication and encryption [13]. Each of the three layers of NB-IoT architecture can be exploited in different ways. To prevent this, data should be encrypted with cryptographic algorithms. Sigfox presents additional security through unique symmetrical authentication key and cryptographic tokens [14].

14.2 Low-power wide-area network technologies for wearable medical devices

LPWAN technologies are enablers for IoT [15] and are based on radio communication at a low bit for wearable devices based on smart sensors or actuators.

LPWAN technology allows long-range data communication, low-power computing, and low cost by reducing the hardware complexity. Most of the LPWAN technologies have a star topology.

Some of the LPWAN technologies are:

- NB-IoT,
- Random phase multi access (Ingenu Inc.),
- LoRa,
- NB-Fi (WAVIoT) [16],
- GreenOFDM (GreenWaves Technologies),
- DASH7 (Haystack Technologies Inc.),
- Symphony Link (Link Labs Inc.),
- ThingPark Wireless (Actility),
- UNB (Telensa, nWave, and Sigfox),
- WAVIoT, and
- LTE Cat-M1 (LTE-M).

LPWAN technology can be used for medical devices connection over long distances and is structured by using star topology, mesh topology, or a mixed star-mesh topology [17].

The comparison between short-range networks (SRN) versus LPWAN is presented in Table 14–1.

In the context of IoT development, wearable devices for personal health monitoring, to use for inside environment (home, hospitals) scenarios the SRN technologies. In addition, for outside environment scenarios are recommended LPWAN technologies available in the range of 11–50 km.

Table 14–1 Comparison between short-range network and LPWAN.[18].

	Short-range network			LPWAN			
	BLE	ZigBee	Wi-Fi	LoRa	Sigfox	NB-IoT	LTE Cat-M1 [19]
Frequency	2.4 GHz	Sub-GHz/ 2.4 GHz	2.4/5 GHz	Sub-GHz	Sub-GHz	LicensedGSM/LTE bands	Licensed LTE bands
ISM	Yes	Yes	Yes	Yes	Yes	No	No
Range	100−400 m	100 m	50 m	15 km	50 km	15 km	11 km
Data rate	<25 Mbps	250 kbps	600 Mbps	50 kbps	1 kbps	250 kbps	1 Mbps
Power	Low	Low	High	Low	Low	Low	Low

Note: BLE, bluetooth low energy.

14.3 Body-centric wireless smart sensors networks topologies

Body area network (BAN), also known as wireless body area networks (WBAN) developed within the last 20 years with the sustain of wireless personal area networks [20]. WBAN may be considered another category of WSNs focused on physiological parameters monitoring and emergency health care services provisioning [21,22], human activity monitoring, accidents prevention (the so-called connected safety applications [23]), artificial organs and prosthetic control [24], or smart textiles and accessories [25].

WBAN applications may use traditional architecture (when only one individual is monitored) or distributed (many individuals are monitored, e.g., in the context of a hospital) [26].

WBAN architectures may be divided into two categories: single-tier and multitier [3,27]. The higher the tier number, higher is the coverage demands. For instance, Tier 1 sights the intra-BAN communication, where the coverage does not exceed 2 m [28]. Tier 2 is the inter-WBAN, which may be considered as an interface with Beyond-WBAN (Tier 3), sending data to Tier 3 directly or through other access points [29]. Finally, Beyond-WBAN is designed to improve communication with the other areas or institutions (e.g., hospitals) [30].

According to Alam and Hamida [30], WBAN architectures are defined as follows: on-body (intra-WBAN), body-to-body (inter-WBAN), and off-body (beyond-WBAN).

Manirabona et al. [31] proposed a four-tier WBAN architecture. Table 14−2 resumes the architectures and briefly describes the functionality of each tier.

Besides other features, the network topology is essential, as it affects the performances of the WBAN [34] in terms of costs, energy consumption, and latency.

Depending on the purpose of the deployment, different topology approaches were proposed, but all the architectures currently implemented contain at least wireless body nodes. These nodes can be sensors, actuators, and personal devices [27].

An efficient approach uses different topologies depending on each tier of the architecture. For instance, Sliman et al. [32] employed a star topology for the Tier 1 (body sensor network)

Table 14–2 Proposed WBAN architectures.

Architecture	Tier 1	Tier 2	Tier 3	Tier 4
Khan et al. [28]	Intra-WBAN Connects body sensor units and body control unit	Inter-WBAN The interface between the body control unit and Internet or cellular networks	Beyond-WBAN WBAN extension to the outside world	-
Alam et al. [30]	On-body Body-worn wireless devices	Body-to-body WBAN interaction	Off-body Wireless access points, servers, and cellular networks.	-
Manirabona [31]	Sensor node Source of information	BAN coordinator Collects sensing data; receives and prompt actions from Tier 4	Data storage and processing Cloud/dedicated/web/local server	Monitor Data access, visualization, analysis; database interactions; recommendations
Sliman et al. [32]	Body sensor network (BSN) Impulse radio UWB sensors	Personal area network BSNs coordination, data gathering from BSN	Wireless hospital sensor network Sensing data collection, processing, and analysis	-

Note: UWB, ultra-wideband.

for its simplicity, while mesh topology is used both in personal area network (Tier 2) for energy saving and quality of service (QoS) and in the wireless hospital sensor network (Tier 3) for ensuring the QoS.

In Fig. 14–1, all state-of-the-art topologies are mentioned. Point-to-point (P2P) and star are single-hop topologies. Alternatively, a two-hop extended star may be used [31]. Also, mesh, tree-based, chain based, and cluster-based are multihop topologies [21].

P2P topology (Fig. 14–2) is the simplest topology and is usually used when the application requires sensor data transmission to a personal device. Ling et al. [34] proposed such a P2P topology application with energy harvesting capability, which is very important for increasing the lifetime of the system.

In star topology (Fig. 14–3), every node is connected to a coordinator (central node) that aggregates the data. Each node sends data within a transmission interval, based on the duty cycle information received from the coordinator node [36].

As single-hop topology, it can carry life-threatening data, as the latency and data loss may be reduced [21,33]. Therefore, it can be used for real-time applications. Moreover, star topologies are energy efficient and more comfortable to deploy (new nodes can be added easily, and undesired nodes can be removed without difficulty), are extensively used in health care. Other advantages reside in centralized management and maintenance.

Point-to-point

Star

Mesh

Tree-based

Chain-based

Cluster-based

Star-mesh hybrid

FIGURE 14–1 WBAN topologies.

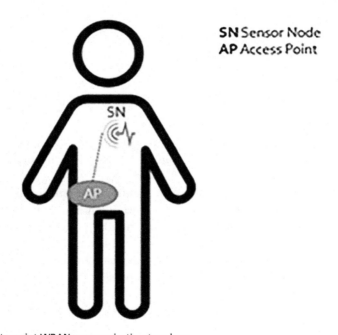

SN Sensor Node
AP Access Point

FIGURE 14–2 Point-to-point WBAN communication topology.

There are several drawbacks, though, mainly because the entire network is dependent on the coordinator node. In case of a failure of the coordinator node, the network is compromised. In addition, the performances of the coordinator influence the network capacity. Also, if the nodes are far, they cannot communicate anymore with the coordinator node.

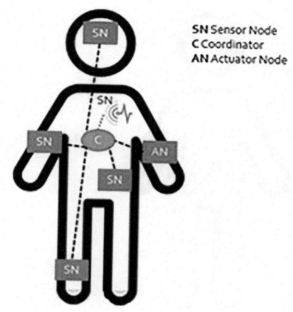

SN Sensor Node
C Coordinator
AN Actuator Node

FIGURE 14–3 Star topology in wireless body area networks.

Mesh topology (Fig. 14–4) is a complex and flexible [37] multihop topology that exhibits a better performance than star topology in terms of energy consumption, also reducing the path losses [34]. Mesh topology may be used both in Tier 1 of WBAN architecture to establish the communication between sensor nodes and coordinator node and in Tier 2 to assure the link between coordinator and data servers [36].

Mesh topology can be deployed successfully in large networks, concerning star topology that is more efficient for a reduced number of nodes, but mesh topology is not suitable for real-time applications [37].

Tree topology is presented in Fig. 14–5. As in the case of mesh topology, it employs multihop communication. The coordinator node is the node that initializes communication [37]. Each child node sends the data to the parent node (which acts as a router), and the parent node forwards the data to the next layered parent node or the coordinator node in order to reach the corresponding parent node of the destination node.

Tree topology is used to limit the power consumption of the network [21]. However, it is not a reliable topology since the failure of a parent node compromises all its child nodes.

Chakraborty [20] reviewed four different arrangements of tree topology (Figs. 14–6–14–9). They used the concept of relay network, widely used in wireless networks, but less studied in WBANs. In relay networks, the range is considerably increased, and the energy consumption is reduced.

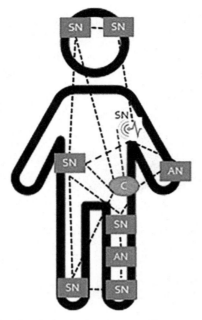

SN Sensor Node
C Coordinator
AN Actuator Node

FIGURE 14–4 Mesh topology in wireless body area networks.

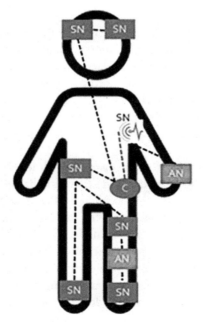

SN Sensor Node
C Coordinator
AN Actuator Node

FIGURE 14–5 Tree topology.

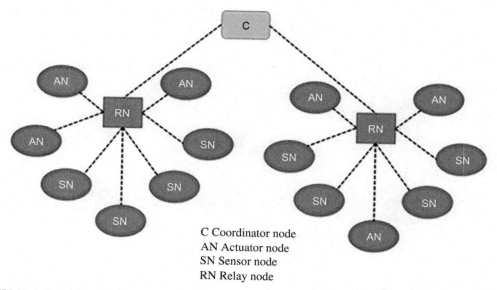

C Coordinator node
AN Actuator node
SN Sensor node
RN Relay node

FIGURE 14–6 Cluster-based tree topology.

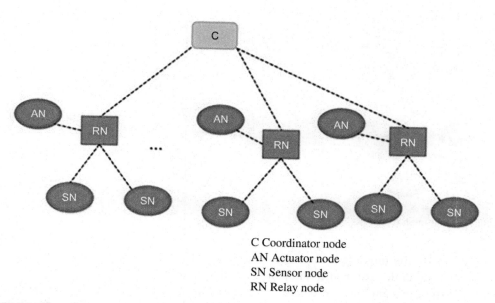

C Coordinator node
AN Actuator node
SN Sensor node
RN Relay node

FIGURE 14–7 Balanced tree topology.

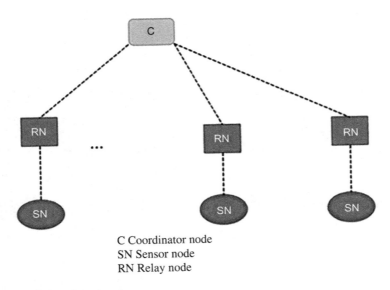

FIGURE 14–8 Relay node-based star topology.

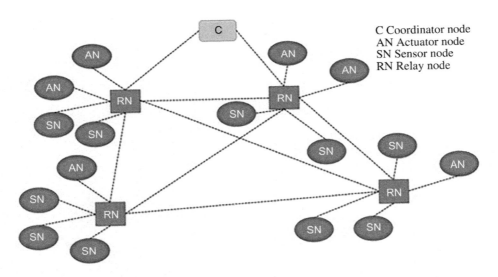

FIGURE 14–9 Star-mesh topology.

In Fig. 14−6, the topology comprises dense clusters of sensor and actuator nodes aggregated by a relay node. The coordinator node manages all relay nodes. This topology can be useful in exhaustive monitoring electroencephalography or critical monitoring (postoperatively) [21]. As in the case of tree-based WBAN, this topology can be unreliable as it is highly dependent on the well-functioning of relay nodes.

A more efficient topology can be found in Fig. 14−7. This one is designed to acquire data from important body sites but can also provide redundant data to increase the reliability of the network. Thus the WBAN based on this structure may be used in monitoring patients with Parkinson's disease [21] and other pathologies that require sensor data from multiple sites.

Another tree-based topology is presented in Fig. 14−8. This structure does not use local aggregation and acts as a star whose range is extended with the aid of relay nodes. It may be used for electrocardiography (ECG), and body temperature and acceleration monitoring [21].

Star-mesh hybrid topology (Fig. 14−9) is aimed to enhance the traditional star topology and is capable of reorganization when a relay node is out of operation. In addition, one could employ star-mesh hybrid topology to allow communication between WBAN and other tiers. The topology can meet the requirements for wearable networks and data fusion [21]. This configuration is suitable for emergency cases or in ECG pathologies when the continuous and reliable monitoring of the patient is required.

14.4 Low-power computing versus data accuracy for low-power wide-area network wearable devices

14.4.1 Low-power computing

Energy saving is a problem of growing importance due to the low-energy utilization and increasing environmental awareness. However, the challenge of energy optimization is to assure the accuracy of the energy forecast model [38].

According to the Global Standards Initiative on the IoT, a device labeled as IoT must meet seven criteria related to sensors: Internet connectivity, processors, energy efficiency, cost-effectiveness, quality, reliability, and security. In order to be attractive for use in commercial applications, IoT devices need to factor in low power consumption, long distance communication, and affordability [18].

Low power consumption, low transceiver chip cost, and extensive coverage area are the main characteristics of the low-power extensive area network (LPWAN) technologies. We expect that LPWAN can be part of enabling new human-centric health and wellness monitoring applications [39].

LPWAN, which stands for low-power wide-area network, has great importance in the IoT domain. Most of the IoT projects are based on the following requirements: extended range, low data rate, low energy consumption, and cost-effectiveness.

Today, many technologies are based on the LPWAN concept. Among the best-known technologies of this type are Sigfox, LoRa, third generation partnership project (3GPP), Weightless, Ingenu, WAVIoT, nWave, Telensa, Cyan's Cynet, Accellus, and SilverSpring's Starfish. Applications such as infrastructure monitoring and metering scenarios, or smart traffic, are the ones that most use LPWAN technology. However, other fields that test the potential of this technology have recently emerged. These fields include smart health care and well-being monitoring.

LPWAN technology makes it possible to increase the distance necessary for the communication of the sensors and the base stations, up to hundreds of meters or even tens of kilometers. On the other hand, the complexity of the network decreases proportionally to this distance.

The main reasons why LPWAN technology is so used nowadays, especially in health care monitoring and well-being applications [40], are the low cost and the low energy consumption for sensors devices.

In technologies such as Sigfox, LoRa, and NB-IoT, the end devices are not being used outside the operation; most of the time they are put in sleep mode, and this can reduce the energy consumption. Although this technique helps to achieve low energy consumption, for NB-IoT technology, this is counterbalanced by the additional energy consumption caused by the synchronous communication and QoS handling, as well as the demand for current of its OFDM/FDMA access modes. NB-IoT technology ensures that the end devices have a shorter lifetime, as compared to Sigfox and LoRa. However, as a trade-off, NB-IoT devices ensure low latency.

Regarding LoRa devices, they are divided into several classes: class A, class B, or class C. Class A is characterized by the fact that the devices of this class start to transmit the packets at any moment in ALOHA fashion; also, the packets will be transmitted using a channel that is selected randomly. After the transmission, there are two receiving windows opened by the sensor node that will be used by the base station to reach the device in the downlink. The devices of class B are different from those of class A because they have additional periodic receive slots. Class C devices increase energy consumption by continually staying in receive mode unless they are transmitting. The devices of class C are used to manage the low-bidirectional latency, to the detriment of increased energy consumption. In addition, there are handover-related signaling procedures, specific to LoRa devices, which help saving energy to the sensor nodes and transmit the packets successfully.

The devices based on the NB-IoT communication protocol use in turn the LTE protocol and also minimize and make its functionalities better. The LTE protocol is the one which broadcast the valid information for all end devices from a cell, and it is maintained to the minimum size and its minimum occurrence while the broadcasting back-end system utilizes the battery power of each end device and achieves the resources. Accordingly, to all mentioned above, the system can be cost-efficient, and the battery consumption small to obtains up to 10 years of battery lifetime with a transmission rate of 200 bytes per day on average [41].

Overall, it is recommended to use Sigfox and class-A LoRa devices for applications that transmit a small amount of data and are not being influenced by the latency. Regarding the applications that need low latency, there are better options such as the LoRa class-C devices and NB-IoT.

14.4.2 Optimizing energy consumption

Increasing the number of connected devices, such as sensors and actuators, has emphasized on low power consumption in wireless communication. Currently, four leading LPWAN technologies are available in the market for IoT devices: LoRa, Sigfox, NB-IoT, and LTE Category M1 (LTE-M). All these technologies are used to connect low-bandwidth devices with low power consumption and low cost, but there are some differences.

The LoRa system operates at the ISM frequency bands, which makes it an attractive solution for the IoT and machine-to-machine systems. Its low power associated with the long-range communication pushes LoRa to the top of the LPWAN technologies. The LoRaWAN characteristics have a great influence in determining the battery lifetime of the sensor node. Using LoRaWAN technology can optimize energy consumption by adapting its main parameters.

NB-IoT, also known as LTE Cat NB1, is an LPWAN technology that works virtually anywhere. It connects simpler and more efficient devices to already established mobile networks and handles small, rare, bidirectional, secure, and reliable data. It provides meager energy consumption, excellent penetration coverage, and lower component costs. At present, 29 companies support NB-IoT (including Vodafone and Deutsche Telekom). The 3GPP ecosystem is fully aligned to a single NB-IoT standard using existing mobile network infrastructures.

Digi is the European version of Digi XBee Cellular designed for Cat NB-IoT. Optimized for Europe's Vodafone networks, this version of Digi XBee Cellular will enable original equipment manufacturers to quickly and easily integrate reliable, low-cost, low-power NB-IoT cells into their projects with additional security incorporated. Digi XBee means a complete system of wireless modules, gateways, adapters, and required software. All of these are designed to accelerate the deployment of wireless applications for worldwide deployments.

14.4.3 Data accuracy

Most of the projects based on the LPWAN technology require the installation of transceivers in buildings or even on the ground, which makes the LPWANs to work in hard conditions, mainly in cities. For this reason, the transmission of data will not always be successful.

Using the NB-IoT protocol allows the user to connect even 100k devices per cell, and there is the possibility for capacity expansion by adding more NB-IoT carriers. This technology uses the quadrature phase-shift key modulation and the method of the SC-FDMA in the uplink. In the downlink, it utilizes the OFDMA. As a result, we have a limit for the rate of data to 200 kbps for the downlink and 20 kbps for the uplink. The message can have a maximum payload size of 1600 bytes [42].

The lowest payload length, 12 bytes, is owned by the Sigfox technology. As a result, to this payload length, we have a limitation of its utilization on different IoT applications, especially for those that have to send a large amount of data. However, the significant advantage of using Sigfox devices is that we can use a single base station to cover an entire city [41].

On the other hand, LoRa devices can send up to 243 bytes of data, and it needs a minimum of three base stations to cover a city because the range is lower than 20 km [43]. In this technology, the network server has many responsibilities, such as assuring the security, diagnostics, filtering the unnecessary packets, and the optimization of data rate for static sensor nodes.

14.5 Algorithms for efficient data processing by low-power wide-area networks

The efficient data processing by LPWAN technologies use several algorithms such as:

- AI algorithms for predictive modeling of the low power wireless communications [44] and
- genetic algorithms for efficient data processing.

The MAC methods used are:

- Carrier-sense multiple access with collision avoidance network algorithm.

If an IoT device wants to transmit data, it must follow the steps below:

1. Listen and check if the wireless network is free or busy (carrier sensing).
2. Start the transmission, based on internal back-off counter.
- Carrier-sense multiple access with collision detection,
- ALOHA network algorithm [45]. The algorithm consists in prioritizing the mode of sending information by node using the following steps:
  1. Send data.
  2. Wait for acknowledgment. If not received, go back to step 1.

14.6 Future perspectives of the low-power wide-area network technologies for medical Internet of things

As the health care industry is in constant change, advances in technology have an essential role in improving health care. At present, it is becoming increasingly crucial for early detection and prediction of diseases that may affect public health, health services and capabilities, and financial budgets. In the future, the development and exploitation of IoT technologies can lead to the overcoming of existing limits. These challenges are resource management, network availability, and localization, security and privacy, interference control, massive number of devices, and hardware constraints.

LPWAN technologies are suitable for applications that require long battery life and infrequent transmission. In terms of long-range communications, the main requirements of health care applications are frequently low-cost communication and services such as remote patient heart monitoring or ECG monitoring. Although most of the LPWAN technologies are not capable of offering the same service quality as cellular protocols, they are optimized for low-cost and high-volume applications. Moreover, for LPWAN technologies such as Sigfox and LoRa, the end device can enter sleep mode as long as the application requires, due to asynchronous ALOHA-based protocols. In terms of cellular protocols, energy consumption is higher because the end device must periodically synchronize with the network at constant intervals. However, in new cellular protocols (e. q. NB-IoT) the synchronization has been substantially reduced.

14.6.1 Network availability and localization

Some of the essential requirements of health care applications are network availability and localization. An essential advantage of cellular technology is that the existing infrastructure can be developed to deliver the service. However, this option is not viable for rural or suburban areas that do not have 4G/LTE coverage but only for dense urban environments. For areas that do not support 4G/LTE coverage, LPWA technologies such as Lora or Sigfox will be a better choice. In cellular systems, localization is usually accomplished with high accuracy and can be achieved by carefully designing and implementing the network. On the other hand, limited bandwidth and absence of direct path make it hard to get error free-localization in LPWA technologies. Thus doing accurate localization using LPWA transceivers only is a challenge.

14.6.2 Resource management

Due to the implementation of many IoT devices, along with an aggressive frequency reuse scheme, severe interference between adjacent cells is generated, thus degrading the system performance. In order to provide a suitable solution for IoT health care, the required throughput, delay, and device density need to be optimized with a novel resource management algorithm. Cellular standards that can operate within the LTE band, such as NB-IoT, will experience interference between cells caused by adjacent cell users. The adjacent Wi-Fi users or other transmissions will affect the other LPWA technologies that operate within the unlicensed spectrum band. In the field of health care applications, the data size is quite small (comprising of few bytes) for applications such as heart rate monitoring, sweating, and blood pressure. For such applications, cellular technologies such as NB-IoT are not an appropriate option because of high control channel overhead cost.

In order to obtain lower delays and increased reliability, LPWAN protocols require changes. Some of these are:

- increasing the subcarrier spacing between the OFDM symbols in order to enable fast and efficient data transmission;
- redesigning physical channels in order to enable early channel estimation,
- providing fast and reliable decoding of data transmission by using convolution codes and block codes for control channels; and
- improving the reliability and availability of signal detection and decoding.

14.6.3 Security and privacy management

In order to meet the needs of IoT applications, the combination of multiple technologies is required. To choose which technology to use for a specific application, the application's requirements, implementation scenarios, and cost should be considered. Both cellular and noncellular LPWA technologies have different advantages and disadvantages. Thus, in the future, they may be implemented together in order to mitigate the disadvantages and combine the advantages [46].

14.6.4 Support a considerable number of devices

In the future, it is expected that LPWAN technologies will connect tens of millions of IoT devices that will transmit data through shared radio resources, which are unfortunately limited. The density of devices located in different geographic areas, as well as the cross-technology interference, hampers limiting resource allocation.

In order to address problems caused by limited resources within LPWAN technologies, different research paths can be approached. These include the use of channel diversity, opportunistic spectrum sensing, adaptive transmission strategies, or nonorthogonal multiple access schemes. If multiple multimode antennas are used at LPWAN base stations, the diversity gain or the data rates of specific IoT devices located in the coverage area of the base station can be significantly increased.

14.6.5 Interference mitigation

In order to increase the performance of LPWAN networks, interference issues should be addressed. To solve this problem, the reduced complexity and energy efficiency of traditional cellular networks could be used, as well as innovative interference compensation techniques, depending on the capability of the IoT devices.

14.6.6 Hardware complexity

In general, requirements on product cost, flexibility, device size, and energy efficiency lead to the use of simplified radio architectures and low-cost radio circuits [38,47−49]. Thus the direct conversion radio architecture of these systems reflects a viable front-end solution for LPWAN networks, as it does not require either external frequency interference filters or image rejection filters [38,47−50].

14.7 Conclusions

The LTE-M is part of Release 13 of the 3GPP standard to reduce power consumption, complexity, and costs, to provide more in-depth coverage to reach difficult locations (e.g., deep inside buildings). The LTE-M standard will improve NB-IoT in terms of bandwidth and has the highest security among LPWAN technologies.

In conclusion, all those technologies can be used in the IoT field. NB-IoT will offer a high QoS with low latency. In addition, if the application needs a high coverage, Sigfox or LoRa technology can be used, with low cost, infrequent communication rate, and very long battery lifetime. They can also be used for reliable communication if the devices work at high speeds.

More than that, LoRaWAN technology is beneficial when it comes to implementing indoor communications, even communication links over 300 meters, for human-centric applications, such as physical activity or location tracking, various vocational well-being and staff management applications, or applications for monitoring the pets.

Acknowledgments

This work has been supported in part by UEFISCDI Romania and MCI through projects ESTABLISH, PARFAIT, and WINS@HI, funded in part by European Union's Horizon 2020 research and innovation program under grant agreement No. 787002 (SAFECARE) and 813278 (A-WEAR).

References

[1] J.P. Queralta, T.N. Gia, H. Tenhunen, T. Westerlund, Edge-AI in LoRa-based health monitoring: fall detection system with fog computing and LSTM recurrent neural networks, in: 2019 42nd International Conference on Telecommunications and Signal Processing (TSP), 1−3 July 2019, Budapest, Hungary, IEEE, 2019.

[2] LoRa Alliance ®, LoRa technology. <https://lora-alliance.org/>, 2019 (accessed 01.04.19).

[3] W.R. Da Silva, L. Oliveira, N. Kumar, R.A. Rabêlo, C.N. Marins, J.J. Rodrigues, An Internet of things tracking system approach based on LoRa protocol, in: IEEE Global Communications Conference (GLOBECOM), 1−7 December 2018,, 2018.

[4] The Things Network. Limitations. <https://www.thethingsnetwork.org/docs/lorawan/limitations.html>, 2019 (accessed 20.04.19).

[5] X. Zhang, M. Zhang, F. Meng, Y. Qiao, S. Xu, S. Hour, A low-power wide-area network information monitoring system by combining NB-IoT and LoRa, IEEE Internet Things J. 6 (1) (2019) 590−598.

[6] K. Mekki, E. Bajic, F. Chaxel, F. Meyer, Overview of cellular LPWAN technologies for IoT deployment: Sigfox, LoRaWAN, and NB-IoT, in: 2018 IEEE International Conference on Pervasive Computing and Communications Workshops (PerCom Workshops), March 2018, pp. 197−202.

[7] G. Dregvaite, R. Damasevicius, Information and software technologies, in: Proceedings of 22nd International Conference, ICIST 2016 (vol. 639), Druskininkai, Lithuania, 13−15 October 2016, Springer, 2016.

[8] K. Al Agha, G. Pujolle, T. Yahiha, Mobile and Wireless Networks, ISTE, 2016.

[9] R. Jaeku, NB-IoT Handbook, 2016.

[10] WAVIoT LPWAN, What is NB-Fi protocol—WAVIoT LPWAN. <https://waviot.com/technology/what-is-nb-fi>, 2019 (accessed 02.04.19).

[11] C. Kolias, G. Kambourakis, A. Stavrou, J. Voas, DDoS in the IoT: Mirai and other botnets, Computer 50 (7) (2017) 80−84.

[12] D.J. Slotwiner, T.F. Deering, K. Fu, A.M. Russo, M.N. Walsh, G.F. Van Hare, Cybersecurity vulnerabilities of cardiac implantable electronic devices: communication strategies for clinicians—Proceedings of the Heart Rhythm Society's Leadership Summit, Heart Rhythm. 15 (7) (2018) e61−e67.

[13] S. Chacko, M.D. Job, Security mechanisms and vulnerabilities in LPWAN, IOP Conf. Ser. Mater. Sci. Eng. 396(1) (2018) 012027.

[14] J. Chua, D. Yang, An examination of LPWAN technology in IoT. <https://www.electronicdesign.com/industrial-automation/examination-lpwan-technology-iot>, 2019 (accessed 20.04.19).

[15] N. Tsavalos, A. Abu Hashem, Low power wide area network (LPWAN) technologies for industrial IoT applications, 2018.

[16] L. Petter, N. Johanna, A study of low-power wide-area networks and an in-depth study of the LoRaWAN standard, 2017.

[17] J. Chua, D. Yang, An examination of LPWAN technology in IoT. <https://www.electronicdesign.com/industrial-automation/examination-lpwan-technology-iot>, 2019 (accessed 02.04.19).

[18] u-blox. LTE Cat M1. <https://www.u-blox.com/en/lte-cat-m1-old>, 2019 (accessed 21.04.19).

[19] B. Antonescu, S. Basagni, Wireless body area networks: challenges, trends and emerging technologies, Proceedings of the 8th International Conference on Body Area Networks, ICST (Institute for Computer Sciences, Social-Informatics and Telecommunications Engineering), 2013, pp. 1–7.

[20] S. Chakraborty, Study on topology control in body area network, in: 2015 Internet Technologies and Applications (ITA), September 2015, pp. 305–311.

[21] M. Li, S. Yu, J.D. Guttman, W. Lou, K. Ren, Secure ad hoc trust initialization and key management in wireless body area networks, ACM Trans. Sens. Netw. 9 (2) (2013) 18.

[22] F. Wu, T. Wu, M. Yuce, An Internet-of-things (IoT) network system for connected safety and health monitoring applications, Sensors 19 (1) (2019) 21.

[23] A. Reichman, Body area networks: applications, architectures and challenges, in: World Congress on Medical Physics and Biomedical Engineering, 7–12 September, 2009, pp. 40–43.

[24] D. Dias, J. Paulo Silva Cunha, Wearable health devices—vital sign monitoring, systems and technologies, Sensors 18 (8) (2018) 2414.

[25] A. Sangwan, P.P. Bhattacharya, Multi-level parameter processed model to optimise clustered distributed WBAN, J. Telecommun. Electron. Comput. Eng. 10 (3) (2018) 47–56.

[26] H. Kaur, N. Bilandi, Topological mechanism for sensor placement in wireless body area network, in: International Conference on Information Technology and Computer Science, July 2015.

[27] K. Shafiullah, K.P. Al-Sakib, A.A. Nabil, Wireless Sensor Networks: Current Status and Future Trends, CRC Press, 2016.

[28] R.A. Khan, A.S.K. Pathan, The state-of-the-art wireless body area sensor networks: A survey. International Journal of Distributed Sensor Networks, 14 (4) (2018) p.1550147718768994.

[29] K. Suriyakrishnaan, D. Sridharan, A review of reliable and secure communication in wireless body area networks, in: Proceedings of Thirteenth IRF International Conference, September 2014.

[30] M. Alam, E. Hamida, Surveying wearable human assistive technology for life and safety critical applications: standards, challenges and opportunities, Sensors 14 (5) (2014) 9153–9209.

[31] A. Manirabona, L.C. Fourati, A 4-tiers architecture for mobile WBAN based health remote monitoring system, Wirel. Netw. 24 (6) (2018) 2179–2190.

[32] J.B. Sliman, Y.Q. Song, A. Koubâa, M. Frikha, A three-tiered architecture for large-scale wireless hospital sensor networks, in: Workshop MobiHealthInf 2009 in conjunction with BIOSTEC, January 2009, p. 64.

[33] N.K. Ray, A.K. Turuk, Handbook of Research on Advanced Wireless Sensor Network Applications, Protocols, and Architectures, IGI Global, 2016.

[34] Z. Ling, F. Hu, L. Wang, J. Yu, X. Liu, Point-to-point wireless information and power transfer in WBAN with energy harvesting, IEEE Access 5 (2017) 8620–8628.

[35] B. Manickavasagam, B. Amutha, P. Sudhakara, Review of wireless body area networks (WBANs), International Conference on Communications and Cyber Physical Engineering, Springer, Singapore, 2018, pp. 645–656.

[36] Z.K. Farej, A.M. Abdul-Hameed, Performance comparison among (star, tree and mesh) topologies for large scale WSN based IEEE 802.15. 4 Standard, Int. J. Comput. Appl. 124 (6) (2015).

[37] F. Ma, H. Zhang, H. Cao, K.K.B. Hon, An energy consumption optimization strategy for CNC milling, Int. J. Adv. Manuf. Technol. 90 (5-8) (2017) 1715−1726.

[38] J. Petäjäjärvi, K. Mikhaylov, M. Hämäläinen, J. Iinatti, Evaluation of LoRa LPWAN technology for remote health and wellbeing monitoring, in: 10th International Symposium on Medical Information and Communication Technology (ISMICT), March 2016, pp. 1−5.

[39] M.M. Alam, H. Malik, M.I. Khan, T. Pardy, A. Kuusik, Y. Le Moullec, A survey on the roles of communication technologies in IoT-based personalized healthcare applications, IEEE Access 6 (2018) 36611−36631.

[40] K. Mekki, E. Bajic, F. Chaxel, F. Meyer, A comparative study of LPWAN technologies for large-scale IoT deployment, ICT Express 5 (1) (2019) 1−7.

[41] K. Mekki, E. Bajic, F. Chaxel, F. Meyer, Overview of cellular LPWAN technologies for IoT deployment: Sigfox, LoRaWAN, and NB-IoT, in: 2018 IEEE International Conference on Pervasive Computing and Communications Workshops (PerCom Workshops), March 2018, pp. 197−202.

[42] The Register, Meet the LPWAN clan: the Internet of things' low power contenders. <https://www.theregister.co.uk/2018/08/15/lpwan_runners_and_riders>, 2019 (accessed 20.07.19).

[43] M. Chen, Y. Miao, X. Jian, X. Wang, I. Humar, Cognitive-LPWAN: towards intelligent wireless services in hybrid low power wide area networks, IEEE Trans. Green Commun. Netw. (2018).

[44] B. Gilboa, Network algorithms for LPWAN, Texas Instruments. <http://www.ti.com/lit/an/sprt734/sprt734.pdf>, 2019 (accessed 21.04.19).

[45] T. Bouguera, J.F. Diouris, J.J. Chaillout, R. Jaouadi, G. Andrieux, Energy consumption model for sensor nodes based on LoRa and LoRaWAN, Sensors 18 (7) (2018) 2104.

[46] T. Hossain, M.A.R. Ahad, T. Tazin, S. Inoue, Activity recognition by using LoRaWAN sensor, in: Proceedings of the 2018 ACM International Joint Conference and 2018 International Symposium on Pervasive and Ubiquitous Computing and Wearable Computers, October 2018, pp. 58−61.

[47] K. Mekki, E. Bajic, F. Chaxel, F. Meyer, A comparative study of LPWAN technologies for large-scale IoT deployment, ICT Express 5 (1) (2019) 1−7.

[48] S. Omelchenko, D. Batura, V. Anisimov, Method for simultaneous confirmation of many messages in low power wide area networks, and use of correction of frequency when transmitting data over UNB LPWAN networks, based on analysis of data obtained on receiving. U.S. Patent Application 15/858,784, 2018.

[49] M.M. Alam, H. Malik, M.I. Khan, T. Pardy, A. Kuusik, Y. Le Moullec, A survey on the roles of communication technologies in IoT-based personalized healthcare applications, IEEE Access 6 (2018) 36611−36631.

[50] A.A.A. Boulogeorgos, P.D. Diamantoulakis, G.K. Karagiannidis, Low power wide area networks (LPWANs) for Internet of things (IoT) applications: research challenges and future trends, 2016, arXiv preprint arXiv:16.

LoRaWAN for smart cities: experimental study in a campus deployment

Rakshit Ramesh[1], Mukunth Arunachalam[1], Hari Krishna Atluri[2], Chetan Kumar S[3], S.V.R. Anand[1], Paventhan Arumugam[2], Bharadwaj Amrutur[1]

[1]ROBERT BOSCH CENTRE FOR CYBER PHYSICAL SYSTEMS, IISC, BENGALURU, KARNATAKA [2]ERNET INDIA, INDIA [3]AIKAAN LABS PVT. LTD, BENGALURU, KARNATAKA

15.1 Introduction

According to the UN report [1], 55% of the world population live in urban areas and it is projected to grow to 68% by 2050. Many countries are adopting smart cities approach to address the urbanization challenge by integrating cyber-physical system technologies with city infrastructure toward improving the overall quality of life in a sustainable manner [2]. Government agencies and municipalities worldwide deploy wide-area network infrastructure to collect data and analyze it to provide advanced applications such as smart energy, smart transportation, smart health care, smart water management, smart waste management, and smart governance. LoRa technology is considered to be one of the enablers that contributes to smart city solutions.

With the advent of LoRa (the radio technology) and long-range wide-area network (LoRaWAN) (the underlying networking protocol) in 2012 by Semtech, two important problems have been addressed, that is, long-range and low-power communication. Owing to LoRa's sub-GHz band of operation (433−1000 MHz) and underlying proprietary chirp spread spectrum (CSS) modulation technique with a maximum link budget of 157 dB and a high sensitivity of −137 dBm, the signals are capable of traveling through buildings and are capable of being demodulated despite heavy fading and degradation. This paves way for fewer base stations/gateways to receive signals and employ smaller antennas.

One of the major advantages of LoRa is its ease of deploying an operational network through easy onboarding of end nodes, LoRa gateways, and establishing connectivity to a network server. This aspect has allowed public aggregators such as The Things Network [3] to cover large swathes of urban environments in cities with a publicly accessible network for radios, so developers can focus more on their applications. Contrasting LoRa with other

LPWAN Technologies for IoT and M2M Applications. DOI: https://doi.org/10.1016/B978-0-12-818880-4.00016-8

multihop mesh networks such as IEEE 802.15.4 (2.4 GHz), issues related to campus-scale deployment has been studied in Ref. [4]. In this paper, the authors pointed out unlike LoRa, higher battery consumption and short range of the mesh networks as the limiting factors in the campus deployment.

While LoRa offers long-range communication, deploying large-scale Internet of things (IoT) network that covers the entire city poses challenges for radio propagation due to the lack of line-of-sight (LoS) communication especially in environments dominated by dense foliage and interspersed buildings. In this chapter, we focus on such radio environment and present our results obtained from the experimentation carried out to study LoRa performance in our campus at Indian Institute of Science (IISc), Bangalore, that mimics this setting. We present the results of our campus experimentation and offer insights and analysis in Section 15.3.

In city-scale IoT deployments where multiple networks of different kinds coexist, there is a need for an interoperable middleware that interfaces to servers servicing devices running on different radio technologies. For a city that would like to make some of its publicly deployed sensors to become discoverable, there arises a need for a city-scale central platform that allows for these resources to be searched and acts as a single point of contact for data ingression. We discuss, in Section 15.4, the integration of our LoRaWAN deployment closely with India Urban Data Exchange (IUDX) [5], which allows for sensor/resource discovery and a seamless way of accessing data.

An often overlooked aspect of network deployment is network maintenance and seamless access to debug and audit certain network parameters. In this regard, we go into detail in Section 15.4.2 on our network management system (NMS) and highlight the importance of having one.

15.2 LoRa, radio, and network

In this section, we give an overview of the technology covering both LoRa's physical layer and LoRaWAN's open standard.

15.2.1 LoRa modulation basics

LoRa corresponds to the physical layer for radio communication, which supports the use of media access control (MAC) protocols such as LoRaWAN, Symphony Link, and MoT: MAC on Time on top of it [6]. We use LoRaWAN as the network protocol with LoRa which is optimized for battery-powered devices and offers long-range communication link in the sub-GHz (400−1100 MHz) band. The network is typically laid out in a star of star topology with the device's radio (node) communicating to a gateway which relays messages to the orchestrator (network server).

Communication between the node and the gateway is spread out on different frequency channels and modulating data rates. LoRa uses a proprietary spread spectrum modulation scheme that is derivative of CSS modulation [7]. CSS-modulated waves have chip signals represented by chirps that spreads the information signal across its bandwidth. These chips are usually up-chirps of a fixed bandwidth usually between 4 and 500 kHz. This scheme allows for a total of six modulation schemes or data rates commonly called as spread factors (SFs), ranging from SF7 to SF12. The data rates of SF12 modulation typically are around 0.2 kbps, whereas data rates of SF7 modulation are around 5 kbps. There is further forward error correction which provides additional coding gain, reducing the overall bit rate. In this section, we discuss some key formulations that play an important role in determining the desired network, especially the packet air time and bit rate.

15.2.1.1 Bit rates

The bit rate for a given SF is given by:

$$R_b = SF * \frac{BW}{2^{SF}} * CR \qquad (15.1)$$

where

R_b = Bit rate offered by the LoRa signal,
SF = Spread factor $\subset$ {7, 8, 9, 10, 11, 12},
BW = Chip signal bandwidth, and
CR = Coding rates $\subset$ {4/5, 4/6, 4/7, 4/8}.

A message is composed of a combination of 2^{SF} symbols that are spread across many frequency levels. There is an inverse correlation between SF and distance at which the signal can be decoded. Clearly, a larger SF, for example, SF12, spreads the signal into 2^{12} different chips when compared to SF7 which spreads the data signal into 2^7 signal chips, making SF12 more robust to interference and fading-based deterioration of the signal. The caveat with long range is the slow bit rate. The effects of this spreading is discussed in Ref. [8].

15.2.1.2 Packet air time

Each SF also determines the air time occupied by the signal to completely transmit a message.

$$T_{sym} = \frac{2^{SF}}{BW} \qquad (15.2)$$

where T_{sym} is the symbol time. With every CSS modulation scheme, there is a need to synchronize the receiver and transmitter radios. This synchronization is achieved by a lock on the packet's prefixed n preamble symbols transmitted. Every packet is transmitted first with a preamble to synchronize the radios; the packet time a symbol takes to transmit is given by:

$$T_{preamble} = (n_{preamble} + 4.25)T_{sym} \qquad (15.3)$$

where

$n_{preamble}$ = Number of symbols constituting the preamble, and
T_{sym} = Symbol time as given in Eq. (15.2).

The extra 4.25 symbols are reserved to indicate the end of a preamble sequence. The maximum number of symbols that constitute the packet payload is given by:

$$n_{payload} = 8 + \max\left(ceil\left(\frac{8PL - 4SF + 28 + 16 - 20H}{CR(SF - 2DE)}\right), 0\right) \qquad (15.4)$$

where

PL = Number of payload bytes,
H = 0 if addition LoRa header is present, 1 if not,
DE = 1 when low data rate optimization is enabled, and
CR = Coding rates $\subset$ {4/5, 4/6, 4/7, 4/8}.

Header H is the low-level header which indicates the type of coding rate used, the payload length, and cyclic redundancy check. DE is a mode intended to correct for clock drifts in SF11 and SF12. Therefore,

$$T_{payload} = n_{payload} * T_{sym} \qquad (15.5)$$

With these, we can now define total packet time as

$$T_{packet} = T_{preamble} + T_{payload} \qquad (15.6)$$

We will see in further sections how these equations affect a network deployment.

15.2.2 Long-range wide-area network protocol

LoRaWAN is the network management scheme commonly used by all LoRa nodes to ensure scalable and sustainable deployments of nodes on LoRa networks, ensuring secure communication and reliable management of available bandwidth and channels [9].

15.2.2.1 Long-range wide-area network nodes

LoRaWAN nodes are broadly classified into three categories based on the power optimization scheme of transceivers employed. Class A devices are bidirectional end-devices which wake up from deep sleep to transmit an uplink packet and then turn off after a fixed reception window where it may receive a downlink packet from the server (ALOHA), making it the most power consumption−optimized class. Class B devices are bidirectional end-devices with scheduled receive slots. This is also a power-optimized category but transmissions/receptions are on fixed time slots instead of random ALOHA slots. This enables periodic/slotted reception on the end node possible in cases where the device has less to transmit but more to receive and actuate upon. Class C devices are the least optimized for power consumption and have their receive windows open always. This category most suits applications where a request−response styled behavior on the downlink is desired.

Since LoRaWAN in most countries operates on the free and unlicensed spectrum, there is great impetus on the fair use of channels by the end nodes and applications. Therefore all channels are duty-cycled and allowed to transmit only a limited number of times in a day. This applies moderately to uplink from the nodes to gateways but a lot more to downlink from gateways to the nodes, since there are fewer (only one fixed downlink channel) for all nodes. It is to be noted that in most deployments, Class A and Class C devices are preferably used because Class B devices require strict synchronization between all the gateways and radio nodes of the system.

15.2.2.2 Long-range wide-area network channel management

LoRaWAN with its available 2 MHz bandwidth (865–867 MHz in India), the most commonly used channel plan is to split this into three pseudo-randomly selected uplink + downlink channels and one fixed downlink channel. Each of these channels are of 125 kHz bandwidth each. Sufficient guard-band intervals are allowed to prevent interchannel interference. There are limits on scalability owing to just 2 MHz of unlicensed spectrum allocated.

The high demand of unlicensed spectrum used by LoRa leads to heavy contention by the devices. To support many devices in LoRaWAN, there is a duty cycle imposition capping the packet transmission frequency. There are three duty cycle regimes that are employed, 1%, 0.1%, and 0.01%, depending on the packet air time taken and the transmit power used. The amount of time between subsequent transmissions is given by:

$$T_{next} = \frac{T_{packet}}{DutyCycle} - T_{packet} \tag{15.7}$$

where T_{packet} can be obtained from Eq. (15.6), implying that for the best range using SF12 and a maximum packet size for this SF being 51 bytes, it would take around 2 seconds for the packet to be transmitted. Therefore the next transmission on the typical 1% duty cycle band would be 3 minutes later. For downlink messages from the server to the node, the duty cycles are even more stricter at 0.01% duty cycle (due to a single fixed downlink band) implying around 10 downlink messages per day (SF10 only).

The channel capacity for LoRa networks is discussed in detail in Ref. [10], where it is shown that for a 500 device system with three channels running on a 1% duty cycle scheme, the maximum throughput is 84 packets per node per hour (each message of 51 bytes). Ref. [8] gives a detailed discussion on outage and coverage probabilities, which indicate the probabilities that a message be received or missed, respectively. It therefore becomes crucial to optimize channel utilization and selection. To address this, LoRaWAN allows for an adaptive data rate (ADR) scheme to be used, which would enable the *network* to determine the optimum SF for the nodes to use. Nodes can further fragment messages to transmit longer infrequent messages.

15.2.2.3 Long-range wide-area network gateway

LoRaWAN gateways are practically dumb packet forwarders in the architecture whose role is to simply forward packets to its affiliate LoRa network server through a secure transport layer security based message queuing telemetry transport connection. Configurations on the gateway need to be made to set the proper channels and TX power lookup tables to ensure compliance with the spectrum policy of the country and to comply with the device's channels. Apart from these, the gateways have no knowledge of the devices in the network.

15.2.2.4 Packet encoding with Protobuf

An important aspect in LoRaWAN communication is the payload size. In a related work [11], the authors proposed a schema for communication with the smart street light, which is based on JSON owing to its properties of self-description and sheer readability. However, due to LoRaWAN packet size limitations, we see that sending a JSON data representation for the sensor values on the uplink and actuation commands on the downlink is bandwidth-inefficient, due to unnecessary usage of characters such as "{" and "," . To minimize the payload size which in turn reduces the bandwidth requirements, a process of data serialization is performed on the payload. The choice of data serialization format for an application depends on factors such as data complexity, need for human readability, speed, and storage space constraints. BSON, YAML, CBOR, Avro, MsgPack, and Protobuf are some commonly used data serialization formats. A comparison of these can be found in Ref. [12]. We use Google's *Protobuf* and its embedded library *nanopb* [13] for encoding and decoding messages. This involves defining a proto file that represents field names with byte "flags" and involves *varint* encoding for most data types. An example proto file we used in our smart street light application is provided in Ref. [14]. At the receiving end, the same proto file is used to reference the memory location of a field in the message body and decode the message to JSON.

15.2.2.5 LoRa network server

The bulk of routing and device management is undertaken by the network server. The network server provides advanced encryption standard based key management to ensure secure communication, device address management to ensure correct *activation* with session management of context and keys, *ADR* to ensure proper channel selection, *duty cycling*, and gateway packet *de-duplication* to ensure multiple packets from the same device coming in from different gateways aren't logged separately. An *application server* is colocated with this network server which deals with device/user management and provides interfaces to make the devices data available to users having appropriate credentials. Fig. 15−1 summarizes LoRa network topology showing sensors at different distances in regions A and B with two well-separated noninterfering gateways, their probable SFs, and respective *next packet transmit times* for an uplink and downlink message.

15.3 Performance in real-world long-range wide-area network deployment scenarios

In this section, we focus on the network deployment specifically at IISc, its performance, means, and metrics to ascertain the network coverage. We briefly mention some of the other LoRaWAN global deployments.

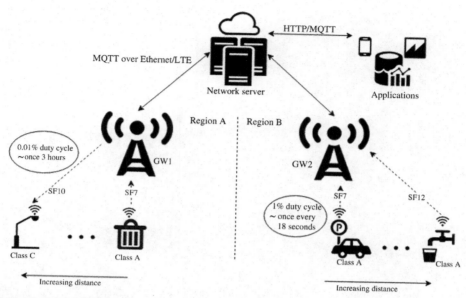

FIGURE 15–1 LoRaWAN network topology.
LoRaWAN network topology showing two gateways covering different regions A and B. Also shown, a few end nodes at different distances with the choice of SF and its effect on the duty cycle.

15.3.1 Campus-wide long-range wide-area network deployment

We conducted an experiment inside the IISc campus having a continuous layer of foliage, housing a few buildings beneath the canopy. Fig. 15—2 shows a satellite terrain view of the campus with region A depicting the area of experimentation of network coverage.

The position of the gateway is indicated and it is clear that only few spots lie in the LoS of the gateway with heavy interference from the thick foliage of the campus. Region B depicted in the figure is the surrounding area in the vicinity of the IISc campus, which is an urban area with mainly residential dwellings and very little tree cover. This offers a stark contrast between the conditions described in previous experiments describing an urban environment (like region B).

15.3.1.1 Experimental setup

For our experiments, one among the two gateways in IISc campus was turned on and this was located at the terrace of a five-storied building at a height of approximately 20 m.

The gateway used here is Kerlink Wirnet 868 MHz station [15] with an antenna of gain 3 dBi and an Ethernet backhaul to the network server. The end node used was IMST's im880b [16] which are equipped with a Semtech SX1272 transceiver and a 3 dBi antenna. The end device was mounted on to a tripod and was taken to different locations in the campus. The transmission power of the end device was set to 14 dBm.

FIGURE 15–2 IISc campus satellite view.
A satellite view of the campus showing the dense forest cover and interspersed buildings. In contrast, a dense urban area is shown, which is typically the subject matter of most studies.

Coding rate was set to the standard 4/5 which is most prevalent. Bandwidth was the standard 125 kHz that is generally prescribed in European and Indian bands. Packet acknowledgment and retransmission were turned off. The test was performed on SFs SF7−SF12 by transmitting 30 packets of size 51 bytes (MAC) to the network server from each SF. Packets thus transmitted were received by the LoRa network server [17] (namely, the gateway) and packet specific information such as the received signal strength indicator (RSSI) and the signal-to-noise ratio (SNR) were logged. A special case with LoRa is that packets can still be decoded when its SNR is below 0 [8]. In this case, $RSSI = RSSI + SNR$.

By the end of transmission of the 30 packets, a packet error rate (PER) could also be derived. The network coverage heatmap for the region of interest (within campus limits) was obtained by statistical cubic interpolation across the RSSI values logged at different regions. The path loss curve was obtained by curve fitting. We have also made the code for this available [18].

15.3.1.2 Measurements on SF7

With reference to Eq. (15.1), we find that on substituting $SF = 7$ makes $R_b = 5468$ bits/second. With reference to Eq. (15.2), we find $T_{sym} = 1.024$ ms. Since 51 MAC bytes were used, we obtain from Eq. (15.4), $n_{payload} = 88$ symbols in total, for which maximum air time would be $T_{air} = 102$ milliseconds. From Eq. (15.7), we obtain time between subsequent transmissions to be $T_{next} = 10$ seconds. This means that the end node can transmit data at the highest bit rate and occupy the lowest time for that band with the *caveat* of reduced range owing to fewer encoded chips. As seen in Fig. 15–2, with the gateway shown as a triangle, regions around it (100 m) have excellent PER rates (nearly 0%) and good RSSI (nearly −80 dBm). Fig. 15–3 shows the network coverage map for SF7. Regions (A) and (B) show the gradient along which signal predominantly propagates, owing to low lying buildings and parting in the foliage to allow for roads. Region (C) faces large PER (96%) and low RSSI (−130 dB). Surprisingly, region (D) faces better PER and higher RSSI. This is because of a tall building in its vicinity which is causing a fringing effect and offering a reflected LoS to the gateway. Region (E) is in the shadow of the network owing to large concentrations of high-rise (4 + storied) buildings in its vicinity. Overall, SF7 offers a maximum range of around 350 m.

15.3.1.3 Measurements on SF12

With reference to Eq. (15.1), we find that on substituting $SF = 12$ makes $R_b = 292$ bits/second. With reference to Eq. (15.2), we find $T_{sym} = 32.768$ milliseconds. Since 51 MAC bytes

FIGURE 15–3 SF7 network coverage map.
Network coverage for SF7 range is limited to a 200 m radius around the gateway.

were used, we obtain from Eq. (15.4) $n_{payload} = 53$ symbols in total, for which maximum air time would be $T_{air} = 2138$ milliseconds. From Eq. (15.7), we obtain time between subsequent transmissions to be $T_{next} = 213$ seconds. This means that the end node can transmit data for the largest distance (due to data being spread over 2^{12} number of chips) with the *caveat* of occupying the channel for a longer period of time. Clearly, regions around it (300 m) have excellent PER rates (nearly 0%) and good RSSI (nearly -100 dBm). Packets transmitted at SF12 are decoded at very low RSSI values. As expected, a significantly larger area can be covered using SF12. Fig. 15-4 shows the network coverage map for SF12. As with SF7, even for SF12, regions (A) and (B) show the gradient along which signal predominantly propagate. We can also see that region (C) now has good coverage (RSSI $= -115$ dB, PER $= 10\%$). Region (D) exhibits interesting behavior; note that there are regions where the PER is 0% but the quality slowly degrades over a very small region. We believe this is because of a tower (20 m) that is in direct LoS with the gateway. The tower now exhibits fringe effect, which causes electromagnetic (EM) waves transmitted from region (D) to reflect off of it and render itself to the gateway, though there is no clear LoS. Region (F) also exhibits good network coverage with PER of less than 10% in some regions. This is because of an occluded LoS from the gateway and the existence of a straight road along the gateway which guides the wave toward the gateway. Overall, the maximum range was observed to be 860 m.

15.3.1.4 Range and packet error rate for SF8–SF11

For the sake of brevity, we have only shown results of SF12 and SF7. Our experiments show that the range of network gradually increases with increase in SF and the PER for peripheral regions in each SF becomes smaller. In SF11 (comparing with Fig. 15–4), it is observed that region (G) is not covered by the network and region € experiences a PER of >80%. In SF8 (comparing with Fig. 15–3), it is observed that region (D) comes under the network coverage and PER of region C decreases to <80%. Likewise, the trends for SF9 and SF10 become increasingly better.

15.3.1.5 Path loss estimation

An important parameter that indicates the nature of the propagation environment is the path loss exponent γ. Considering the Friis equation for free-space propagation of an EM wave from [19],

$$P_r = P_t \frac{G_r G_t}{L} \left(\frac{\lambda}{4\pi} \right)^2 \left(\frac{1}{d} \right)^\gamma \tag{15.8}$$

where

  P_r and P_t = Received and transmitted power, respectively;
  G_r and Gt = Gain of receiver and transmitter antenna, respectively;
  L = System loss (attenuation) such as insertion loss, and matching loss;
  d = Distance between the receiver and transmitter antenna;
  $\lambda = c/f$, where f = 865 MHz; and

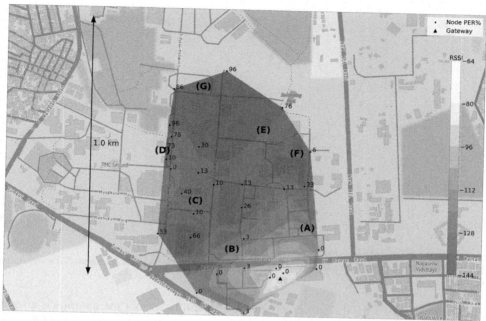

FIGURE 15–4 SF12 network coverage map.
Network coverage for SF12 range is limited to 800 m and good PER rates for areas closer than 500 m.

γ = The path loss exponent.

Converting to decibels, and substituting for our test setup $G_t = G_r = 3$ dBi, insertion loss of 0.5 dB for antenna and 0.5 dB matching loss at worst, $P_t = 14$ dBm, $f = 865$ MHz, we get

$$P_r(dBm) = 19 - 10\gamma log_{10}(d) \tag{15.9}$$

We now empirically find the path loss exponent γ by fitting the above equation (using the Levenberg–Marquardt method) to the data we collected as shown in Fig. 15–5. For this experiment, we chose only SF12 measurements, since path loss is not affected by the modulation scheme.

With $\gamma = 5.616$, we can clearly see how much of a harsh environment the campus is for LoRa signals. This is mainly because of heavy scattering of signals due to trees and very little reflections. A similar experiment over a smaller region was conducted in Ref. [4] where path loss exponents for less-dense regions in the IISc campus are provided.

15.3.1.6 Campus deployment: key observations
As a result of our experiments, we arrive at the following conclusions.

- The best range LoRa radios (14 dBm TX Power, 3 dBi antenna gain) can offer is 0.8 km in campus with thick foliage.

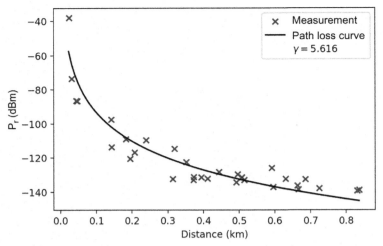

FIGURE 15–5 Received power (dBm) versus distance.
Estimated received power through empirical curve fitting. We find that $\gamma = 5.616$ for the campus (having heavy foliage cover).

- Path loss exponent γ for the campus is 5.616. This indicates that the campus is a very harsh environment for signals to propagate. Keeping this in mind, we increase the deployment density of gateways around "quality-of-service (QoS)-Shadow" regions to offer better coverage.

We provide below a few guidelines that a network service provider would need to keep in mind while deploying their network.

- To determine how many gateways a region requires, a thorough field survey with an end node needs to be made for all SFs. Usually, one would go with the manufacturers' claims that LoRa affords 10 km range but we have found this to not be true in urban settings.
- It is possible that a region may face lesser PERs despite no clear LoS to the gateway, and despite being further away radially than a region with higher PERs. This is because of the topology of that region, reflected LoS components, etc. at that region.
- Placement of the node/gateway plays an important role in network coverage. Gateway height should be as high as possible making sure that a clear LoS is available to all nodes. There should be no obstacle in the near field of node or gateway (along the LoS). Node and gateway antennas should be oriented in such a way that their field of maximum radiation align with each other.

15.3.2 Note on scalability and drawbacks of long-range wide-area network under dense foliage scenario

Our experimentation in the campus has given us key insights on scalability and performance aspects of LoRaWAN in dense foliage radio environments within a city. Our experimentation, as was mentioned in Section 15.3.1.3, has shown that a radio range of around 500 m is the expected radio range with 10% PER with the LoRa device operating at SF12. This crucial result throws light on the scalability of LoRaWAN in terms of cost, performance, and operational management. For a city-scale deployment, the range of around 500 m between device and gateway demands more gateways to be deployed, and the SF of SF12 results in less number of devices per gateway for a given duty cycle [10]. All this leads to higher installation costs and operational expenditure.

For a scalable deployment with regions of urban foliage, careful network planning involving optimal gateway placement that minimizes the cost while improving the radio coverage and the selection of appropriate applications that can tolerate low duty cycle operation are essential.

15.3.3 Other global long-range wide-area network deployments

As mentioned earlier in Section 15.3, below are some of the other LoRaWAN global deployments.

15.3.3.1 Bologna, Italy

The network deployed in Bologna [20] for server applications such as environment monitoring was focused on two districts, Saragozza and Navile, with one gateway at each location. Radios used had 3 dBi gateway receiver antenna gain and 2 dBi transmitter antenna gain with signals transmitted at 14 dBm TX power. At Saragozza, a region with fairly low-height buildings and near LoS from the position of the gateway positioned at 71 m above ground level, it was observed that the maximum range obtained was in the order of 1−2 km for SF12.

15.3.3.2 Paris, France

The network was deployed in a suburb of Paris [21] with fairly low lying buildings with clear LoS between gateway and end nodes. The radios again had similar antenna gain parameters to the ones used in Bologna. It was observed that a maximum range of around 2.5 km was obtained in a LoS path.

15.3.3.3 Bangkok, Thailand

The experiment [22] was composed of one end-device, one gateway, and one server with MQTT protocol. The gateway and end device were from Libelium with the gateway antenna having a 5 dBi gain and end device with a 4.5 dBi antenna. The results of the experiment show that the range is only up to 2 km in outdoor rural area and 55−100 m in an indoor urban environment.

15.3.3.4 Lille, France

The network was deployed at the Scientific Campus of the University of Lille in the North of France [23]. The end device used here was a Libelium Waspmote with antenna gain of 4.5 dBi and the gateway used was a Kerlink Wirnet 868 MHz station with gain of 3 dBi and was situated at the first floor of Building in the campus. The results of the experiment showed that LoRaWAN provides good performances over the major part of the Campus over a distance of 1.2 km. Poor signals are due to the presence of high-rise buildings, which disturb the local quality of data transmission.

15.4 Internet of things middleware for smart cities

In the case of a smart city IoT network deployment, LoRaWAN has to coexist with networks of various kinds such as SigFox and Zigbee. In addition, network servers of this kind don't take into consideration discovery of other LoRa networks and the geographical and network management contexts of constituent LoRa nodes which could facilitate in network planning and management. Network servers also cannot report quality of the sensors interfaced by the radio which would aid in sensor prognostics.

In this regard, our implementation details a second server, the smart city middleware which interfaces with network servers of different protocols such as LoRa, Zigbee, and NB-IoT and plugs the void between radio management and sensor/device management. It is required that these servers make their data access interfaces known and available to ensure seamless and interoperable data brokering as shown in Fig. 15−6.

15.4.1 Aspects in Internet of things network deployments

Typical low-bandwidth network implementations present a siloed approach of an IoT system, wherein all devices follow one kind of radio technology, for example, LoRa. For instance,

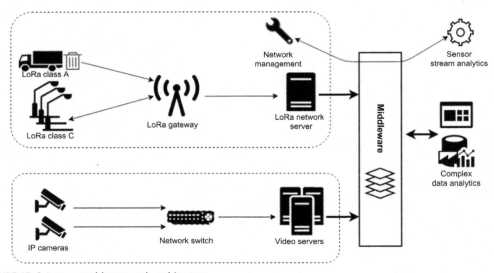

FIGURE 15–6 Interoperable network architecture.
An interoperable network architecture showing how devices running on different network modalities can be accessed from a common middleware.

a device on Wi-Fi cannot communicate with another device on LoRa. This prohibits applications running on one platform to consume sensor information from devices on another platform. To address this limitation, we have developed a middleware framework, IUDX [5], that enables applications to work with devices on heterogeneous networks. IUDX offers features such as data queuing over a Pub-Sub broker (publish-subscribe), data storage, security, and APIs to make application development simple. It specifies modalities of interactions, that is, the query language, semantics, and keyword definitions that are well defined and regulated. In the context of LoRaWAN, the process typically involves a thorough description of the LoRaWAN-based resources including static attributes such as the resource's ID, application server, resource location, and provider of the resource, and data attributes such as the sensor values, sensor units, and the sensor data range. Additional GIS information of the topography of the region where the nodes are deployed can be obtained through discovery features made available through a catalog.

IUDX expects payloads to purely be in JSON format; therefore it is important to have a component that does this translation from Protobuf. Onboarding to IUDX then involves writing a software module *adapter* that maps application program interfaces (APIs) from one particular type of network server to the IUDX's APIs.

An adapter is essentially a software entity residing in the middleware that is a client to both the middleware and the foreign platform (e.g., private LoRa network server). Its role is to consume sensor data from the foreign platform and publish it into IUDX and similarly extract data from IUDX and publish it to the devices on the foreign platform. Further, it has the ability to process this data and make it usable by either platforms. Fig. 15–7 shows an adapter that does the Protobuf to JSON conversion for messages originating from the street light on the foreign platform side and JSON to Protobuf conversion for street light actuation commands from applications on IUDX. It can even download the encode/decode descriptions via entries made in a catalog on IUDX at the time of device registration and generate the required classes to perform data conversion.

15.4.2 Long-range wide-range area network operation and management

As mentioned earlier, IUDX makes it possible to dispatch device maintenance teams by integrating sensor stream analytics (which could detect a faulty sensor) to the LoRa network management suite. In addition, sensor refresh (upload) rates can be controlled on an on-demand basis with inputs from the sensor stream analytics. The operational and management aspects of LoRaWAN include the ability to configure the LoRa network to meet end application demands, monitor the network status and performance, perform quick fault diagnosis, and minimize service disruption with predictive maintenance. To achieve these essential objectives, ISO recommends the FCAPS model [24] - Fault management (F), Configuration management (C), Accounting management/Administration, Performance management (P), and Security management (S). Fig. 15–8 shows a gateway management infrastructure considering the above mentioned FCAPS model.

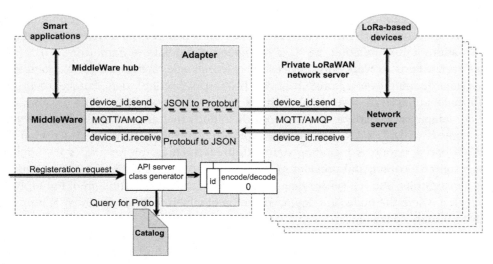

FIGURE 15–7 Adapter architecture overview.
Solving the interoperability problem through adapters, which acts as a client to both the middleware and the network server, converting packet format where necessary.

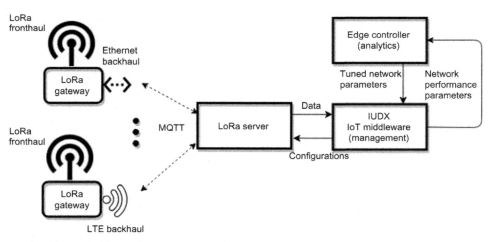

FIGURE 15–8 Gateway infrastructure management.
IoT network infrastructure supporting network agnostic gateway infrastructure management.

15.4.3 Configurable parameters

For configuring and optimizing the performance, LoRaWAN offers various configurable parameters such as SF, coding rate, bandwidth, operating channel, transmit power, and an ADR mode for devices and gateways.

In our implementation, the configuration is done at the LoRa network server over RESTful APIs. The server then pushes the configuration to the LoRa gateways and subsequently to node over a secure MQTT channel.

15.4.4 Network management

Performance management enables building a reliable system by monitoring performance parameters in real time and predict failures, thereby aiding in minimizing the downtime and maintaining the QoS of the end applications. This section outlines the mechanisms and methods for collecting system-level operational telemetry from various LoRa entities, processing, analyzing, and visualizing of operational telemetry data, and publishing it to authorized consumers.

The main parameters that determine the performance of LoRa are RSSI, SNR, SF, packet delivery ratio. Monitoring these parameters help the network operators to provide reliable and timely data to the end applications and thereby ensure QoS. The operational telemetry can further be processed at NMS.

The NMS performs system telemetry operation by (1) monitoring various parameters of interest pertaining to devices, communication network infrastructure, and systems; (2) alerting; (3) report generation; (4) dash-boarding, (5) data visualization through time-series plots, and so forth. For LoRaWAN, it is also important to take spacial and temporal information into consideration. The operational telemetry while assisting in the routine troubleshooting and making performance measurements also assists in predictive maintenance and resource planning with the help of analytics.

Due to current lack of standardized and comprehensive management framework, LoRa operations and management system is expected to work with a vendor specific framework. In our implementation, we used commercial edge controller [25] software with customization to process the operational telemetry data received at the LoRa gateway. The gateway further pushes the telemetry data (as header section of the data payload) to LoRa server. The LoRa server offers RESTful API to retrieve the telemetry using which we developed a dashboard and visualization systems using LoRa telemetry data. As shown in Fig. 15−8, the presence of an interoperability layer such as IUDX streamlines such integration.

The fault management in a network deployment includes fault detection, fault isolation, and fault resolution of network devices and their connectivity. For accomplishing this, it is important for an IoT network service provider to have continuous access to their remotely located gateways which might be LoRa gateways, Zigbee border routers, and so on, In a practical scenario where a LoRa gateway is connected to a 4G/long-term evolution network backhaul, which is often assigned a dynamic IP address, it is hard to access the gateway via secure shell (SSH). To circumvent this problem, we have developed a network management service, whereby the administrator will still be able to SSH to the LoRa gateway seamlessly by establishing a managed reverse SSH tunnel to the administrator's server. Architectural details and code for this are available in Ref. [26].

15.5 Summary

In this chapter, we covered the performance and operational management aspects of a LoRaWAN deployment for supporting smart city applications in an environment comprising regions of moderate-to-dense foliage and buildings as can be typically found in large campuses, parks, and so forth.

Through extensive physical testbed experimentation done in a large academic campus that represents such an environment, we evaluate the performance of LoRaWAN by considering key metrics such as PER, and radio range for various LoRa SFs. We believe this study would be useful to the network service providers in the design and deployment of LoRaWAN in the scenario considered and can be directly scaled up for city-wide deployments.

In this context, we briefly presented studies conducted in some of the large-scale LoRaWAN deployments found in the literature. As LoRaWAN offers network services to the data layer in the overall architecture of a smart city application, we introduce an interoperable data exchange platform, IUDX, an ongoing effort undertaken by one of the smart city initiatives in India. We discussed the operational management activities that need to be carried out to ensure smooth functioning of the LoRaWAN that meets the QoS requirements of smart city applications.

References

[1] un.org, World urbanization prospects: the 2018 revision. <https://www.un.org/development/desa/publications/2018-revision-of-world-urbanization-prospects.html> (accessed 01.03.19).

[2] rg.smartcitiescouncil.com, Smart cities readiness guide: the planning manual for building tomorrow's cities today. <https://rg.smartcitiescouncil.com> (accessed 01.03.19).

[3] The Things Network, The things industries. <https://www.thethingsnetwork.org/> (accessed 20.02.19).

[4] N. Rathod, et al., Performance analysis of wireless devices for a campus-wide IoT network, in: 13th International Symposium on Modeling and Optimization in Mobile, Ad Hoc, and Wireless Networks (WiOpt), May 2015, pp. 84−89. <https://doi.org/10.1109/WIOPT.2015.7151057>.

[5] iudx.org.in, Indian urban data exchange. <https://www.iudx.org.in/explore-iudx/>, 2019 (accessed 01.03.19).

[6] J. Pena Queralta, et al., Comparative study of LPWAN technologies on unlicensed bands for M2M communication in the IoT: beyond LoRa and LoRaWAN, in: The 14th International Conference on Future Networks and Communications (FNC), 19−21 August, 2019, Halifax, Canada.

[7] Semtech Corporation, AN1200.22 LoRa modulation basics. <https://www.semtech.com/uploads/documents/an1200.22.pdf>, 2015 (accessed 07.11.19).

[8] O. Georgiou, et al., Low power wide area network analysis: can LoRa scale?, IEEE Wirel. Commun. Lett. 6 (2017) 162−165. <https://doi.org/10.1109/LWC.2016.2647247>.

[9] LoRa Alliance. <https://www.lora-alliance.org> (accessed 20.02.19).

[10] F. Adelantado, et al., Understanding the limits of LoRaWAN, IEEE Commun. Mag. 55 (2017) 34−40. < https://doi.org/10.1109/MCOM.2017.1600613 >.

[11] S. Abhay , et al., Schemas for IoT interoperability for smart cities, in: poster presented at BuildSys@SenSys, 2017.

[12] B.S. Petersen, et al., Smart grid serialization comparison, in: Computing Conference, Jul. 2017, pp. 1339–1346. <https://doi.org/10.1109/SAI.2017.8252264>.

[13] Embedded Library for Protobuf, Nanopb, Petteri Aimonen. <https://jpa.kapsi.fi/nanopb/>, 2011.

[14] An example Protobuf schema file. <https://raw.githubusercontent.com/rbccps-iisc/applications-street-light/master/proto_stm/txmsg/sensed.proto>.

[15] Kerlink Wirnet Station IoT outdoor LoRaWAN gateway. <https://www.kerlink.com/product/wirnet-station/>.

[16] iM880B-L—Long range radio module. <https://www.wireless-solutions.de/products/radiomodules/im880b-l>.

[17] LoRa Server, forum.loraserver.io. <https://www.loraserver.io/overview/architecture/>.

[18] Open source RSSI heatmap generator. <https://github.com/rraks/sigcatch>.

[19] T.S. Rappaport, Mobile radio propagation: large-scale path loss, Wireless Communications, Principles and Practice, 2nd ed., Prentice Hall, India, 1996, pp. 105–120.

[20] G. Pasolini, et al., Smart city pilot project using LoRa, in: 24th European Wireless Conference, Catania, Italy, May 2018, pp. 62–67.

[21] A. Augustin, J. Yi, T. Clausen, W.M. Townsley, A study of LoRa: long range & low power networks for the Internet of things, Sensors 37 (2016) 1466.

[22] N. Vatcharatiansakul, et al., Experimental performance evaluation of LoRaWAN: a case study in Bangkok, in: 14th International Joint Conference on Computer Science and Software Engineering (JCSSE), 2017. <https://doi.org/10.1109/JCSSE.2017.8025948>.

[23] M. Loriot, et al., Analysis of the use of LoRaWAN technology in a large-scale smart city demonstrator, in: 2017 Sensors Networks Smart and Emerging Technologies (SENSET), 2017. <https://doi.org/10.1109/SENSET.2017.8125011>.

[24] M.3010 Principles for a telecommunications management network, ITU-T, 02/2000, International Telecommunications Union. <https://www.itu.int/rec/dologin_pub.asp?lang = e&id = T-REC-M.3010-200002-I!!PDF-E&type = items>, 2000.

[25] AiCon Edge Controller, Aikaan Labs Pvt. Ltd. <https://aikaan.io/products/#ai_con>.

[26] Remote Gateway Manager. <https://github.com/rraks/RemoteGatewayManager>.

16

Exploiting LoRa, edge, and fog computing for traffic monitoring in smart cities

Tuan Nguyen Gia, Jorge Peña Queralta, Tomi Westerlund

TURKU INTELLIGENT EMBEDDED AND ROBOTIC SYSTEMS (TIERS) GROUP, UNIVERSITY OF TURKU, TURKU, FINLAND

16.1 Introduction

LoRa is one of the most prominent wireless technologies in the low-power wide-area network (LPWAN) family. LoRa is a patented energy-efficient wireless communication protocol that achieves very low-power and very long-range transmissions, of over 10 km in line-of-sight, trading-off data rate, and time-on-air. LoRa's duty cycle is limited by regional regulations because it operates using unlicensed sub-GHz radio bands, mostly on the 433, 868, and 915 MHz frequency bands. Taking into account this and the low-transmission bandwidth, LoRa is naturally most suitable for applications where transmissions are sparse in time and payloads are relatively small. This includes metering applications of noncritical parameters such as weather meters, air quality or other environmental meters, and also in animal tracking, smart agriculture, and connected farms [1–3].

As LoRa only provides the definition of the physical layer in data transmission, different solutions for the link and network layer have been developed. The most widely used media access control (MAC) protocol over LoRa is low-range wide-area network (LoRaWAN), an open standard defined by the LoRa Alliance keeping in mind the energy-efficient nature of LoRa and the fact that many devices relying on LoRa are battery-powered [4]. LoRaWAN enforces the limitations of LoRa data rates and duty cycles in its specification, and therefore its most relevant application fields are those that do not require real-time transmissions. LoRaWAN has the advantage of being open, and global LoRaWAN-based networks have been already deployed, such as The Things Network [5]. In a smart city, deploying an open network that can be used by both public and private organizations or individuals can enable a more rapidly growing number of applications to appear, benefiting the city's economy. Open LoRaWAN networks have been city-wide deployed in Amsterdam or Bristol [6].

LPWAN Technologies for IoT and M2M Applications. DOI: https://doi.org/10.1016/B978-0-12-818880-4.00017-X

LoRa was designed clearly targeting battery-powered devices enabling years, or even decades, of battery life for metering applications with time-sparse low-data uploads. Nonetheless, its long-range transmission, ease of setup, and low cost of deployment can be leveraged for specific applications that might require real-time, low data rate transmissions. The most advanced proprietary protocol that overcomes the duty cycle limitations of LoRaWAN is Symphony Link [7,8]. Symphony Links use frequency hopping and different available frequencies to avoid duty cycle limitations and increase the data rate. In this chapter, we propose a similar approach to enable real-time applications in smart cities that are not possible with LoRaWAN but still benefit from the very long-range and energy-efficient nature of the LoRa modulation. Moreover, if compared to other LPWAN solutions that also enable higher data rates and continuous transmissions, such as narrowband Internet of Things or LTE Cat M1, LoRa does not have the overhead of a mobile network operated by a third party in terms of SIM cards and recurrent payments. This reduces significantly the infrastructure deployment cost for an independent network. At the same time, the same infrastructure can be leveraged for enabling open LoRaWAN or other networks in the same environment (Fig. 16−1).

The IoT can be described as an end-to-end platform where physical and virtual objects are interconnected and communicate with each other. A traditional IoT platform broadly contains the following items: (1) connected objects, which can be divided into sensor nodes or actuators, depending on whether they are data producers or they operate following instructions from end-users or cloud-based applications; (2) local gateways, which provide local networks and connectivity for sensor nodes and actuators; and (3) cloud servers and services. Sensor nodes in IoT systems sense data via one or several sensors and then transmit the collected data to a gateway via a short-range wireless communication protocol such

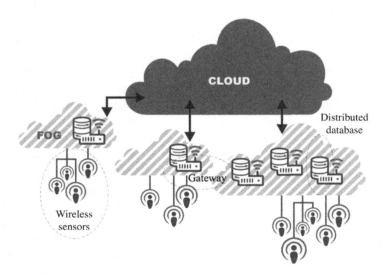

FIGURE 16–1 Fog computing.
Illustration of the fog computing concept, as an intermediate layer between end-devices and cloud servers [9].

as Bluetooth Low Energy (BLE), Wi-Fi, or ZigBee [10−14]. Depending on the applications, a specific protocol is chosen. For instance, Wi-Fi-based sensor nodes can be applied for high data rate applications such as 20-channel EEG or 12-channel EMG in which each channel often collects thousands of bytes per second. On the other hand, BLE or nRF is used for IoT applications focusing on energy efficiency such as fall detection [15−17]. In these IoT applications, the gateway forwards the data sent from sensor nodes to cloud via the Internet for global storage and further processing. IoT applications are able to provide advanced services such as real-time monitoring and big data analysis [18,19]. For example, end-users such as factory managers can easily monitor a manufacturing line in a factory anywhere at anytime or factory managers will receive instant messages when the production line has some problems. Due to the large number of benefits, IoT is ubiquitously applied in many applications and fields such as smart home, education, health monitoring, traffic monitoring, and control [20,21].

Traffic management is one of the key aspects that city administrators need to work with on a daily basis [22]. Most of the largest cities in the world have problems with traffic jams and congested roads, and new regulations on traffic circulation limitations are starting to be implemented around the globe to tackle this problem. Therefore, it is imperative to have a proper understanding of traffic flow and enough data in order to develop effective solutions and regulations. Existing solutions for traffic management include (1) video feeds from traffic cameras installed near traffic lights; (2) Wi-Fi or Bluetooth sensors that record unique IDs of devices that connect to them and send this information to the cloud where data are aggregated and used to rebuild the path traveled by these devices; or (3) vehicle counting sensors in intersections or traffic lights [23−25]. The data gathered in situations (1) and (3) are able to provide information about traffic flow in a given set of points of the city, while the path of individual vehicles remains unknown. In the case of situation (2), since only a relatively small percentage of the number of vehicles is measured, the traffic flow has to be estimated. The authors have proved, however, that reconstructed individual vehicle paths are a representative sample of the total population if the penetration of the measured sensors is at least 25%−30% [26].

These and other examples of services for smart cities mostly base their operation on transmitting acquired data via short-range and relatively high-bandwidth technologies such as Wi-Fi and Bluetooth, or 3G/4G mobile networks [22,27,28]. This is, in turn, combined with powerful cloud services that store the raw data and perform data analysis in real-time or in batches. However, this architecture threatens to collapse the spectrum in dense urban areas and overload both the network and cloud servers. At the same time, most of these technologies require relatively large amounts of energy for short-range transmission [29].

By leveraging the advantages of an edge computing architecture, the network load is reduced to the minimum, reducing the required number of gateways that must be placed across the city. This reduces the installation cost significantly, as compared to traditional approaches that use Wi-Fi or Bluetooth and require a much larger number of gateways. Taking into account that most of the acquired data is processed at the edge devices, we also reduce the network load and computational load on the cloud servers, lowering further the overall system cost.

The integration of LoRa into an edge fog—based IoT platform can be an appropriate solution for the above-mentioned challenges. This architecture takes advantage of LoRa's long-range communication, low-cost infrastructure investment, and energy efficiency for remote areas where Internet connection is not widely available. This, together with other wireless protocols such as Wi-Fi, 6LoWPAN, and ZigBee for a short-range communication where high data rate can be utilized, and proper data compression in between, enables a much wider range of use cases with network capacity being no longer a limitation.

In this chapter, we propose an advanced edge-fog-IoT architecture for traffic management and monitoring applications. The architecture helps to overcome the challenges of the existing IoT-based architecture and also enables advanced services to be deployed at different layers in the network. The proposed architecture consists of sensor devices, smart edge devices, smart LoRa-based gateways with fog computing, cloud-based services, and end-user terminal applications.

In addition, a new lightweight algorithm that enables traffic density estimation through real-time image processing of traffic cameras feeds is presented. This algorithm has the advantage of providing a good estimation of traffic density with relatively light computation methods. Therefore, it is a good candidate to run in resource-constraint edge devices.

The rest of this chapter is organized as follows. Section 16.2 introduces related work in IoT for traffic management in smart cities and integration of LoRa with edge and fog computing. Section 16.3 presents the concepts of edge and fog computing and edge artificial intelligence (AI). Section 16.4 introduces different LPWAN technologies and compares them for different applications. Section 16.5 then presents a proposed system architecture for integrating edge and fog computing with LPWAN technology to expand the range of application scenarios for long-range and low-power wireless technologies by compressing the data with AI at the edge. In section 16.6, we implement the proposed architecture for the specific use case of traffic monitoring in a smart city and propose a compression algorithm that runs at the edge. Section 16.7 discusses the challenges and future direction for both research and industry. Finally, section 16.8 concludes the work.

16.2 Related work

The development of the IoT has been one of the main pillars behind the introduction of the smart city concept [24,28,30—32]. Mitton et al. proposed in 2012 the integration of the rising paradigm that the IoT was at the moment, with cloud services and software as a service platforms. They defined a high-level modular architecture that offered adaptability to a wide variety of sensor data. The authors used an abstraction structure based on the sensor web enablement (SWE) standard defined by the Open Geospatial Consortium in 2011, but that overcame SWE limitations and was extended to include a larger number of Internet clients. Their implementation was based on the Contiki operating system and was aimed at developing advanced techniques for data filtering and aggregation in the use case of a smart city.

Zanella et al. published in 2014 their paper *Internet of Things for Smart Cities*, with a proof of concept deployment in an island in the city of Padova, Italy [33]. The authors implemented a system for gathering data from sensors placed on street light poles. The data that were collected included information about descriptive parameters of the environment, such as temperature, CO or benzene levels, relative humidity, noise levels, or illuminance. The sensors were connected to the Internet using the 6LoWPAN standard, with unique IPv6 addresses assigned to each node and wireless sensor network gateways that acted as 6LoWPAN border router and provided the interface to connect the resource-limited, low-power link with traditional WAN. The paper presents an exhaustive and detailed overview of the different challenges and critical points that need special attention in the deployment of an integrated urban IoT solution for a smart city.

A variety of sensors is since being deployed in cities across the world to provide city administrators with more in-depth information of the environment and the interaction of citizens with a city's infrastructure. Another example of a smart city solution was presented by Pla-Castells et al., in which they perform traffic characterization using Bluetooth wireless sensors [25]. The gathered data were made available to the public via an application programming interface (API) and licensed under Creative Commons.

In the past few years, more solutions have been presented that use LoRa for low-rate, low-power, long battery life applications in smart cities, as opposed to previous solutions based on Wi-Fi, Bluetooth, or GSM/3 G/LTE [34−39]. Pasolini et al. have shown that the range of LoRa in a dense urban environment is about 1−2 km, with the gateway deployed in a favorable position at 71 m above average ground level [35]. The authors have run different simulations to estimate the bit error rate and the percentage of packets successfully received at the gateway for different configurations. A mobile edge computing architecture has been tested by Chen et al. using unmanned aerial vehicles to gather environmental data at different points of a city [36]. In this case, sensors are not static, instead the drones move toward points of interest to gather data.

Even more closely related to our proposed application is the work carried out by Truong [40]. The authors have demonstrated how edge computing can bring the capabilities of sensor nodes transmitting using LoRa very close to those that transmit using 4G networks or local area networks. This is possible due to the movement of the data analysis from the cloud to the edge nodes. Rather than sending raw data to cloud servers directly, it can be analyzed and compressed at the edge of the network, reducing the minimum required data rate by several orders of magnitude. The author focuses on the middleware application layer to provide a framework that enables data sharing and availability at the edge.

Traffic management is arguably one of the most challenging problems that city administrators have to deal with in a large metropolis due to the increasingly large number of vehicles and the small changes that have occurred in terms of infrastructure over decades or centers in the oldest areas of the city. At the moment, traffic jams seem to be unavoidable in such large urban areas. For instance, the 2016 Tomtom traffic index shows that travel time is increased by an extra 41% in Shanghai, China, during rush hours, which this figure will likely increase in the near future [41]. Traffic management targets have partially shifted from avoiding congestions to managing them more efficiently and distributing traffic dynamically. When traffic is not properly managed, it might cause significant consequences to the city environment. For example, an increase in traveling time or the amount of CO_2 emissions reduces the quality of life. This has the potential to indirectly or directly threaten human health, whether through higher incidence of respiratory diseases or even in the case of an ambulance being trapped in a traffic jam during an emergency.

Taking the above considerations into account, many systems have been introducing to overcome traffic congestion and manage the traffic more efficiently. In Ref. [42], authors present a customized LPWAN-based system for traffic management in a smart town. The system leverages different concepts and techniques from traffic theory, machine learning, and LoRa in order to improve traffic manageability. Although the proposed system is implemented and tested in a small town, it is scalable for suiting to a larger scale application such as smart cities. In another application [43], authors introduce a vehicle monitoring system using LoRa and a custom IoT platform. The system is able to detect a vehicle location and measure a traveling distance together with different environmental parameters such as temperature, humidity, and air quality. In summary, a sensor node placed at a vehicle can acquire data (e.g., CO_2 or O_3) from different sensors and transmit the data to a LoRa-based gateway which then forwards the data to a cloud server for global storage and data processing. In Ref. [44], authors present a system for autonomously controlling traffic lights to avoid traffic congestions. The system consists of multiple sensor nodes that are placed on street lanes and LoRa-based gateways. Sensor nodes collect information related to vehicles passing by (e.g., the number of vehicles) and transmit the collected data via LoRa to gateways, which then send the data to cloud. Based on the collected information, cloud-based applications and services can send commands to control traffic lights in a more efficient manner. Sensor nodes are able to transmit the data over long distances, up to 15 km, which helps reduce the investment and deployment costs while providing a good quality service.

16.3 Edge and fog computing

With the rise of interest around IoT development, and supported by a stronger growth of the industrial IoT, new technologies have emerged to solve the problem of dealing with a rapidly increasing amount of sensor data that threaten to take existing network infrastructure capability to its limit and collapse cloud servers [45]. Cloud-centric data analysis, processing, and storage are not a scalable solution as sensor data can easily surpass the capacity of a local network. This is the case of multiple high-definition cameras or real-time bio-signal acquisition devices. This high amount of data transfer is often even unnecessary in many cases where there is no need to store raw data but just the result of its analysis. In those cases, a more viable and practical solution is to perform part of the data processing at the local network level, so that only the critical information is transferred to the cloud services. This enables several orders of magnitude of data compression, which reduces the network load and allows more efficient use of cloud servers. Both edge computing and fog computing are computing paradigms that refer to the distribution of computational intelligence at different layers in the network, and moving data analysis and compression capabilities as near as possible to where the data originate. By analyzing local data at edge nodes, we can distribute the computing load across the network and keep only the level of detail necessary at each layer.

Even though IoT can offer many advantages, it still has some limitations. For instance, a local network's capacity may be overused when the number of sensor nodes and data are large. This might cause delays or uncontrolled data loss. In real-time applications, latency and undelivered data can produce unexpected results. In safety-critical scenarios, this can develop into serious consequences. For instance, engineers can be injured when a corobot does not properly function and cannot be stopped in real-time. In case of health monitoring, data loss can cause an incorrect disease analysis. Therefore, new technology ought to be developed, which help overcome some of the limitations of the IoT. The target can be addressed with the assistance of edge computing and fog computing, which are essentially a series of extra layers between sensor nodes and gateways, or gateways and cloud servers, respectively [9,46,47]. In other words, instead of using the traditional IoT architecture having three layers (sensor nodes, gateways, and cloud), a new system architecture consisting of sensor nodes, edge devices, fog gateways, and cloud servers is proposed. The new architecture inherits the benefits of edge and fog computing for bringing computation and other advanced services to the edge of the network. For instance, an edge fog—based architecture helps reduce energy consumption of sensor nodes by shifting computation from sensor nodes to edge devices/gateways. In addition, the architecture helps reduce the network load by preprocessing data before transmission.

16.3.1 Edge AI: artificial intelligence at the edge of the network

AI and, more concretely, machine learning algorithms often rely on powerful servers that are able to run computationally expensive data analysis and processing. Cloud servers can easily adapt to the necessities of machine learning algorithms and methods. However, in a hybrid IoT architecture where the computational load is distributed, more computationally efficient algorithms need to be designed to run at the edge and fog layers. In this chapter, we present a lightweight algorithm that can process several frames per second when running in computationally constrained and cost-effective single-board computers such as Raspberry Pi, Onion Omega, Banana Pi, Odroid, or Intel Up boards.

Before introducing our proposed algorithm, we review trends in the integration of AI techniques with edge and fog computing. Tang et al. introduced a big data analysis architecture that integrated fog computing and aimed at smart cities [22,48]. In particular, they applied their architecture to the use case of monitoring the state of pipelines. The authors used fiber optic sensors and sequential learning algorithms to be able to predict situations threatening the proper operation and safety of pipelines. Their proposed architecture consisting of multiple Fog-enabled layers is able to support neighborhood-wide, community-wide, and city-wide levels with quick response time and high computational performance.

Multiple research efforts have been recently put into reducing the computational footprint of deep learning algorithms and making them *lightweight*. For example, Kim et al. showed how to enable state-of-the-art accuracy in multicategory object detection task without compromising computational resources by integrating several AI techniques [49]. The authors claim to reduce the computational time of their classification approach to as little as 12.3% of the computational cost compared to ResNet-101, the winner on VOC2012. Even though they tested the proposed neural network efficiency for object classification, they state that the same design principle can be potentially used in a much wider range of applications, including semantic analysis and face recognition.

Zhang et al. have performed an extensive analysis of AI platforms with different edge devices and hardware [50]. The analyzed software includes TensorFlow, Caffe2, MXNet, PyTorch, and TensorFlow Lite. The authors have measured parameters such as latency, memory footprint, and power consumption. The experimental edge-devices were MacBook FogNode Jetson TX2. They concluded that Tensorflow was the fastest framework on more powerful CPUs, while Caffe2 is faster when running small-scale model; MXNet is the fastest package when running on Jetson TX2; PyTorch is more memory efficient on CPUs and MXNet is the most energy-efficient package on FogNode and Jetson TX2. The study is mainly based on the comparison of the AlexNet and SqueezeNet neural network models. The authors also noticed that in edge devices, loading a pretrained model often takes more time than running the model itself. Therefore, this can be optimized, which might be ignored in cloud-based machine learning where this has little to no effect.

16.4 Low-power wide-area network technology

Semtech's LoRa specification and definition provide a communication modulation protocol that can be mapped to the first layer of Open Systems Interconnection model—the physical layer. Different solutions have been proposed and developed to map second and third layers—the link and network layers. The most popular MAC protocol for wide area networks is LoRaWAN, which is designed to operate with either LoRa or FSK modulation [4]. In this section, we briefly overview the LoRaWAN protocol specification and compare it with other alternatives in the LPWAN spectrum.

16.4.1 LoRa for the physical layer

LoRa transmissions can be adapted with different modulation parameters, as well as the transmitted power. The parameters that have the most effect on packet length, time-on-air, and link budget are the bandwidth, coding rate, and spreading factor (SF). The bandwidth affects the data rate and sensitivity. A larger bandwidth enables more data to be sent but reduces sensitivity. In average, halving the bandwidth corresponds to $3-4$ dBm gain in the link budget. The SF defines the number of chirps that are used to modulate each symbol before transmission. A single symbol is modulated with 2^{SF} chirps. When SF increases one unit, the length of the transmitted signal and its time-on-air double. At the same time, it increases the link budget by about 2−3 dBm. The exact gain depends on the bandwidth. The coding rate represents the number of extra symbols used for forward error correction. It directly affects robustness to interference, and in an ideal situation, it would be changed to adapt to channel conditions. It has no effect on the link budget. A higher CR decreases the probability of error during decoding. A LoRa packet consists of a preamble, a header if in explicit mode, the payload, and the cyclic redundancy check bits, and its structure is shown in Fig. 16−2.

16.4.2 Long-range wide-area network

At its most basic level, LoRaWAN is a network protocol that targets battery-powered devices by design. Therefore there is an emphasis in keeping energy consumption as low as possible. In contrast with single-link radio-based communication, LoRaWAN-connected devices do not build a connection with a specific gateway but instead always broadcast the data. Several gateways can receive the transmission, which is then handled by one of them according to a cloud-based classification. All gateways are connected through back-end servers. This enables connected devices to be either static or mobile. This naturally leads to a star-based topology at the network level, with each gateway handling a potentially large number of devices. Because of the duty cycle and data rates limits of LoRaWAN, LoRa gateways are able to handle many more devices than Bluetooth or Wi-Fi routers. Moreover, signals modulated with different LoRa parameters such as bandwidth and SF are orthogonal; therefore the maximum number of concurrent device connections that a LoRaWAN access point can handle is significantly larger than Wi-Fi routers.

Because LoRa and LoRaWAN target low-power usage, transmissions must be as short as possible and receiving windows be opened as seldom as possible. Transmission time, or payload time-on-air, directly depends on the amount of data to be transmitted and the LoRa modulation parameters. If longer transmission range is required, then the time-on-air needs to be significantly increased. Because of this, uplink transmission is predominant in LoRaWAN, with most transmissions having no acknowledgment at all. This means that when a sensor node transmits data, it has no way of knowing whether any LoRaWAN access point has received the transmission unless an acknowledgment is specifically requested. At the same time, this also has an impact on the protocol specification, with most of the network capacity reserved for uplink transmissions versus downlink. LoRaWAN takes this into

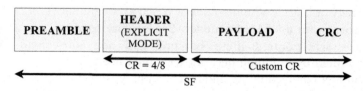

FIGURE 16−2 LoRa packet.
Illustration of a LoRa packet structure and the effect of the modulation factors in different parts of the packet.

account by defining different classes of communication depending on the energy consumption constraints of different end-devices. The protocol tackles this by defining three different classes of connected end-devices with an incremental number of features: (1) Class A devices, with a short receiving window opened after each transmission; (2) Class B devices, with scheduled receiving windows; and (3) Class C devices, with full bidirectional communication. In practice, open LoRaWAN networks mostly implement only support for class A devices. This is because class B and C devices require all gateways to be synchronized and to operate with a common set of features in their firmware. Because any individual or organization can deploy a LoRaWAN access point and connect it to an open network, it is impractical to assume control over all gateways.

16.4.3 Symphony Link

Symphony Link is a proprietary network protocol that uses the LoRa modulation as its base. Developed by Link Labs, Symphony Link specifically addresses some of the limitations of LoRaWAN, mostly in terms of duty cycle and data rates [7,8]. Its main target is industrial or private use, as tight control over the network infrastructure is necessary to eliminate the regulated duty cycle limitations and ensure proper synchronization of end-devices and gateways. Because of this, Symphony Link is able to leverage the long-range and low-power nature of LoRa transmissions while enabling real-time communication. This significantly increases the number of LoRa use cases in case of private deployments. Moreover, the specification provides built-in support for repeaters, which are able to extend the range of single gateways. However, the gateway capacity remains constant, so the total amount of connected devices does not increase. Finally, Symphony Link allows for multicast messages to be transmitted from one gateway to many end-devices at the same time. This is particularly useful for over-the-air updates or similar one-to-many broadcasts.

The Symphony Link specification has several differences when compared to the LoRaWAN standard. While LoRaWAN provides different device classes for different end-users or applications, in a Symphony Link network all devices operate under a common standard. In order to optimize the LoRa modulation parameters, both end-devices and access points perform interference scans periodically to select the most appropriate communication channels. This enables energy saving and reduces the probability of packet-loss by choosing the channels with most available capacity. This can also be done in a LoRaWAN network, even though it is a functionality reserved for the end-user to implement. In a Symphony Link network, access points choose a 500 kHz channel (or 125 kHz in Europe) and transmit beacons with a 0.5 Hz frequency. Beacons are used to keep access points and connected devices synchronized. Both beacons and any subsequent communication exchange are encrypted by a unique network ID, which needs to be configured in the end-devices before the first connection. As frequency hopping is used to avoid duty cycle limitation, access points periodically broadcast information about the number of channels being used during transmission for the frequency hopping scheme. They can be configured as 1, 8, or 64 channel access points, and repeaters and end-nodes adapt to this setup. Access points also transmit information about the quality of service, which enables the ordering of messages by priority in the end-devices. These can choose which packets to transmit in situations where the network capacity might be near saturation and more priority needs to be given to some information. Moreover, nodes can dynamically adapt the LoRa SF and transmitting power to save energy based on the received beacon signal power. Even though this can be done in LoRaWAN, the lack of frequent downlink messages for standard Class A devices makes the estimation less accurate and increases the probability of data loss. In Symphony Link, all messages are acknowledged.

16.4.4 MoT: MAC on time

Many research efforts are being put into the development of LoRa-based MAC protocols that address some of the limitations of LoRaWAN, while providing an open solution. Hassan and Hossam have developed a MAC protocol that is based on LoRa for the physical layer, with a focus on reducing energy consumption, enabling scalable networks with fair usage, and maximizing capacity [51]. The proposed protocol, namely MAC on time (MoT), guarantees packet delivery and improves bandwidth utilization when compared to LoRaWAN. The authors overcome the problem of synchronization with a centralized scheduling system that controls uplink windows with instructions sent to end-devices via packet acknowledgment. Therefore, end-nodes do not decide by themselves the time between transmissions, often equivalent to the time they stay in idle or sleep mode. This allows for centralized power consumption management, network load management, and optimization of the network capacity. The authors have tested the performance of the protocol in multiple simulations, where MoT has proved to be able to provide 4−15 times more network capacity and better latency than LoRaWAN.

16.5 System architecture

We propose an extension of traditional IoT system architectures to leverage the advantages of LoRa while taking into account its limitations in terms of duty cycle and data rate. By integrating LoRa with edge and fog computing, we can benefit from the latter two's capabilities in data analysis and compression. This converges to a five-layer architecture that consists of (1) end-devices, (2) smart edge gateways, (3) fog layer with LoRa access points, (4) cloud servers, and (5) end-user terminal applications. The proposed architecture is illustrated in Fig. 16−3.

16.5.1 Device layer

In the system architecture, the device layer plays the main role in terms of integration and interaction with the real world. When sensor nodes of the device layer do not properly function due to, for instance, a hardware failure or disconnection with edge devices/gateways, the whole system might stop working. Therefore, it is necessary to design and implement the

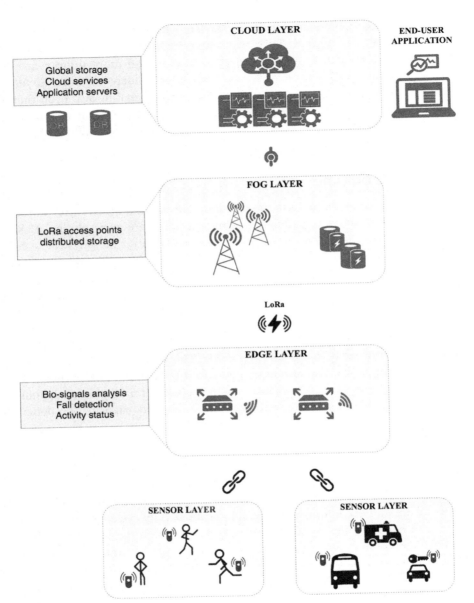

FIGURE 16–3 Proposed system architecture.
Proposed four-layer hybrid sensor-edge-fog-cloud system architecture with LoRa wireless link used as the edge-fog bridge.

device layer carefully. The device layer can consist of sensing nodes and actuating nodes or hybrid nodes which can both sense data and control some actuators. Depending on the application scenarios, one or several sensor node types can be used. Sensing nodes are responsible for collecting different data from a wide variety of sensors. In a traffic monitoring and management system, data such as the number of cars passing by a given street or intersection, the approximate number of vehicles in an area, or the public transportation flow can be acquired via some sensors such as piezoelectric sensors, inductive loop sensors, magnetic sensor, acoustic sensor, or passive infrared sensors. Each sensor has its own advantages and limitations. For instance, passive infrared sensors might not be suitable for highways with many lanes. In addition to sensing nodes collecting the primary data, sensing nodes acquiring contextual data should be applied. For example, sensing nodes can collect contextual environmental data such as temperature, humidity, CO_2 level, and other parameters related to pollution emissions. Even though the information might not be directly related to traffic management systems, they are helpful for the traffic system administrators, city planners, and regulation makers. Actuating nodes can be used to control actuators such as fans, traffic lights, or water pump systems. For instance, they can turn on a water pumping system to emit steam to increase humidity level when the air is too dry.

These nodes are often built from a combination of energy-efficient sensors/actuators, microcontroller, and wireless communication modules. In summary, sensors and a wireless communication module are connected to a microcontroller via SPI protocol because SPI is more energy-efficient than other wire communication protocols such as I2C and UART. The microcontroller is energy-efficient with different sleep modes. The communication protocol nRF is used in nodes because the protocol is energy-efficient and capable of having many-to-many communication.

16.5.2 Edge layer

Edge layer consists of edge devices that are built from several components such as a powerful microcontroller with high-capacity memory or single-board computers. Edge devices might be either battery-powered or connected to a power source, depending on the specific application scenario. Single-board computers usually need to be connected to a power source as they can only operate for a period of a few days at most if powered with a battery. In the case of battery-powered edge devices, these must be energy-efficient. In this case, edge device can be equipped with nRF communication modules to collect the data transmitted from sensing nodes. Due to the capability of supporting high data rates (i.e., 250 kbps in theory and 150 kbps in practice), an edge device with a single nRF module can support more than 120 sensing nodes as each sensing node often gathers less than 1 kbps data. When the number of sensing nodes increases dramatically, more nRF modules can be added into an edge device. The collected data is then processed for eliminating noises and saving bandwidth. For instance, the data can be both encrypted and compressed before being sent to Fog gateways via LoRa. Loss-less and lossy compression algorithms can be used. However, lossy compression algorithms are more preferred as their compression rate is high (e.g., around 100 times). Although lossy decompression cannot fully recover the original data, the decompressed data (i.e., the number of cars passing by a street, the number of cars waiting at a crossroad, temperature, and humidity) are still good enough for traffic management systems.

The selection of microcontroller-based edge devices or gateways based on single-board computers is made depending on the computational needs of each specific application. For applications where only standard data compression and encryption are necessary, a microcontroller-based edge gateway might be sufficient. In use cases where more intensive data analysis is preferred, the power of single-board computers might be necessary. This is the case of data processing based on computationally expensive image processing algorithms or neural networks for object classification or semantic analysis. However, lightweight AI models can run in microcontrollers. For instance, a face recognition algorithm has been developed for Espressif's ESP32 microcontroller [52]. This is possible, of course, by trading-off frame resolution (QVGA, 320 × 240 pixels) and the number of frames per second analyzed. Nonetheless, widely used machine learning frameworks such as TensorFlow now offer edge-focused versions (TensorFlow Lite), which are able to run in microcontrollers or less powerful CPUs [50,53,54].

16.5.3 Fog layer

Fog layer encompasses interconnected fog-assisted gateways that are placed at particular places and able to communicate with each other and share certain information. These gateways use power from a socket wall and they are built with hardware much more powerful computationally wise than edge devices. Therefore, fog-assisted gateways are capable of performing heavy computation and offering advanced services such as distributed storage, data processing, and push notifications. The detailed information of Fog services is mentioned in some of our previous works [10−13,55]. Fog-assisted gateways are equipped with LoRa modules for receiving data from edge devices. In addition, these gateways are connected to cloud servers via Ethernet or high-capacity wireless links. This includes mobile networks, such as 4G or 5G, which are useful for mobile or remote access points.

In the proposed system architecture, LoRa access points, or gateways, are the central element in the fog layer. These gateways are connected to back-end servers that, for instance, prioritize which gateway is used for a given downlink message to a specific end-device. Moreover, for some applications, services available at fog computing can be essential, such as distributed storage. This is the case of collaborative SLAM, where local maps are stored at the fog layer so that they are always available even if the edge gateway to which an end-node is connected changes [56].

16.5.4 Cloud layer

The cloud layer consists of servers and different services that support the system and provide a bridge between the IoT devices and end-users for management or monitoring. Time-series data of sensor state, and the results of the data analysis and compression performed at the edge and fog layers are stored in cloud servers where global storage is available. Part of the data analysis can also be performed at the cloud. However, this should be only data that do not require time-critical or safety-critical responses, as unexpected increases in communication latency or the possibility of data loss might develop into unexpected behavior.

Cloud servers are also essential in terms of bridging the IoT platform with end-users and administrators. Even if the edge and fog layers perform most or all of the data analysis, servers are necessary to host cloud-based web and mobile applications and all related data. Typical cloud server providers are Amazon AWS, Google Cloud, or Digital Ocean. The cloud layer can also be deployed with private servers.

16.5.5 Terminal layer

The terminal layer consists of any type of connected devices that are used by end-users for interacting with the IoT platform. These include any devices able to open or execute native or web applications. Cloud-based applications are the public side of the IoT platform and allow end-users to monitor and control the system, analyze historical data, or overview the platform's state and effectiveness.

In the case of a traffic management system for a smart city, end-users are typically city administrators, and terminal devices are used to monitor traffic and receive alerts of traffic accidents, traffic congestion, or other important events.

16.6 LoRa and mobile edge computing: a use case for traffic monitoring

The complete system from sensor nodes to cloud servers has been implemented. Sensor nodes are made from an 8-bit ATmega328P AVR microcontroller, a nRF24L01 wireless communication module and some other sensors such as a MQ-135 CO_2 sensor or a BME280 temperature and humidity sensor. The data collected by sensors are sent to the microcontroller which runs some basic filtering threshold-based methods to eliminate incorrect values. The microcontroller sends the data to nRF24L01 forwarding the data via nRF to edge devices. Three sensor nodes are deployed in our experiments.

Edge devices are made from a combination of a Raspberry Pi 3 Model B, an RGB camera, an nRF24L01 module, and a LoRa module. The nRF24L01 module is used to collect data from sensor nodes while images acquired directly at the edge are processed and analyzed together with other sensor data. Only the result of the analysis is sent over the LoRa network to the fog layer (Fig. 16−4).

Because traffic monitoring requires near real-time data transmission, a basic LoRa configuration is not suitable for this application. In the same sense, LoRaWAN cannot be used in practice as it limits the data rate and duty cycle by design. Therefore, we propose the use of a frequency hopping technique to avoid the duty cycle restrictions and enable real-time transmission of traffic status from all end-devices. A LoRa concentrator is a module that is able to receive concurrent LoRa signals from different devices if the modulation parameters

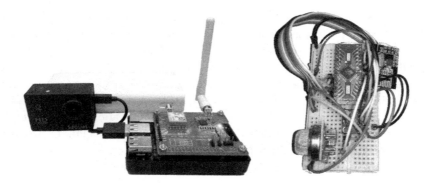

FIGURE 16–4 Edge device and sensor node.
The edge device, on the left, consists of a Raspberry Pi with a LoRa shield, a wide-lens camera, and backup battery. The sensor node has smoke, benzene, alcohol, temperature, and humidity sensors. The sensor node and edge device communicate with via an nRF module, which is not visible for the edge device in this image.

are different. For instance, a LoRa concentrator might be able to simultaneously decode up to eight signals if they have been transmitted with different bandwidths and/or different SFs. Moreover, because signals modulated with different SFs are orthogonal, the quality of signal is not affected by the existence of multiple simultaneous transmissions.

The backbone of the fog layer is a series of fog-assisted gateways consisting of LoRa access points and powerful computers (i.e., having Intel Core i7, 16 GB of RAM, and 2 TB of hard-disk). The fog-assisted gateways are able to perform heavy computation and temporarily store data in distributed manner. LoRa access points have been configured to take into account the frequency hopping scheme that end-devices use to transmit consecutive messages to avoid the limitations imposed by regulations on the duty cycle.

Cloud servers and terminal applications were already implemented in our previous works [47]. Therefore, they are customized (e.g., updating new user interface) and reused in the proposed system.

The implemented system can be installed in several locations across a city. For instance, edge device with one or multiple cameras can be installed in an intersection or a road outside the city, with a power supply nearby. Then, sensor nodes can be placed in the vicinity of the edge device. In the case of nRF module used in this implementation, its range varies from 30 to 50 m, but other versions have a range of up to 1 km in line of sight. Sensor nodes can either be battery-powered or connected to the mains, depending on the specific power consumption requirements and the type and amount of data gathered.

While sensor nodes gather information about environmental parameters such as temperature, humidity, and concentration of different gases, the main data source is a camera on the edge device. By processing the video feed directly at the edge, we can extract only a single feature, the current traffic density, which is sent over LoRa. Raw images can be stored at the edge device for a certain period of time, in case end-users might request them. However, these cannot be transmitted over LoRa as the bandwidth is not sufficient.

In order to perform real-time traffic density estimation in resource-constrained edge devices, we have developed a lightweight image processing algorithm that does not count individual cars in the image, but instead uses the combination of multiple iterations of a cascade classifier to estimate how full of vehicles the image is. The algorithm has been implemented using Python3 and OpenCV4.0 and relies on the OpenCV built-in *CascadeClassifier*. The algorithm works as follows. First, a region of interest is selected from a sample image where the vehicle density will be estimated. In a two-way road, there can be either a single region of interest including both directions, or two separate regions for each of the directions. We refer to this step as *Lane Segmentation* in the algorithm overview shown in Fig. 16−5. After this, fixed noise is obtained by running the cascade classifier in a sample image without any vehicles. Because cascade classifiers are based on finding a set of matching features, some items present in the video feed such as traffic signals or some parts of near buildings might be wrongly identified as vehicles. The next step is to run the built-in cascade algorithm with different scale factors and calculate the intersection of the identified objects. By using this, rather than focusing only on areas where there is a concentration of detected objects, we are not able to identify individual cars but instead we are able to also obtain parts of cars that might be covered by others when the traffic is congested. The obtained density distribution is then integrated to obtain the current traffic density. The obtained value is then compared with a series of previously calculated values under different road conditions, and a traffic density level is chosen accordingly (Fig. 16−6).

Fig. 16−7A and B show two separate lane segmentation from the same original image, one for each direction. The same images shows in graphs (C) and (D) the accumulation of rectangles with detected objects, as compared to a traditional vehicle finder algorithm

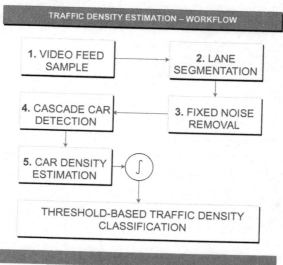

FIGURE 16–5 Traffic density estimation algorithm workflow.
The traffic density estimation algorithm uses lane segmentation and merges several iterations of a car cascade detector in order to obtain an estimation of car density.

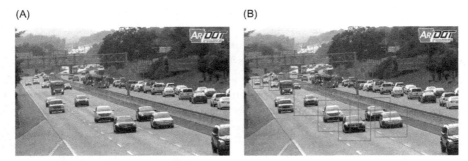

FIGURE 16–6 Traffic camera and cascade-based car detection algorithm.
Figure (A) shows a snapshot of a traffic camera in Arkansas, United States, and figure (B) shows a traditional cascade-based detection algorithm only able to detect some of the cars. Data from *https://www.kasu.org/post/arkansas-begin-streaming-nearly-100-traffic-cameras-online*.

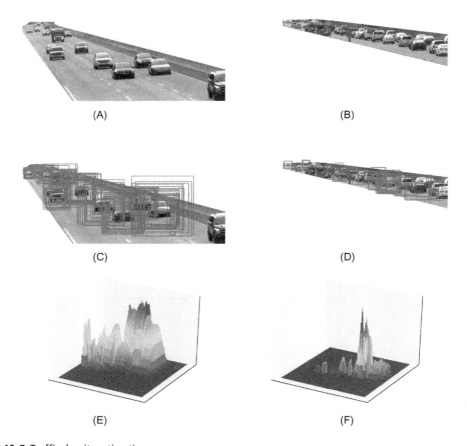

FIGURE 16–7 Traffic density estimation.
Estimation of traffic density in each of the two lanes based on integration of several levels of cascade classifiers. Figures (A) and (B) show two different lanes extracted from Fig. 16–6. Figures (C) and (D) show the superposition of the cascade classifier at different iterations. Finally, figures (E) and (F) show the density distributions obtained by intersecting the classified areas.

based on the same cascade classifier that is shown in Fig. 16−6. The 3D distribution resulting of accumulating the intersection of this detected objects is shown in Fig. 16−7E and F. We have tested the performance of the algorithm for different types of roads and times of the day. Fig. 16−8 shows the density distribution before integration for (A) a congested road outside a city, (B) a traffic jam in a big metropolitan avenue, and (C) an evening road in the outskirts of another town.

(A)

(B)

(C)

FIGURE 16−8 Traffic estimation in different situations.
Estimation of traffic density using the presented algorithm in three different situations: Figure (A) shows a congested road outside a city (https://www.pexels.com/photo/cars—*traffic*—1328359/), figure (B) shows a traffic jam in a big artery of a metropolis (https://pxhere.com/en/photo/1214630), and figure (C) shows a normal road in the outskirts of Belfast. The images on the right side show the density distributions before integration. *Source: (C) Reproduced from Evening traffic, Belfast (November 2015) cc-by-sa/2.0—© Albert Bridge—geograph.org.uk/p/ 4727179.*

The lane segmentation can be either done in an automatic way or manually. A manual approach has the benefit of choosing an area of interest for each situation, but an automatic approach is also possible given a sample without any vehicles or with a reduced number of objects. In the same way, the different levels of traffic density can be dynamically calculated by obtaining a daily minimum, maximum, and the mean and standard deviation, or manually assigned based on a series of sample images. The integration output is hardly comparable between different scenarios as the distance of the camera with respect to the road, its view field, and other aspects of the camera positioning significantly affect the behavior of the cascade classifier. Traffic density data are then merged with data from sensor nodes, compressed, and encrypted before being sent using LoRa. Because of the real-time need of the system, we use raw LoRa and implement a frequency hopping scheme and dynamically adjust the LoRa modulation parameters in order to overcome duty cycle limitations. Since we are not using LoRaWAN or other network protocols, the messages include edge device identification and are encrypted based on a preconfigured key using AES-256.

16.6.1 Performance evaluation

A lightweight algorithm can provide a good estimation of traffic flow or vehicle density with real-time image processing in a resource-constrained edge device. Therefore, while there is a minimum level of accuracy required in order to obtain meaningful information, the focus of the algorithm is on reducing latency as much as possible.

Table 16−1 shows the run-time needed on an Intel(R) Core(TM) i5-6200U CPU for a state-of-the-art algorithm implemented with OpenCV and neural networks based on the YOLO and SORT algorithms [57,58], which is compared to the run-time needed by our algorithm. Even though the performance of both algorithms is not comparable, as our proposed algorithm only provides a rough estimate of *how full* a given area of the road is of vehicles, we have measured and compared their latency to see how they would perform on an edge device. In that sense, the traffic counter algorithm has very high accuracy and is able to track individual vehicles through consecutive video frames. However, even in a relatively powerful Intel i5 CPU with four cores, the algorithm is able to analyze a single frame per second. If it is run on a more resource-constrained platform, the performance can be significantly reduced and it would become the bottleneck of real-time operation. The proposed algorithm, instead, only takes about 150 milliseconds to analyze one frame. Therefore, multiple frames per second can be processed even in a less powerful CPU such as that of a Raspberry Pi. The

Table 16–1 Traffic counter and density estimation latency.

	Loading model (ms)	Frame analysis (ms)
YOLO + SORT	121	985
Density estimation	5	144

Notes: Latency comparison between a NN-based vehicle counter and the proposed traffic density estimation algorithm. Algorithms have been tested using an Intel(R) Core(TM) i5-6200U CPU @ 2.30 GHz.

(A)   (B)

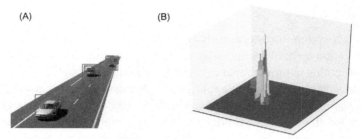

FIGURE 16–9 Traffic density estimation in a highway. (A) Image after lane segmentation and (B) density function before integration.

model loading time is also significantly lower even though this has less impact on performance as it is done only once at boot (Fig. 16−9).

In practice, the main limitation for real-time operation is the LoRa link between the edge and fog layers. Therefore, having 1−2 frames per second processed at the edge device is sufficient for a practical application, and the proposed algorithm performs fast enough to meet this requirement.

16.7 Discussion

When designing a system based on LoRa, it is essential to consider application requirements and regulations. For instance, although LoRa-based modules are capable of transmitting up to a data rate of 250 kbps in theory, the actual data rate in a typical transmission is much lower in practice. Depending on the scenarios and application environment, the successful transmission range varies from one kilometer to a few kilometers (e.g., 2−5 km). For instance, the transmission range in a city with several multi-story buildings between receiver and transmitter can be of 1−2 km [60].

Similar to other wireless communication protocols, LoRa is vulnerable in terms of packet injection attacks. **For example, a third party can play** anonymous sensor nodes near gateways, mimic **LoRa sender nodes and inject a large amount of data in the network**. These data might be collected by LoRa gateways. As a result, actual data sent by sensor nodes cannot reach the gateways.

LoRa itself does not contain any secured mechanism by definition. It is thus necessary to apply some secured method for LoRa network. For example, LoRaWAN applies AES-128 for packages transmitted over LoRa. The benefit of AES-128 is that it is required fewer resources (e.g., hardware, and energy consumption budget) to implement and run AES-128 at sensor nodes and gateways. However, the AES-128 cannot be considered as a highly secured mechanism. Applying complex security mechanisms can cause an increase in energy consumption due to the heavy computations of the mechanisms. Therefore, system administrators need to choose an appropriate security mechanism depending on the application.

16.8 Conclusion

We have proposed a system architecture that integrates LoRa as a long-range and low-power wireless communication technology and edge and fog computing as solutions for reducing network load and distributing computational capabilities. The results show that the proposed system architecture can be suitable for different applications. Moreover, we have presented a lightweight image processing algorithm that can be used to estimate traffic density in a smart city.

We have reviewed the concepts of edge and fog computing and introduced their benefits in terms of network load reduction and computational load distribution when compared to a traditional cloud-centric IoT architecture. We have also presented different technologies in the LPWAN spectrum and shown the advantages and disadvantages of each of them from the point of view of data rate, range, openness, or ease of deployment.

The proposed system architecture is flexible enough to accommodate a wide variety of sensor nodes, while providing complex information such as traffic flow without the need of heavy computation or transmitting high-quality images to cloud servers. We have implemented the proposed architecture with an edge device and a sensor node and shown the performance of our lightweight image processing algorithm for traffic density estimation. Finally, we have compared the latency of our algorithm for single frame processing with state-of-the-art solutions in vehicle counters.

Future work will include the incorporation of more types of sensors into the same platform, and a focus will be put on specific compression techniques that can be used at the edge devices to further reduce the network load and take more advantage of the LoRa link. We will also further improve the traffic density estimation algorithm and perform benchmarking with other similar solutions.

References

[1] U. Raza, P. Kulkarni, M. Sooriyabandara, Low power wide area networks: an overview, IEEE Commun. Surv. Tutor. 19 (2017) 855–873.

[2] C. Swedberg, IoT aims to track free-ranging reindeer in Finland, in: RFID Journal. <https://www.rfidjournal.com/articles/view/17106/>, 2019 (accessed 26.06.19).

[3] Ahoy Systems, LoRa smart street light solutions. <http://www.ahoysys.com/lora-street-light.php>, 2019.

[4] The LoRa Alliance Technical Committee. LoRaWAN™ 1.1 Specification, LoRa Alliance, 2017.

[5] The Things Network—Coverage. <https://www.thethingsnetwork.org/map> (accessed 03.01.2019).

[6] S. Battle, B. Gaster, LoRaWAN Bristol, Proceedings of the 21st International Database Engineering and Applications Symposium, ACM, 2017.

[7] Link Labs. A comprehensive look at low power, wide area networks, 2016.

[8] B. Ray, Link Labs, Low power wide area network technology. Symphony Link™ vs. LoRaWAN™, 2018.

[9] T.N. Gia, M. Jiang, A. Rahmani, T. Westerlund, P. Liljeberg, H. Tenhunen, Fog computing in healthcare Internet of things: a case study on ECG feature extraction, 2015 IEEE International Conference on Computer and Information Technology; Ubiquitous Computing, Communications; Dependable, Autonomic, Secure Computing; Pervasive Intelligence and Computing, IEEE, 2015.

[10] T.N. Gia, N.K. Thanigaivelan, A. Rahmani, T. Westerlund, P. Liljeberg, H. Tenhunen, Customizing 6LoWPAN networks towards Internet-of-things based ubiquitous healthcare systems, 2014 NORCHIP, IEEE, 2014.

[11] M. Jiang, T.N. Gia, A. Anzanpour, A. Rahmani, T. Westerlund, S. Salantera, et al., IoT-based remote facial expression monitoring system with sEMG signal, 2016 IEEE Sensors Applications Symposium (SAS), IEEE, 2016.

[12] M. Ali, T.N. Gia, A. Taha, A. Rahmani, T. Westerlund, P. Liljeberg, et al., Autonomous patient/home health monitoring powered by energy harvesting, GLOBECOM 2017—2017 IEEE Global Communications Conference, IEEE, 2017.

[13] T.N. Gia, M. Ali, I.B. Dhaou, A. Rahmani, T. Westerlund, P. Liljeberg, et al., IoT-based continuous glucose monitoring system: a feasibility study, Procedia Comput. Sci. 109 (2017) 327–334.

[14] V.K. Sarker, M. Jiang, T.N. Gia, A. Anzanpour, A. Rahmani, P. Liljeberg, Portable multipurpose biosignal acquisition, wireless streaming device for wearables, 2017 IEEE Sensors Applications Symposium (SAS), IEEE, 2017.

[15] I. Tcarenko, T.N. Gia, A. Rahmani, T. Westerlund, P. Liljeberg, H. Tenhunen, Energy-efficient IoT-enabled fall detection system with messenger-based notification, International Conference on Wireless Mobile Communication and Healthcare, Springer, 2016.

[16] T.N. Gia, M. Jiang, V.K. Sarker, A. Rahmani, T. Westerlund, P. Liljeberg, et al., Low-cost fog-assisted health-care IoT system with energy-efficient sensor nodes, 2017 13th International Wireless Communications and Mobile Computing Conference (IWCMC), IEEE, 2017.

[17] T.N. Gia, V.K. Sarker, I. Tcarenko, A. Rahmani, T. Westerlund, P. Liljeberg, et al., Energy efficient wearable sensor node for IoT-based fall detection systems, Microprocess. Microsyst. 56 (2018) 34–46.

[18] T.N. Gia, M. Jiang, Exploiting fog computing in health monitoring, Fog and Edge Computing: Principles and Paradigms, John Wiley & Sons, Inc, Hoboken, NJ, 2019.

[19] T.N. Gia, I.B. Dhaou, M. Ali, A.M. Rahmani, T. Westerlund, P. Liljeberg, et al., Energy efficient fog-assisted IoT system for monitoring diabetic patients with cardiovascular disease, Future Generation Computer Systems, Elsevier, 2019.

[20] J. Peña Queralta, T.N. Gia, H. Tenhunen, T. Westerlund, Collaborative mapping with IoE-based heterogeneous vehicles for enhanced situational awareness, IEEE Sensors Applications Symposium (SAS), IEEE, 2019.

[21] S.R. Moosavi, T.N. Gia, E. Nigussie, A. Rahmani, S. Virtanen, H. Tenhunen, et al., Session resumption-based end-to-end security for healthcare Internet-of-things, 2015 IEEE International Conference on Computer, Information Technology; Ubiquitous Computing, Communications; Dependable, Autonomic, Secure Computing; Pervasive Intelligence and Computing, IEEE, 2015.

[22] Z. Li, R. Al Hassan, M. Shahidehpour, S. Bahramirad, A. Khodaei, A hierarchical framework for intelligent traffic management in smart cities, IEEE Trans. Smart Grid 10 (2019) 691–701. Available from: https://doi.org/10.1109/TSG.2017.2750542.

[23] V. Kostakos, T. Ojala, T. Juntunen, Traffic in the smart city: exploring city-wide sensing for traffic control center augmentation, IEEE Internet Comput. 17 (2013) 22–29. Available from: https://doi.org/10.1109/MIC.2013.83.

[24] K. Su, J. Li, H. Fu, Smart city and the applications, in: 2011 International Conference on Electronics, Communications, Control (ICECC), 2011. <https://doi.org/10.1109/ICECC.2011.6066743>.

[25] M. Pla-Castells, J.J. Martinez-Durá, J.J. Samper-Zapater, R.V. Cirilo-Gimeno, Use of ICT in smart cities. A practical case applied to traffic management in the city of Valencia, in: 2015 Smart Cities Symposium Prague (SCSP), 2015. <https://doi.org/10.1109/SCSP.2015.7181559>.

[26] J. Martínez, R.V. Cirilo, A. García, F. Soriano, Influence of percentage of detection on origin-destination matrices calculation from Bluetooth, WiFi Mac address collection devices, in: Proceedings of International Simulation Conference, 2015.

[27] A. Sharif, J. Li, M. Khalil, R. Kumar, M. I. Sharif, A. Sharif, Internet of things—smart traffic management system for smart cities using big data analytics, in: 2017 14th International Computer Conference on Wavelet Active Media Technology and Information Processing (ICCWAMTIP), 2017. <https://doi.org/10.1109/ICCWAMTIP.2017.8301496>.

[28] S. Musa, Smart cities—a road map for development, IEEE Potentials 37 (2018) 19–23. <https://doi.org/10.1109/MPOT.2016.2566099>.

[29] W. Shi, J. Cao, Q. Zhang, Y. Li, L. Xu, Edge computing: vision and challenges, IEEE Internet Things J. 3 (2016) 637–646. <https://doi.org/10.1109/JIOT.2016.2579198>.

[30] o, K.S. Trivedi, Combining cloud and sensors in a smart city environment, EURASIP J. Wirel. Commun. Netw. 2012 (2012) 247.

[31] A. Caragliu, C. Del Bo, P. Nijkamp, Smart cities in Europe. J. Urban Technol. 18 (2011) 65–82.

[32] H. Ahvenniemi, A. Huovila, I. Pinto-Seppa, M. Airaksinen, What are the differences between sustainable and smart cities? Cities 60 (2017) 234–245. <http://www.sciencedirect.com/science/article/pii/S0264275116302578>.

[33] A. Zanella, N. Bui, A. Castellani, L. Vangelista, M. Zorzi, Internet of things for smart cities. IEEE Internet Things J. 1 (2014) 22–32. <https://doi.org/10.1109/JIOT.2014.2306328>.

[34] V. A. Stan, R. S. Timnea, R. A. Gheorghiu, Overview of high reliable radio data infrastructures for public automation applications: LoRa networks, in: 2016 8th International Conference on Electronics, Computers and Artificial Intelligence (ECAI), 2016.

[35] G. Pasolini, C. Buratti, L. Feltrin, F. Zabini, R. Verdone, O. Andrisano, et al., Smart city pilot project using LoRa, in: 24th European Wireless Conference, 2018.

[36] L. Chen, H. Huang, C. Wu, Y. Tsai, Y. Chang, A LoRa-based air quality monitor on unmanned aerial vehicle for smart city, in: 2018 International Conference on System Science and Engineering (ICSSE), 2018.

[37] J. J. Chen, J. E. Chen, V. Liu, L. Fairbairn, L. Simpson, C. Wang, et al., A viable LoRa framework for smart cities, in: 22nd Pacific Asia Conference on Information Systems (PACIS 2018), AIS Electronic Library (AISeL), 2018. <https://eprints.qut.edu.au/118239/>.

[38] J.G. James, S. Nair, Efficient, real-time tracking of public transport, using LoRaWAN and RF transceivers, 2017 IEEE Region 10 Conference, TENCON 2017, IEEE, 2017.

[39] M. Centenaro, L. Vangelista, A. Zanella, M. Zorzi, Long-range communications in unlicensed bands: the rising stars in the IoT and smart city scenarios. IEEE Wirel. Commun. 23 (2016) 60–67.

[40] H. Truong, Enabling edge analytics of IoT data: the case of LoRaWAN, 2018 Global Internet of Things Summit (GIoTS), IEEE, 2018.

[41] Tomtom Traffic Index, Shanghai China. <https://www.tomtom.com/engb/trafficindex/city/shanghai>, 2016.

[42] S.B. Seo, D. Singh, Smart town traffic management system using LoRa, machine learning mechanism. <http://sites.ieee.org/futuredirections/tech-policy-ethics/november-2018/smart-town-traffic-management-system-using-lora-and-machine-learning-mechanism/>, 2018.

[43] C.L. Hsieh, Z.W. Ye, C.K. Huang, Y.C. Lee, C.H. Sun, T.H. Wen, et al., A vehicle monitoring system based on the LoRa technique, Int. J. Transp. Veh. Eng. 11 (2017) 1100–1106.

[44] R.F.A.M. Nor, F.H.K. Zaman, S. Mubdi, Smart traffic light for congestion monitoring using LoRaWAN, 2017 IEEE 8th Control and System Graduate Research Colloquium (ICSGRC), IEEE, 2017.

[45] C. Tseng, F.J. Lin, Extending scalability of IoT/M2M platforms with Fog computing, in: 2018 IEEE 4th World Forum on Internet of Things (WF-IoT), 2018.

[46] F. Bonomi, R. Milito, J. Zhu, S. Addepalli, Fog computing and its role in the Internet of things, Proceedings of the First Edition of the MCC Workshop on Mobile Cloud Computing, ACM, 2012.

[47] A.M. Rahmani, T.N. Gia, B. Negash, A. Anzanpour, I. Azimi, M. Jiang, et al., Exploiting smart e-Health gateways at the edge of healthcare Internet-of-things: a fog computing approach, Future Gener. Comput. Syst. 78 (2018) 641−658.

[48] B. Tang, Z. Chen, G. Hefferman, S. Pei, T. Wei, H. He, et al., Incorporating intelligence in fog computing for big data analysis in smart cities, IEEE Trans. Ind. Inform. 13 (2017) 2140−2150.

[49] K. Kim, Y. Cheon, S. Hong, B. Roh, M. Park, PVANET: deep but lightweight neural networks for real-time object detection, 2016.

[50] X. Zhang, Y. Wang, W. Shi, pCAMP: performance comparison of machine learning packages on the edges, USENIX Workshop on Hot Topics in Edge Computing (Hot-Edge 18), USENIX Association, Boston, MA, 2018. Available from: https://www.usenix.org/conference/hotedge18/presentation/zhang.

[51] G. Hassan, S. H. Hossam, MoT: a deterministic latency MAC protocol for mission-critical IoT applications, in: 2018 14th IWCMC, 2018.

[52] Espressif, ESP-WHO: face recognition with ESP32, OV2640. <https://github.com/espressif/esp-who>, 2018.

[53] M. Abadi, P. Barham, J. Chen, Z. Chen, A. Davis, J. Dean, et al., Tensorflow: a system for large-scale machine learning, in: 12th USENIX Symposium on Operating Systems Design and Implementation (OSDI16), 2016.

[54] Alsing O., Mobile object detection using TensorFlow lite and transfer learning, 2018.

[55] J. Peña Queralta, T. N. Gia, H. Tenhunen, T. Westerlund, Edge-AI in LoRa based healthcare monitoring: a case study on fall detection system with LSTM recurrent neural networks, in: 2019 42nd International Conference on Telecommunications and Signal Processing (TSP), 2019.

[56] V. K. Sarker, J. Peña Queralta, T. N. Gia, H. Tenhunen, T. Westerlund, Offloading SLAM for indoor mobile robots with edge-fog-cloud computing, in: 1st International Conference on Advances in Science, Engineering and Robotics Technology (ICASERT-2019), 2019.

[57] J. Redmon, A. Farhadi, YOLO9000: better, faster, stronger, in: arXiv Preprint arXiv:1612.08242, 2016.

[58] A. Bewley, Z. Ge, L. Ott, F. Ramos, B. Upcroft, Simple online and realtime tracking, in: 2016 IEEE International Conference on Image Processing (ICIP), 2016. <https://doi.org/10.1109/ICIP.2016.7533003>.

[59] G. Lopez, Python traffic counter. <https://github.com/guillelopez/python-traffic-counter-with-yolo-and-sort>, 2018.

[60] R. Madoune Seye, B. Ngom, B. Gueye, M. Diallo, A study of LoRa coverage: range evaluation, channel attenuation model, in: 2018 1st International Conference on Smart Cities and Communities (SCCIC), 2018.

17

Security in low-power wide-area networks: state-of-the-art and development toward the 5G

Radek Fujdiak[1,2], Konstantin Mikhaylov[1,3], Martin Stusek[1], Pavel Masek[1], Ijaz Ahmad[3,4], Lukas Malina[1], Pawani Porambage[3], Miroslav Voznak[2], Ari Pouttu[3], Petr Mlynek[1]

[1]BRNO UNIVERSITY OF TECHNOLOGY, BRNO, CZECH REPUBLIC [2]TECHNICAL UNIVERSITY OF OSTRAVA, OSTRAVA, CZECH REPUBLIC [3]UNIVERSITY OF OULU, OULU, FINLAND [4]VTT TECHNICAL RESEARCH CENTRE OF FINLAND, ESPOO, FINLAND

17.1 Introduction

Low-power wide-area (LPWA) technologies are the most recent addition to the portfolio of the massive machine-to-machine (MM2M) applications for the Internet of things (IoT). As one can see from the term itself, the low energy consumption and the broad coverage are the two distinctive features of the LPWA technologies. Also, the cost of the solution and ease of the deployment, both for the devices and the infrastructure components, are of particular importance.

Despite appearing on the commercial market only a few years ago, the LPWA technologies are now seen as the key driver for the IoT market growth in the coming years. The number of the LPWA chipsets shipped yearly is estimated to exceed 100 million units [1] and the number of the machines connected with these technologies is expected to demonstrate the compound annual growth rate of 109% and reach 1 billion by 2023 [2].

Given this tremendous growth, in a very short time, there will be hundreds, if not thousands of LPWA networks-based IoT devices deployed around us to address the diverse applications and use cases. However, can we rely on these systems? Moreover, are they safe and secure to use? Also, if not, how can we fix them? These questions are especially crucial due to the fast development cycle and short laboratory-to-market path for many LPWA technologies and commercial LPWA-based applications and solutions.

Therefore, in what follows we approach these issues by providing a comprehensive overview and the detailed discussion of the critical security aspects for LPWA technologies in general and the security solutions underlying the three dominant state-of-the-art (SotA)

LPWAN Technologies for IoT and M2M Applications. DOI: https://doi.org/10.1016/B978-0-12-818880-4.00018-1

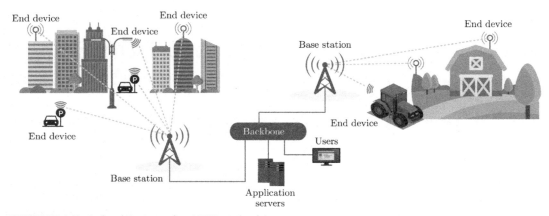

FIGURE 17–1 Typical architecture of an LPWA network.

LPWA technologies, namely long-range wide-area network (LoRaWAN), Sigfox, and narrowband-Internet of things (NB-IoT) [2], in particular. The discussion is followed by the analysis and discussion of the ways how the security of the LPWA solutions may be improved in the future.

17.1.1 Low-power wide-area architecture

Albeit the LPWA technology market is still very fragmented and composed of several dozen different technologies, the technical solutions underlying many of them have much in common. Specifically, many LPWA technologies feature the architecture similar to that illustrated in Fig. 17−1.

A conventional LPWA network features the star-of-stars topology with three major actors:

- The machines, which we term end devices (EDs)[1] and which are typically represented by resource-limited radio-enabled machines such as, sensors and actuators. Among the most common limitations for LPWA EDs are their limited processing capabilities (due to the push for ED cost minimization) and their energy budget (for the EDs powered with batteries). Also, to address the scalability constraints, the LPWA network operators often limit the amount of uplink or downlink traffic for particular EDs (e.g., in the form of monthly uplink/downlink data traffic, the share of time the ED can operate in the channel, or the number of uplink/downlink packets transferred per a unit of time).

---

[1] For the sake of consistency throughout the chapter we use the terms "end device" or "device" to disambiguate an IoT device, and "gateway (GW)" to denote the first element of network infrastructure, with which the ED communicates using a nonlicensed band LPWA technology. In the case of licensed LPWA technologies, we used more common terms user equipment (UE) and evolved NodeBs (eNB) instead of ED and GW, respectively. The terms "end-user" and "subscriber" are used interchangeably to denote an external human or machine prosumer of the IoT data.

The EDs typically communicate only with the GW, while communication between the individual EDs or between an ED and a third-party system is typically not supported.

• The LPWA network core is typically composed of one or multiple GWs, which are connected through an Internet protocol (IP)-based backbone link to a server or cloud, which manages the LPWA network. Typically, a wired interface (e.g., Ethernet or optical fiber cable) is used as the physical layer for backbone, albeit there are also commercial solutions featuring the broadband wireless backbone. Also, the LPWA network core may encapsulate other elements such as the dedicated authorization server or the server storing the data and managing the access to the data by external systems and services. These may also be integrated with the network-managing server or, in some cases, even the GW. Most often, the elements of the LPWA network core are powered from mains and thus are not significantly restricted for their energy consumption.

• The third component of an LPWA application is the various external systems and services —end-users and subscribers, which communicate with the LPWA network core to obtain the data sent by EDs in the uplink or to inject the data that need to be delivered to EDs in the downlink. The communication between end-users and LPWA core are typically carried using the IP-based interfaces and using special application programming interfaces.

17.1.2 Low-power wide-area technology—security and challenges

From the discussion above, it becomes clear that an LPWA system has rather complex architecture and several different interfaces, all of which have to be secured. Specifically:

• The over-the-air interface between EDs/UEs and GWs/eNBs,
• The wired/wireless interface used to interconnect the components of the core LPWA network (e.g., the GWs/eNBs and the data server), and
• The interfaces between the LPWA core network and the end-user application servers/ services.

Note that the two latter interfaces typically feature fully functional IP-based communication and thus can rely on the well-established security and authorization protocols developed for the Internet. Therefore, in what follows, we will primarily focus on the security aspects of over-the-air (OTA) LPWA communication.

Depending on their main purpose, the potential attacks on LPWA can be divided into several major subgroups:

• Data-focused attacks: These are the attacks focusing on accessing the data circulating in LPWA networks or on injecting the counterfeit data in the LPWA networks.
• Denial of service (DoS) attacks: The main purpose of these attacks is to intentionally obstruct or even completely prevent the data transfers in the LPWA network. Jamming, hijacking of destroying the EDs, or making the GWs use their limited resources (e.g., the radio spectrum) inefficiently are just a few illustrative examples of these.

- Monetary-focused attacks: Another possible target for an attacker might be to inquire the maximum monetary losses for LPWA network operators and/or ED owners. To give an example, blocking the home network, thus forcing an ED to use roaming is just one possibility. Alternatively, an attacker may attempt to get his/her data sent at no cost or putting the costs to another user.
- Hardware-exploitation attacks: These are the attacks intended to obtain control over the elements of LPWA networks in order to reallocate them for other tasks (e.g., cryptocurrency mining, spying, or exploiting them to launch a DoS attack).
- Hybrid attacks: These attacks attempt to approach several goals at a time.

The two former attack types are rather common and can take place in any wired or wireless network (e.g., the Internet). The monetary attacks are also possible only for the systems involving billing. Although attacks of this type are also feasible in the other wireless networks (e.g., cellular or commercial Wi-Fi, to give an example), we are not aware of any studies investigating the tolerance of LPWA technologies to this type of attack. Given the limited capabilities of many LPWA devices, gaining control over the hardware operation (i.e., the hardware-exploitation attack) of the individual devices may not be the most appealing goal for an attacker. However, if there will be a possibility to get control over thousands or even millions of devices through a single attack—it will be certainly exploited. Another potential goal for hardware-exploitation attack is the elements of LPWA network backbone infrastructure, especially the ones possessing substantial processing power.

Among the hybrid attacks, which focus on several of the discussed above goals, the energy depletion attacks are especially relevant for IoT and LPWA networks. These attacks focus on increasing the consumption of the IoT devices and LPWA network infrastructure, thus reducing the lifetime of the network and increasing the operation costs. These attacks are much less common since they make sense only for the systems with limited energy availability. Previously, the energy attacks have got limited attention in the context of the wireless sensor networks (WSNs). Nonetheless, the WSNs differ from LPWA networks, and energy depletion LPWA network attacks have been almost not investigated to this day (the only paper partially addressing this problem for LoRaWAN known to the authors is [3]).

17.2 Security features of state-of-the-art low-power wide-area technologies

In what follows, we discuss the main security features of the three SotA LPWA technologies. We assume that the basics of the technical solutions and the key terminology for the three LPWA technologies focused (i.e., Sigfox, LoRaWAN, and NB-IoT) are already familiar to the readers. Should this information be needed, it can be found from the initial chapters of this book or from various external sources.

For each of the technologies, we first introduce the critical security credentials and detail the way these are produced and distributed. Next, we focus on the identity protection approaches utilized within each technology. The discussion of the authentication mechanisms for ED, network, individual messages, and even the end-users follow.

Next, we focus on the security procedures related to the actual data transfer. We detail the mechanisms intended to ensure data integrity and confidentiality, which protect data against unauthorized changing and access, respectively. To characterize the network operators' role and capability to access the data, we detail the end-to-middle (E2M) and end-to-end (E2E) security solutions next.

Finally, we go through other relevant parameters such as forward secrecy (protection of past sessions against compromises of credentials happening in future), replay protection (robustness against replay attacks), delivery reliability (mechanisms preventing data loss), prioritization (possibility to prioritize selected messages), updatability (possibility to change the security credentials or even the algorithms), network monitoring and filtering (capabilities of the network to detect abnormal behavior or an attack), algorithm negotiation (support of more than one security algorithm), class break resistance (possibility of an ED to keep secrecy of its communication if the credentials of another ED are compromised), certification processes (specific procedures related to ED connection to the LPWA network, the commercial availability of devices and infrastructure elements), and IP support (which provides interoperability, but may introduce some new vulnerabilities).

17.2.1 Sigfox

The Sigfox security design is based mainly on symmetric cryptography. Moreover, some security features, such as encryption, are on-demand and not provided by default. This section compiles results of documentations analysis, best practices, and even reverse engineering to generate the complete picture of the main security features of Sigfox technology [4–6].

- *Credential and their provisioning.* The Sigfox technology uses three main credentials: device identification (*ID*, 4 bytes(B)), porting authorization code (*PAC*, 16 B), and key (sometimes referred also as network access key, *NAK*, 16 B). There are three different ways how credentials are delivered, namely via (1) Sigfox Central Registration Authority (CRA), (2) secure element (SE) providers, and (3) the Sigfox build platform (used exclusively for the products in development). The procedures for (1) and (2) are shown in detail in Figs. 17-2 and 17-3, respectively. Since (3) is not intended to be used for any commercial devices, we do not address it here.

  The principal difference between (1) and (2) is the location, where the security credentials are stored. In the former case, the credentials are kept in local memory (e.g., in radio transceiver or a location accessible by the processor of the IoT device). In the latter case, the credentials are stored in a specialized chip, such as STSAFE-A1SX [7].

  The *ID* acts as the unique identifier of an ED in the Sigfox network. The 4 B ED ID results in the address space of 2^{32}, that is, 4,294,967,296 unique addresses. The Sigfox report from May 2018 [8] estimates the number of deployed Sigfox devices to reach 3 million, which is below 0.1% of the total maximum limit. Note that the device *ID* is typically neither encrypted nor protected.

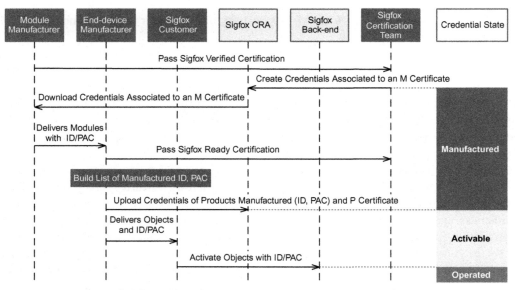

FIGURE 17–2 An overview of Sigfox credentials provisioning processes without secure element.

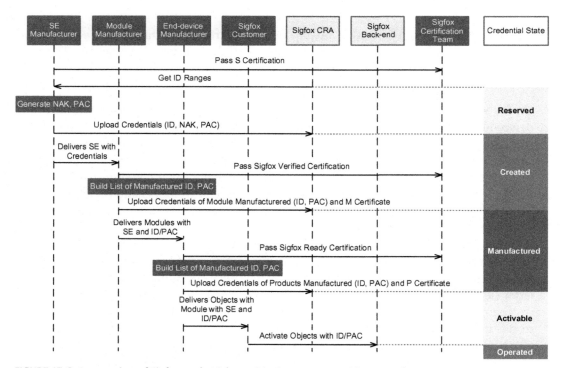

FIGURE 17–3 An overview of Sigfox credentials provisioning processes with secure element.

The *NAK* is the key used to encrypt all the communication, which is known only to the manufacturers/CRA/SE providers. The *PAC* is a one-time code used to confirm ED ownership in the process of its registration as a part of its assignment to a particular device group. The *PAC* is regenerated after each registration and delivered to the group owner and must be kept confidential, since knowing it, one can virtually "hijack" the ED by connecting it to a third-party device group.

In addition to *ID*, *NAK* and *PAC*, two other keys are specified in the Sigfox documentation—the authentication key (*Ka*, 16 B) and the encryption key (*Ke*, 16 B). By default, *Ka* is the same as the *NAK*. The difference between the two is that the *NAK* is used mostly in the registration processes, while *Ka* is employed in the cryptography processes. The *Ke* key is used to encrypt the frames exchanged between the devices and the Sigfox core network, if encryption service is activated. The *Ke* is derived by applying Advanced Encryption Standard AES_{ECB}-128 to the *NAK*.

- *Identity protection.* The Sigfox solution for identity protection is based on combination of unique *ID* and *PAC*. *ID* is typically transferred via air in nonencrypted form. Therefore, the ED identity is protected primarily by *PAC*. The *ID* of the ED is static and does not change through the lifetime of the device. The *PAC* changes only when a device is registered to a new group. Note that there is neither temporary mobile subscriber identity (TMSI) service/protocol nor any equivalent managing identity protection.
- *Authentication (device, network, message, and subscriber).* Each Sigfox device has a unique identifier *ID*, which is not protected. Modification of *ID* is not supported either. To register an ED at the Sigfox cloud service, one has to possess the device *ID* and a valid *PAC*.

  The network is identified by unique private *NAK*, which is known only to the manufacturers/CRA/SE and has to be put in the device for it to communicate with the network. Note that no mechanism for changing the *NAK*, should it ever be compromised, is available. The authentication of each message is provided via AES_{CBC}-128(*Ka*, Data).

  The cloud authenticates the subscriber via an email and a password, which must have at least eight characters containing at least one lower case (26 valid characters), one upper case (26 valid characters), one digit (10 valid characters), and one symbol (32 valid characters). Since the login is not protected against multiple tries, the brute-force attack is possible. Considering, for example, attacker having the maximal distributed power equal to that of distributed.net (1049.20 gigakeys(GKs)/s) [9], the attack would need at maximum 94^8/1049.20 GKs = 96.83 minutes to break the weakest possible password. Increasing the password to 10 characters would increase the time to break the password to over 1.6 years, while brute-force breaking of a 12-character password would require over 14,000 years.
- *Data integrity.* The data integrity is ensured by the 2 (in uplink)/1 (in downlink) bytes of cyclic redundancy check (CRC) produced by the polynomial function. For the uplink the polynomial function is $x^{16} + x^{12} + x^5 + 1$, result is XOR-ed with 0xFFFF. For the downlink the polynomial function is $x^8 + x^5 + x^3 + x + 1$ and the result is put in CRC field.
- *Data confidentiality.* By default, the Sigfox radio messages between an ED and a GW are not encrypted. The secured IP connection is established only between a GW and an

network server (NS) through virtual private network/secure sockets layer (VPN/SSL) (over Eth/DSL/4G/long-term evolution (LTE) depending on the availability and Sigfox infrastructure) and between the NS and a subscriber via HTTPs protocol. However, Sigfox introduced new service in Q4/2017 for the EDs to provide E2M security. This service is available at extra cost on request. Moreover, the service must be supported by both the local Sigfox operator and by the ED. This service implies the use of AES_{CTR}-128 encryption with *Ke* derived from *NAK* (*Ka*). The requirement for Advanced Encryption Standard (AES) is to support encryption of up to 2^{64} blocks of data (approximately $9.22 \cdot 10^{18}$ messages) before the key exchange need arises. Given the typical limit of 140 Sigfox messages per day, the need for key exchange would arise in 1.80×10^{14} years. However, should the key get compromised, no mechanism for changing it is available.

- *E2M security and end-to-end security*. The E2M security is provided via integrity, authentication, and encryption function, as discussed above. The E2E security is not provided and needs to be implemented by the application developer.
- *Forward secrecy*. Sigfox communication protocol does not use any forward secrecy or single-session protection.
- *Replay protection*. To fight a packet replay attack, the message counter value is encrypted and sent in each message. The counter field is 12 bit wide and can encode 4095 values. For an ED sending 140 messages per day, the counter would repeat in approximately 29.25 days. Except the counter, two timestamps are used. The former ($T M_0$) is added to each received message by the GW. The latter ($T M_1$) is generated once Sigfox NS receives the message. The two timestamps are used to compute the NS-GW delivery latency to protect the message against replay attack between the GW and the NS. However, the "stop-and-replay" attack (i.e., jamming the EDs, changing the message, and resending in selected time) is still feasible.
- *Reliable delivery*. Some Sigfox operators claim to achieve the 99% reliability [10]. This is achieved by combining the message repetition and random carrier selection. Moreover, Sigfox also employs cooperative reception, allowing multiple base stations (on average—three) receive each message, to make a transmission more reliable. As a result, the Sigfox solution implements time, frequency, and spatial diversity.
- *Prioritization*. Sigfox technology does not offer any packet prioritization mechanism.
- *Updatability*. The limited downlink capabilities (typically only four messages of 8 bytes per day) offer minimal capabilities for any update. Moreover, the update procedures are out of the scope of the Sigfox specification and must be handled through the application layer.
- *Network monitoring and filtering*. Each GW implements preliminary uplink message validation and CRC check. However, this is the NS, which handles the final message processing (computes and validates the data authenticity, checks the sequence number and timestamps, removes the duplicates) and billing.
- *Algorithm negotiation*. Sigfox EDs use predefined algorithms, and there is no algorithm negotiation mechanism.

- *Class break resistance.* There is no secret/private key sharing between different EDs. All EDs use unique device *ID* and keys. Nevertheless, the same cryptography algorithms are used by each device.
- *Certified equipment.* Sigfox is a proprietary solution, which operates in the unlicensed bands. However, there are strict policy and regulations in place for deploying new EDs and GWs. The Sigfox control authority must always certify an ED or any other device entering the network. The Sigfox EDs are openly available on the market. The network infrastructure (e.g., GWs) is provided by Sigfox only to the selected partners.
- *IP network* Sigfox implements non-IP data delivery (NIDD) over the air. The radio frames are further encapsulated in the IP packet by the GW and delivered to NS via VPN/SSL, which should provide sufficient security in this part of the network.

17.2.2 Long-range wide-area network

Likewise Sigfox, LoRaWAN security solution is based on symmetric cryptography. The following discussion is based on the analyses of the recently released documentation for LoRaWAN specifications 1.1 [11,12] and 1.0.3 [13], and LoRaWAN back-end specification [14]. Note that in what follows, we primarily focus on the public LoRaWANs, in which the ED owner and the network service providers are different parties.

- *Credential and their provisioning.* For communicating in the network, a LoRaWAN device requires to have a unique device address (DevAddr) and two keys: the network session key (NwkSKey), securing the communication between the ED and the NS, and the application session key (AppSKey), encrypting the application payload. The LoRaWAN specification offers two ways to obtain and distribute these keys. The former option is named activation by personalization (ABP) and implies that credentials are provided to the ED and NS offline during production deployment.

  An alternative is the over-the-air activation (OTAA), enabling to generate the needed keys in the process of connecting an ED to the network. Note that before specification 1.1 both AppSKey and NwkSKey were derived using the same application key (AppKey), which, thus, has to be known to the network operator. Specification 1.1 introduces another AES-128 root key, called NwkKey, which is used to derive the FNwkSIntKey, SNwkSIntKey, and NwkSEncKey session keys. This key is shared with a network operator to manage the Join procedure and to derive the network session keys, while the AppKey is kept private. Note that the storage of the keys is outside of the LoRaWAN specification's scope.
- *Identity protection.* The identity protection in LoRaWAN is partial and bases only on the DevAddr. The device addresses are static for ABP-devices but may be changed for OTAA-EDs. More advanced solutions for identity protection, such as TMSI, are currently missing.
- *Authentication (device, subscriber, and network).* The subscriber authentication is out of LoRaWAN specification scope. For an ABP device, no explicit authentication neither for devices nor for network is supported.

Each OTAA-ED node has to have the 64-bit globally unique identifier called device identifier (DevEUI), the application identifier (AppEUI) which uniquely identifies the application in the context of the LoRaWAN network, and the unique 128-bit AppKey. The ED sends the join request message with AppEUI, DevEUI, random DevNonce, and adds the message integrity code (MIC) computed using AppKey. The NS checks the MIC and generates keys. The mutual authentication is ensured by the knowledge of the AppKey on both sides.

The specification 1.1 improves OTAA procedure by modifying JoinAccept MIC to prevent the replay attack. Further, all nonces are not random numbers, but the counters. Also, from specification 1.1, the OTAA may be managed by the dedicated join server. The mutual authentication is still based on the secrets, shared between the devices and the join server. The knowledge of secrets is proved by computing and checking the cipher-based message authentication code (CMAC) functions (i.e., MIC).

- *Data integrity.* It is ensured by the 32-bit MIC produced by the CMAC algorithm using the 128-bit AES encryption. The 4-byte MIC is calculated from a MAC payload, and the NwkSKey key shared between the NS and the ED. This code is added after the MAC payload. To avoid a packet replay attack, a frame counter is used. However, the payload could be flipped due to the fact that AES-counter mode (AES-CTR) mode does not support data authentication.
- *Data confidentiality.* Data encryption is ensured by the 128-bit AES encryption in the CTR mode. Application payloads are encrypted using the AppSKey. For pre-LoRaWAN 1.1 networks, the AppSKey needs to be provided to the NS, but for networks supporting later specification versions, this key can be kept confidential.
- *E2M Security and E2E Security.* The LoRaWAN 1.0 provides only E2M security due to the key sharing. However, LoRaWAN 1.1 provides full E2E Security. The networks composed of both LoRaWAN 1.0 and 1.1 EDs have to use the security solution of LoRaWAN 1.0.
- *Forward secrecy.* LoRaWAN does not imply any forward secrecy or single-session protection.
- *Replay protection.* The replay protection is handled by using message counters. The specification version 1.1 improves OTAA (Join procedure) by modifying JoinAccept MIC to prevent the replay attack. Similarly to Sigfox, the timestamps of packet reception by GW and NS can be used to prevent replay attack in the backbone.
- *Reliable delivery.* Most often the LoRaWAN packets are sent with no repetitions or acknowledgments, even though both these mechanisms are supported.
- *Prioritization.* Currently, neither quality of service (QoS) nor prioritization is supported.
- *Updatability.* The session keys of OTAA-EDs can be updated by making and ED rejoin the network. The preshared master key AppKey cannot be updated using conventional LoRaWAN procedures.
- *Network monitoring and filtering.* Likewise for Sigfox, each GW does the preliminary message check, and the NS handles the final filtering and validation.
- *Algorithm negotiation.* LoRaWAN EDs use predefined algorithms, and no algorithm negotiation mechanism is in place.

- *Class break resistance.* There is no secret/private key sharing between EDs. Each device uses a unique identifier and keys.
- *Certified equipment.* LoRaWAN is a standardized solution based on proprietary modulation, which runs in an unlicensed band. The LoRaWAN Alliance certifies the equipment. Note that both the EDs and the GWs/NSs are openly available for purchase.
- *IP network.* LoRaWAN use NIDD over the radio layer. Communication between a GW and NS is via VPN/SSL.

17.2.3 Narrowband-Internet of Things

The security mechanisms of NB-IoT resemble that of the LTE system, from which it had originated. In what follows the comprehensive discussion of NB-IoT security mechanisms encompassing UE authentication, nonaccess stratum (NAS) security and access stratum (AS) security setup is provided.

- *Credential and their provisioning.* NB-IoT authentication procedures are based on the subject (e.g., the UE) first asserting a particular identity so that the other party (e.g., the network) can then verify that the subject's credentials match that identity. There is an underlying assumption, therefore, that the identifier is (1) unique and (2) permanently mapped to a particular subject. If this is not the case, then any authentication of that identity might potentially be subverted.

  One example of a potentially subvertible identifier is the International Mobile Equipment Identity (IMEI) used in Third Generation Partnership Project (3GPP) networks, including the NB-IoT. The registration protocol ensures that the network receives the IMEI provided by the mobile device. In theory, the IMEIs are unique, but in reality, false and duplicate IMEIs do appear. In contrast, the International Mobile Subscriber Identity (IMSI) embedded in 3GPP Subscriber Identity Module (SIM) or Universal SIM (USIM) represents a reliable, unique identifier. The certification programs are intended to ensure that both the IMSI itself and the associated subscriber authentication key Ki are securely provisioned and stored.
- *Identity protection.* Some protocols include privacy-preserving measures to minimize the use of permanently allocated identifiers, which could be intercepted and correlated with device activity over time. An example of this is the TMSI allocated by 3GPP networks to address the mobile device instead of the IMSI, which is only used when a device connects to the network.
- *Authentication (device, subscriber, and network).* The process of authentication of UE in NB-IoT/LTE network is conducted using the authentication and key agreement (AKA) procedure. The whole process can be divided into three individual steps, which are depicted in Fig. 17−4. Further, all integrity/ciphering keys, which are relevant for this procedure, are listed in Table 17−1.

  When a UE attempts to access the network for the initial attach, the *Attach Request* message containing IMSI, UE network capability (i.e., the supported security algorithms) and $KSI_{AS\ ME} = 7$ (Key Set Identifier) are delivered to the mobility management entity

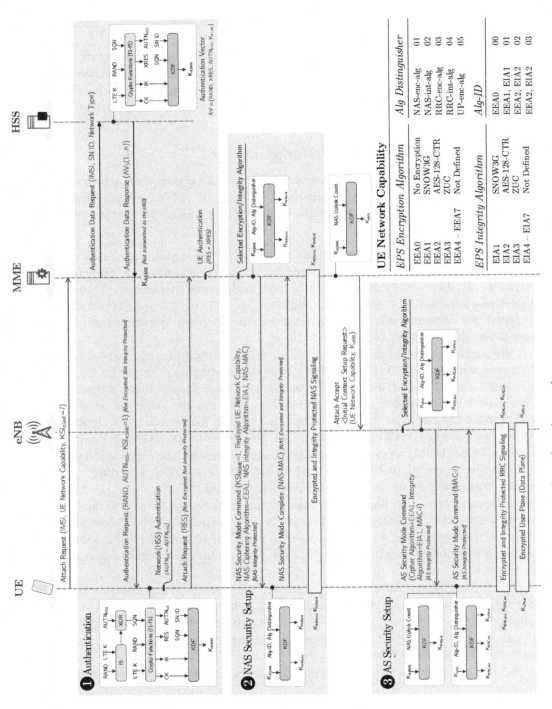

FIGURE 17-4 An overview of NB-IoT communication and security procedures.

Table 17–1 Comparison of LTE/NB-IoT security keys.

Key	Length	Location	Derived from	Description
K	128 bits	USIM, HSS/AuC	—	EPS master key
CK	128 bits	USIM, HSS/AuC	K	Cipher key
IK	128 bits	USIM, HSS/AuC	K	Integrity key
$K_{AS\ ME}$	256 bits	UE, MME, HSS	CK, IK	MME base key
K_{eNB}	256 bits	UE, eNB, MME	$K_{AS\ ME}$	eNB base key
$K_{NAS\ int}$	128/256 bits	UE, MME	$K_{AS\ ME}$	Integrity key for NAS message between UE and MME
$K_{NAS\ enc}$	128/256 bits	UE, MME	$K_{AS\ ME}$	Encryption key for NAS messages between UE and MME
K_{RRCint}	128/256 bits	UE, eNB	K_{eNB}	Integrity key for RRC messages on SRB between UE and eNB
K_{RRCenc}	128/256 bits	UE, eNB	K_{eNB}	Encryption key for RRC messages on SRB between UE and eNB
K_{UPenc}	128/256 bits	UE, eNB	K_{eNB}	Encryption key for user IP packets on DRB between UE and eNB

(MME). The $KSI_{AS\ ME} = 7$ indicates that no access security management entity key ($K_{AS\ ME}$) is available on the UE. When MME receives an access request, the user is identified using his IMSI and request to home subscriber server (HSS) is issued. As a response, the HSS generates a set of authentication vectors (AVs) = {RAND, XRES, $AUTN_{HS\ S}$, $K_{AS\ ME}$}. The derivation of all keys in crypto functions blocks in this step is based on AKA procedure utilizing MILENAGE or TUAK algorithms. The former is a suite of mathematical functions $f1, \ldots, f5$, based on symmetric-key primitives, specifically the AES-128. The TUAK was proposed as an alternative to MILENAGE by the ETSI SAGE committee. Unlike its competitor, TUAK relies on the truncation of Keccak's internal permutation [15,16].

The $K_{AS\ ME}$ generation consists of two steps for both UE and HSS. For the UE, a cipher key (CK) and integrity key (IK) are derived from the LTE key (K), a 48 bits sequence number (SQN), and the random number (RAND). The LTE key is stored in the USIM and cannot be changed, whereas the HSS generates the SQN and RAND. SQN, however, is not transmitted directly, but it is concealed in $AUTN_{HS\ S}$. UE, in order to derive back the SQN, passes the RAND and LTE K on the input of $f5$ (part of the MILENAGE algorithm) and the output is then XOR-ed with $AUTN_{HS\ S}$, which produces the required SQN value.

Further, in the second step, CK and IK are utilized to generate $K_{AS\ ME}$, which is crucial for the following procedures. The $K_{AS\ ME}$ is derived from CK, IK, SQN, and serving network ID (SN ID) via the secure hash algorithm (HMAC SHA-256) which is also used in all key derivation function blocks in the rest of the system [17]. The same procedure of $K_{AS\ ME}$ generation is also conducted on the HSS. However, in this case, LTE key and IMSI are provisioned by the authentication center (AuC) [18].

Now the authentication tokens $AUTN_{UE}$ generated by the UE and $AUTN_{HS\ S}$ received from the MME are compared. If the authentication succeeds, RES is dispatched to the

MME, where the expected authentication result (XRES) received from the HSS is compared with the RES from the UE. After this process, UE and network have authenticated each other, and they both share the same $K_{AS\ ME}$ key, though the real key has not been directly transferred over the air. The actual value of $K_{AS\ ME}$ is transferred only between HSS and MME in AVs. However, more than one AV containing $K_{AS\ ME}$ can be generated in the response. Thus the $K_{AS\ ME}$ is replaced with $KSI_{AS\ ME}$ in *Authentication Request* exchanged between MME and UE. The $KSI_{AS\ ME}$ value then serves as an index of $K_{AS\ ME}$ in a lookup table stored on both UE and HSS. The $KSI_{AS\ ME}$ consists of three bits, thus values $1-7$ ($000-111$ in binary) can be used [19].

The AVs are further passed to the MME and stored. The MME selects one of them to perform an authentication process with the UE. Next, 128 bits random number RAND and authentication token $AUTN_{HS\ S}$ are forwarded from the MME to UE, where authentication result (RES), authentication token ($AUTN_{UE}$), and $K_{AS\ ME}$ are computed utilizing the EPS AKA algorithm [20,21].

Next step in the authentication procedure is the NAS security setup. In this procedure, NAS security keys to be used during NAS signaling are derived from $K_{AS\ ME}$. NAS security setup consists of *Security Mode Command* and *Security Mode Complete* messages exchange between the MME and the UE. The whole process starts when the *Security Mode Command* message from the MME is delivered to the UE. During the first step, the MME selects NAS security algorithms (Alg-ID, Alg Distinguisher)[2] and derives integrity ($K_{NAS\ int}$) and encryption ($K_{NAS\ enc}$) keys from $K_{AS\ ME}$. Next, MME applies the $K_{NAS\ int}$ to the *Security Mode Command* message producing the message authentication code (NAS-MAC). The authentication code is generated utilizing the evolved packet system integrity algorithm (EIA) algorithm selected by the eNB. Later on, the MME conveys the message which includes the selected security algorithm (NAS ciphering algorithm—EEAx, NAS integrity algorithm—EIAx) and NAS-MAC to the UE. Since the UE is not aware of the selected encryption algorithm yet only message integrity is protected.

When the *Security Mode Command* message is delivered, the integrity of the message is verified by the UE applying the same integrity algorithm used by eNB. At the same time, the UE utilizes the NAS integrity/encryption algorithm to generate security keys ($K_{NAS\ int}$ and $K_{NAS\ enc}$) from the $K_{AS\ ME}$. Later on, the *Security Command Complete* message is encrypted (utilizing the EEA algorithm selected by eNB) with the $K_{NAS\ enc}$ and authentication code NAS-MAC is generated utilizing the $K_{NAS\ int}$ and attached to the encrypted message. Finally, the encrypted and integrity protected message is forwarded to the MME. Once the integrity of the received message is verified and the message is decrypted using the security keys ($K_{NAS\ int}$ and $K_{NAS\ enc}$), the NAS security setup is completed [18,20].

When the NAS security setup is finished, AS security setup procedure between UE and eNB begins. However, since the $K_{AS\ ME}$ is not transferred directly to the eNB, it

---

[2] In the example depicted in Fig. 17−4, both EEC and EIC algorithms are set to 1, but in a real-world application, the value is selected by MME.

cannot derive the K_{eNB}. MME thus generates K_{eNB} from $K_{AS\ ME}$ and forwards it to the eNB via *Attach Accept* message. When the K_{eNB} is derived, the AS security setup may continue. The AS security ensures the integrity and encryption of radio resource control (RRC) messages. Similar to the NAS security, the AS security setup procedure consists of the round trip of RRC signaling messages (*Security Mode Command* and *Security Mode Complete*) and begins when the *Security Mode Command* is delivered to the UE. On the eNB, AS security algorithm (Alg-ID, Alg Distinguisher)[3] is selected and used for IK (K_{RRCint}) and encryption key (K_{RRCenc}) generation. Both keys are derived from K_{eNB} and are used for ensuring RRC signaling messages integrity and encryption, respectively. The third key (K_{UPenc}) is also derived from K_{eNB} and used to encrypt data transfers for user plane. Then K_{RRCint} is applied to the *Security Mode Command* message a message authentication code for integrity (MAC-I) utilizing the selected EIA algorithm. Now the *Security Mode Command* message containing the selected AS security algorithm and the MAC-I is transmitted to the UE. When the UE receives the message, its integrity is verified by applying the AS integrity algorithm selected by the eNB. Further, AS integrity/encryption algorithm is used to generate AS security keys (K_{RRCint}, K_{RRCenc}, and K_{UPenc}). Finally, *Security Command Complete* message, containing MAC-I, is generated using RRC IK, and the message is forwarded toward the eNB. When the eNB successfully verifies the integrity of the received message, the AS security procedure is completed [18].

- *Data integrity and control integrity.* There are four algorithms for confidentiality and integrity protection in NB-IoT: NULL (no security algorithm used), SNOW3G (128-bit stream cipher) [22,23], AES (128-bit block cipher in CTR mode) [24,25], and ZUC (Zu Chongzhi) (128-bit stream cipher) [26,27]. For integrity protection, EAIx commands are used, where the x stands for the code of selected algorithm.

Turning on the security feature allows the systems to use the secret keys discussed above to execute encryption/decryption. This allows to protect the confidentiality and integrity of the RRC messages between UE and the eNB, as well as of the NAS messages between UE and the background services. The protection is based on the communication layer and is transparent to users. In essence, the security is built on the predistributed secret key stored in the SIM card.

Network devices can disable the NAS security feature. Then after reaching the eNB, the transmission of the message will be in clear text. Adversaries can use the specialized devices to intercept the data from the base station to the core network. Because of those shortages in communication layer, some applications that demand high-level security use E2E security enhancement on the application layer. The enhancement is based on certificates. It can achieve certificate issuing, secret key agreement similar to TLS [28], and E2E secure communication.

---

[3] In the example depicted in Fig. 17–4, both EEC and EIC algorithms are set to 1, but in a real-world application, the value is selected by eNB.

- *Data confidentiality.* NAS commands control data confidentiality. Also, the same three-cipher algorithms based on symmetric cryptography are used. The selection of the algorithm is driven by EEAx parameter of the command.
- *E2M security and E2E security.* LTE security [29] is based on the AKA procedure, which allows both the UE and the eNB to achieve mutual authentication and to generate session ciphering (CK) and integrity (IK) keys. Different AKA procedures are implemented in the LTE security architecture to support UE access to the evolved packet core (EPC) via non-LTE access networks. When a UE connects to the EPC over the evolved universal mobile telecommunications system terrestrial radio access network, the AKA procedure is performed between the UE and the MME. However, when a UE connects to the EPC via a non-3GPP access network, the authentication is done between the UE and an authentication, authorization, and accounting (AAA) server. If the UE has no preconfigured information, the non-3GPP access network is considered untrusted, and the UE traffic needs to pass through a trusted evolved packet data gateway connected to the EPC by establishing an IPsec tunnel using the Internet Key Exchange Protocol version 2 (IKEv2). If there is preconfigured information, the non-3GPP access network is considered trusted, and the UE and the AAA server will utilize the extensible authentication protocol-authentication and key agreement (EAP-AKA) or improved EAP-AKA (EAP-AKA´) [30].
- *Forward secrecy.* Not implemented by default.
- *Replay protection.* Key hierarchy is used for ciphering and integrity protection of NAS signaling messages between the and the MME.
- *Reliable delivery.* Let us consider a device that sends information to a base station. When a device sends a message, it has no way to know that this message has been correctly received. That is why in many communication protocols, a second message is sent from the base station back to the device, to inform that everything went fine. If one cannot afford to lose the data, the bidirectional communication between the base station and the device is crucial. In NB-IoT, this idea is automatically applied in the MAC layer with what the 3GPP name as the hybrid automatic repeat request (HARQ). To enable low-complexity UE implementation, NB-IoT allows only one HARQ process in both downlink and uplink and allows longer UE decoding time for both narrowband physical downlink control channel and narrowband physical downlink shared channel. Asynchronous, adaptive HARQ procedure is adopted to support scheduling flexibility [31].
- *Prioritization.* NB-IoT uses an S1-based connection between the radio network and the EPC. The connection to the EPC provides NB-IoT devices with support for roaming and flexible charging, meaning that devices can be installed anywhere and can function globally. The ambition is to enable certain classes of devices to be handled with priority to ensure that emergency-situation data can be prioritized if the network is congested.
- *Updatability (device/firmware and Keys/Sec. algorithms).* Out of the box, NB-IoT support OTA firmware updates. However, this functionality is only supported in IP

transmission mode. Symmetric-key management of NB-IoT communication follows the GBA mechanism. Corresponding protocols include 3GPP TS 33.220, 3GPP TS 33.228, 3GPP TS 33.246, and so on. The relatively mature key agreement follows AKA protocol [32]. However, it only handles communication security and is not applied to the secret keys [33].

- *Network monitoring and filtering.* Network can be monitored in real time by the internal processes of the mobile network. As the core of the networks is IP-based, the traffic can be also monitored using firewalls.
- *Algorithm negotiation.* Before the communication over the network can be protected, both the UE and the network need to exchange information on what security algorithms they can use [34]. Evolved packet system part of the network can support multiple algorithms and define three classes of security algorithms [35]: (1) 128-EEA1 and 128-EIA1 based on 3GPP TS33.401 specification [36], (2) 128-EEA2 and 128-EIA2 build on AES [37], (3) 128-EEA3 and 128-EIA3 based on ZUC [38]. The first two classes of the algorithms above need to be supported by all entities in the network, that is, UEs, eNBs, MMEs, etc. The third class is optional for implementation. Furthermore, EPS can be additionally extended to support more algorithms as technology evolves. The algorithms are negotiated at two levels: (1) between UE and the base station (AS level) and (2) between UE and the core of the network (i.e., the MME-NAS level). The network selects the algorithms based on the security capabilities of UEs and the preconfigured list of allowed security algorithms for the entities in the network, that is, MMEs or eNBs. Messages are not protected before security algorithms are agreed, and protection of the signaling has been set up. If UE detects a mismatch between the security capabilities it sent to the network and the configuration received from the network, the UE cancels the attach procedure. In case the security capabilities match, two "Security Mode Command" procedures are used to indicate the selected algorithms and to enable ciphering, and integrity with replay protection. First "Security Mode Command" procedure is handled for the AS, and the second one—for the NAS. From the network side, the MME does select the NAS-level algorithms, and the eNB is responsible for selection of the AS-level algorithms, including the user plane algorithm. If needed, the NAS-level algorithms can be changed through the NAS Security Mode Command procedure. On the other hand, it is not possible to change the AS-level algorithms using the "AS Security Mode Command" procedure.
- *Class break resistance.* No private keys are shared between devices; thus NB-IoT technology is class break–resistant.
- *Certified equipment.* The NB-IoT equipment needs to pass certification. The UE devices are openly available, while the eNBs are available only from a few manufacturers around the globe.
- *IP network.* Both IP and non-IP data can be sent for NB-IoT via NAS using the signaling radio bearers (SRB). This mechanism is designed to facilitate small data transfer, reduce the signaling overhead, and reduce the energy consumption for UE. Small data transfer can further be optimized using NIDD.

17.3 Future vision: Internet of things and 5G core network—security overview

LPWA technology has many attractive features for the growing developments of IoT deployments across multiple sectors such as logistics and transportation, utilities, smart cities, and agriculture. The 4C (capacity, consumption, cost, coverage) model which describes the key characteristics of LPWA network justifies its appropriateness for IoT and M2M applications that require to transmit small chunks of data over long ranges with last longing battery life. The reduced complexity and high scalability in LPWA technologies will also facilitate the less human intervention for the IoT applications of the future.

The tendency of empowering different LPWA technologies in numerous application areas depends on their individual needs. For instance, emerging LPWA applications in manufacturing are identified as machine auto-diagnosis, asset control, and location reporting, monitoring, item tracking, etc. NB-IoT modules can be used to achieve high-precision monitoring and operations within factory premises. The single cloud of Sigfox makes this technology beneficial for various pancontinental tracking applications. LoRaWAN can address resource management in smart agriculture and various utility applications. Likewise, each technology has its advantages in the IoT applications and their deployments. Therefore it is foreseen that 5G wireless mobile communication will lead to a connected world of humans and devices while providing global LPWA solutions for IoT applications. This is not yet entirely clear how exactly will this look like and which technologies will form the 5G basis and which ones will adjoin later in the process of 5G evolution. Nonetheless, this is worth to look at how security for 5G is seen now.

The 5G core network comprises the most important network control elements, mobility, user information, and charging elements and functions. The core network of 4G or LTE consisted of elements such as MME, Policy and Charging Rule Functions, HSS, etc. In some of the core network elements, the data and control parts are bundled together, as shown in Fig. 17—5. One of the major shifts that occur in 5G is cloudification of the core network elements through separating the network control functions from the data forwarding planes. This cloudification logically centralizes the core network elements into high-end servers enabling cost-effective scalability, service provisioning, and availability. The detailed architecture and its elements with description are available in the latest 3GPP specifications [39,40]. The 5G core network is IP-based, ensures QoS and quality of experience, and is more dynamic due to novel technological concepts such as cloud technologies, software-defined networking (SDN), and network function virtualization. However, it will bring forth some potential security challenges, especially critical for the low-power IoT devices.

The Next Generation Mobile Networks consortium has provided several key insights into the possible security challenges in 5G in the form of recommendations, as described in [41]. Few of the main security challenges highly related to LPWA network are

1. *Flash network traffic*: It is projected that the number of end-user devices, for example, IoT ED, will grow exponentially in 5G that will cause significant changes in the network

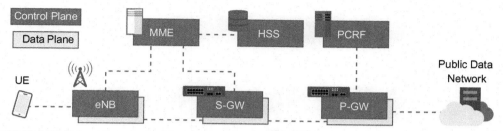

FIGURE 17–5 4G EPC architecture simplified, showing control and data planes.

traffic patterns either accidentally or with malicious intent. Having said that, large swings and burst in traffic will be very common.

2. *DoS attacks*: DoS and distributed DoS attacks can exhaust various network resources such as energy, storage, and computing. Sporadic requests or specifically crafted requests generated toward the network in huge number (e.g., by a massive number of compromised IoT EDs or nonauthentic subscribers) can be highly challenging and can possibly bring the network to a halt.

3. *Security of radio interface keys*: In previous wireless network generations, including 4G, the radio interface encryption keys are generated in the home network and sent to the visited network over insecure links, causing a clear point of exposure of keys.

The first two attacks are particularly very challenging and interrelated. Due to the massive number, it will be tough to differentiate between legitimate requests and malicious requests meant for resource exhaustion attacks. Moreover, most of the signaling involves the core network elements, which are now either physically or logically centralized. Hence, signaling oriented DoS attacks will be one of the critical challenges. This will be more challenging since LPWA IoT devices may not have enough resources to protect the content from integrity or man-in-the-middle attacks through proper encryption or hashing. Signaling oriented challenge in 4G, highlighted in [42], has been difficult to counter due to the penetration of IP traffic in cellular networks. However, 4G networks have mostly distributed control planes where a security loophole in a system will cause local damage (e.g., DoS attack on a control point , e.g., a gateway). In contrast, the core elements in 5G (shown in Fig. 17–6) are centralized; thus security challenges or loopholes will have more adverse consequences since more control points of the network are centralized into singular nodes.

It is worth noting that LPWA IoT devices will mostly comprise low-power embedded systems. In a large-scale analysis of firmware of low-power embedded devices, the authors in [43] show that most of the firmware is ripe with security vulnerabilities. Therefore, low-power embedded systems are highly vulnerable to be masqueraded for security attacks. Since, the domain of IoT is developing and evolving very vast and still yet to be explored in the context of security, the further challenges cannot all be easily identified and responded. Sensitive systems that are supposed to be highly secure can be exposed to security vulnerabilities due to combining and using insecure IoT for different functionalities, specifically

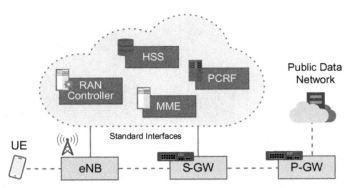

FIGURE 17–6 Simplified network architecture of 5G, showing control and data planes.

when the Internet of hacked things is on the rise. For example, in 2015, 2.2 million BMWs were infected, where the infection allowed remote unlocking of the car. Similarly, 1.4 million Chryslers had vulnerability in their dashboard computers, which allowed hackers to steer the vehicle, apply brakes, and control the transmission [44]. Security loops in such critical systems can cause damages directly to a human.

Connecting infected systems to a network might expose the network to security loopholes [45]. One example is using the compromised devices to launch insider attacks or DoS attacks on the system these devices operate in, for instance, the 5G core network [46]. Resource-constrained devices in significant numbers will require processing and storage in the cloud. The cloud systems will serve diverse and significant number of services and possibly shared through virtualization among different stakeholders. Since the 5G core network is cloudified, IoT will bring many challenges into the signaling plane in the cloud. In LTE, the HSS has been the main point of attacks under the guise of requests for authentication and authorization [45]. 3GPP suggests that IoT devices should periodically update the security credentials; however, the frequent update will increase the burden on the control plane, making it easily prone to resource exhaustion attacks. In such scenarios, compromised IoT devices can induce vulnerabilities into the whole system.

On the other hand, it is worth mentioning that the 5G core network is supposed to be highly resourced with strict access control procedures. For example, the core network elements, such as MME, are now represented as network functions in software. MME is represented as Access and Mobility management Function (AMF) and Session Management Function (SMF), with clearly stated protocols and reference points for interaction among them, as highlighted in the 3GPP specification release 15 [40]. This solves the scalability issues and enables dynamically scaling the resources based on need from highly resourced cloud infrastructures, making resource exhaustion least likely. Furthermore, to effectively handle the signaling, two approaches are discussed by 5GPP [47]. First, using lightweight AKA protocols for massive IoT communication. Second, using group-based AKA protocols to group IoT devices together, which will minimize the individual signaling traffic [47]. Hence, there are a number of group-based authentication schemes for IoT, such as authentication for NB-IoT [48]. The authentication scheme groups IoT devices with similar attributes together and selects a group leader. The group leader aggregates sensitive information and sends it to the core network, which verifies each node independently. The proposed mechanism also preserves identity privacy besides minimizing the signaling involved in authentication in the core network. The same kind of group-based authentication is proposed for vehicular IoT in [49] using the concepts of SDN.

Moreover, secure core network or secure network control points are highly important for smooth, optimized, efficient, and secure work of LPWA IoT devices. Furthermore, robust load balancing mechanisms in the core network will still be needed due to the emergence of big data through IoT to ensure timely resources for authentication and authorization of resources to IoT devices. Secure core network or the network control plane overlooking the behavior of connected things and stats of network components with the capability to remotely monitor the entire ecosystem can increase the security. For instance, the SDN-enabled centralized control plane that can overlook and control the entire network sees the stats of the traffic passing through each node can significantly improve the network security. Compromised LPWA IoT nodes sending excessive data can be recognized at the data plane by using monitoring applications in the SDN application plane. A simple monitoring application in the SDN application plane that gathers statistics from the data plane can help recognize malicious activity within the network by comparing the statistics against predefined thresholds for different services or registered devices. Hence, the centralized monitoring as enabled by the centralized core network in 5G can highly improve the security of not only the network connecting IoT EDs but also that of the IoT EDs themselves (Fig. 17−6).

Acknowledgments

The work was supported by the Academy of Finland 6 Genesis Flagship (grant no. 318927), National Sustainability Program under grant no. LO1401 (by using the infrastructure of SIX Center), Ministry of Education, Youth and Sports under grant no. LM2015070, and LPWAN evolution project of the University of Oulu. The work of Radek Fujdiak was financed by the Ministry of Interior under grant no. VI20172019057. The work of Konstantin Mikhaylov was also supported by the mobility project MeMoV (no. CZ.02.2.69/0.0/0.0/16_027/00083710) of Czech Ministry of Education, Youth and Sport, funded by European Social Fund. The work of Ijaz Ahmad was also supported by the Jorma Ollila Grant. The work of Lukas Malina was financed by the Ministry of Industry and Trade under grant no. FV40340.

References

[1] T. Niwa, LPWA market is still in early stage. <https://iotbusinessnews.com/2018/10/11/68577-lpwa-market-is-still-in-early-stage/>, 2018 (accessed 14.05.19).

[2] E. Pasqua, LPWAN emerging as fastest growing IoT communication technology. <https://iot-analytics.com/lpwan-market-report-2018-2023-new-report/>, 2018 (accessed 14.05.19).

[3] K. Mikhaylov, R. Fujdiak, A. Pouttu, V. Miroslav, L. Malina, P. Mlynek, Energy attack in LoRaWAN: experimental validation, in: Proceedings of 14th International Conference on Availability, Reliability and Security, ARES19, 26−29 August 2019, Kent, Canterbury, UK, ACM, 2019, 175, 1−6.

[4] Sigfox, Sigfox technical overview: technical specification, 2017.

[5] Sigfox, Sigfox security white paper: technical specification (revision 34), 2017.

[6] Sigfox, Sigfox system description: technical specification (revision 04), 2017.

[7] STMicroelectronics, STSAFE-A1SX: Data brief. <https://www.st.com/resource/en/data_brief/stsafe-a1sx.pdf>, 2017 (accessed 14.05.19).

[8] Sigfox, 3 million devices connected milestone: blog post, <https://www.sigfox.com/en/node/775>, 2018 (accessed 14.05.19).

[9] Distributed.net, The largest computer on earth. <http://cowpie.distributed.net/rc5-proxyinfo.html>, 2019 (accessed 14.05.19).

[10] Rakon, Thinxtra, & Sigfox, Sigfox: your questions answered. <https://www.rakon.com/component/docman/doc_download/499-rakon-thinxtra-sigfox-your-questions-answered>, 2017 (accessed 14.05.19).

[11] LoRa Alliance, LoRaWAN™1.1 specification, 2017, white paper.

[12] LoRa Alliance, LoRaWAN™1.1 regional parameters technical specification (final release), 2017.

[13] LoRa Alliance, LoRaWAN™1.0.3 specification, 2018.

[14] LoRa Alliance, LoRaWAN™ Backend Interfaces 1.0 technical specification (final release), 2017.

[15] J. Liu, Y. Yu, F.-X. Standaert, Z. Guo, D. Gu, W. Sun, et al., Small tweaks do not help: differential power analysis of MILENAGE implementations in 3G/4G USIM cards, in: Proceedings of 20th European Symposium on Computer Security, ESORICS15, 21−25 September 2015, Berlin, Germany, Springer-Verlag, Berlin, Heidelberg, 2015, pp. 468−480.

[16] S. Alt, P.-A. Fouque, G. Macario-rat, C. Onete, B. Richard, A cryptographic analysis of UMTS/LTE AKA, in: M. Manulis, A.R. Sadeghi, S. Schneider (Eds.), ACNS 2016: Applied Cryptography and Network Security, Springer-Verlag, Berlin, Heidelberg, 2016, pp. 18−35.

[17] F. Sharevski, Mobile Network Forensics: Emerging Research and Opportunities: Emerging Research and Opportunities, IGI Global, Hershey, PA, 2019.

[18] NMC Consulting Group, LTE security II: NAS and AS security. <https://www.netmanias.com/en/post/techdocs/5903/lte-security/lte-security-ii-nas-and-as-security>, 2015 (accessed 14.05.19).

[19] M. Ouaissa, A. Rhattoy, M. Lahmer, Analysis of authentication and key agreement (AKA) protocols in long-term evolution (LTE) access network, in: A. Kalam, S. Das, K. Sharma (Eds.), Advances in Electronics, Communication and Computing, Springer-Verlag, Berlin, Heidelberg, 2018, pp. 1−9.

[20] NMC Consulting Group, LTE security I: LTE security concept and LTE authentication. <https://www.netmanias.com/en/post/techdocs/5902/lte-security/lte-security-i-concept-and-authentication>, 2013 (accessed 14.05.19).

[21] NMC Consulting Group, LTE network architecture: basic, <https://www.netmanias.com/en/post/techdocs/5904/lte-network-architecture/lte-network-architecture-basic>, 2013 (accessed 14.05.19).

[22] A. Bikos, N. Sklavos, Architecture design of an area efficient high speed crypto processor for 4G LTE, IEEE Trans. Dependable Secure Comput. 15 (5) (2018) 729−741.

[23] B. Wang, L. Liu, A flexible and energy-efficient reconfigurable architecture for symmetric cipher processing, in: Proceedings of 2015 IEEE International Symposium on Circuits and Systems, ISCAS15, 24−27 May 2015, Lisbon, Portugal, IEEE, Piscataway, NJ, 2015, pp. 1182−1185.

[24] J. Blömer, J.-P. Seifert, Fault based cryptanalysis of the advanced encryption standard (AES), in: Proceedings of International Conference on Financial Cryptography, FC03, 27−30 January 2003, Guadeloupe, French West Indies, Springer-Verlag, Berlin, Heidelberg, 2003, pp. 162−181.

[25] J. Daemen, V. Rijmen, The Design of Rijndael: AES-the Advanced Encryption Standard, Springer, Berlin, Heidelberg, 2002.

[26] H. Wu, T. Huang, P.H. Nguyen, H. Wang, S. Ling, Differential attacks against stream cipher ZUC, in: Proceedings of International Conference on the Theory and Application of Cryptology and Information Security, ASIACRYPT12, 2−6 December 2012, Beijing, China, Springer-Verlag, Berlin, Heidelberg, 2012, pp. 262−277.

[27] L. Zhang, L. Xia, Z. Liu, J. Jing, Y. Ma, Evaluating the optimized implementations of SNOW3G and ZUC on FPGA, in: Proceedings of 11th IEEE International Conference on Trust, Security and Privacy in Computing and Communications, TrustCom12, 25−27 June 2012, Liverpool, UK, IEEE, Piscataway, NJ, 2012, pp. 436−442.

[28] E. Rescorla, The transport layer security (TLS) protocol version 1.3. <https://tools.ietf.org/html/rfc8446>, 2018 (accessed 14.05.19).

[29] 3rd Generation Partnership Project (3GPP), 3GPP TS33.401. 3GPP system architecture evolution (SAE); security architecture. <https://portal.3gpp.org/desktopmodules/Specifications/SpecificationDetails. aspx?specificationId = 2296>, 2015 (accessed 14.05.19).

[30] J. Navarro-Ortiz, S. Sendra, P. Ameigeiras, J. Lopez-Soler, Integration of LoRaWAN and 4G/5G for the industrial internet of things, IEEE Commun. Mag. 56 (2) (2018) 60−67.

[31] Y.P.E. Wang, X. Lin, A. Adhikary, A. Grovlen, Y. Sui, Y. Blankenship, et al., A Primer on 3GPP narrow-band Internet of things, IEEE Commun. Mag. 55 (3) (2017) 117−123.

[32] C.-M. Huang, J.-W. Li, Authentication and key agreement protocol for UMTS with low bandwidth consumption, in: Proceedings of 19th International Conference on Advanced Information Networking and Applications, AINA'05, 28−30 March 2005, Taipei, Taiwan, IEEE, Piscataway, NJ, 2005, pp. 392−397.

[33] A.G. Martín, R.P. Leal, A.G. Armada, A.F. Durán, NB-IoT random access procedure: system simulation and performance, in: Proceedings of 2018 Global Information Infrastructure and Networking Symposium, GIIS, 23−25 October 2018, Thessaloniki, Greece, IEEE, Piscataway, NJ, 2018, pp. 1−5.

[34] D. Forsberg, G. Horn, W.-D. Moeller, V. Niemi (Eds.), LTE Security, John Wiley & Sons, Chichester, West Sussex, UK, 2012.

[35] Y. Zhang, J. Chen, H. Li, J. Cao, C. Lai, Group-based authentication and key agreement for machine-type communication, Int. J. Grid Util. Comput. 5 (2) (2014) 87−95.

[36] 3rd Generation Partnership Project (3GPP), 3GPP TS35.216. Specification of the 3GPP confidentiality and integrity algorithms UEA2 and UIA2. <https://portal.3gpp.org/desktopmodules/Specifications/ SpecificationDetails. aspx?specificationId = 2396>, 2018 (accessed 14.05.19).

[37] NIST, Advanced Encryption Standard (AES) [Federal Information Processing Standards Publication 197]. <https://nvlpubs.nist.gov/nistpubs/FIPS/NIST.FIPS.197.pdf>, 2001 (accessed 14.05.19).

[38] 3rd Generation Partnership Project (3GPP), 3GPP TS35.221. Specification of the 3GPP confidentiality and integrity algorithms EEA3 and EIA3. <https://portal.3gpp.org/desktopmodules/Specifications/ SpecificationDetails. aspx?specificationId = 2399>, 2018 (accessed 14.05.19).

[39] 3rd Generation Partnership Project (3GPP), 3GPP TS23.002: Technical Specification Group services and system aspects; network architecture, Release 8, <https://portal.3gpp.org/desktopmodules/ Specifications/SpecificationDetails. aspx?specificationId = 728>, 2007 (accessed 14.05.19).

[40] 3rd Generation Partnership Project (3GPP), 3GPP TS23.501:Technical Specification Group services and system aspects; system architecture for the 5G system, Release 15. <https://portal.3gpp.org/desktopmo-dules/Specifications/SpecificationDetails.aspx?specificationId = 3144>, 2018 (accessed 14.05.19).

[41] Next Generation Mobile Networks Alliance, NGMN 5G white paper. <https://www.ngmn.org/fileadmin/ ngmn/content/images/news/ngmn_news/NGMN_5G_White_Paper_V1_0.pdf>, 2015 (accessed 14.05.19).

[42] R. Bassil, A. Chehab, I. Elhajj, & A. Kayssi, Signaling oriented denial of service on LTE networks, in: Proceedings of 10th ACM International Symposium on Mobility Management and Wireless Access, MobiWac'12, 24–25 October 2012, Paphos, Cyprus, ACM, New York, NY, 2012, pp. 153–158.

[43] A. Costin, J. Zaddach, A. Francillon, & D. Balzarotti, A large-scale analysis of the security of embedded firmwares, in: Proceedings of 23rd USENIX Security Symposium, USENIX Security, 20–22 August 2014, San Diego, CA, Berkeley, CA, pp. 95–110.

[44] M. Liyanage, I. Ahmad, A.B. Abro, A. Gurtov, M. Ylianttila (Eds.), Comprehensive Guide to 5G Security, John Wiley & Sons, Chichester, West Sussex, UK, 2018.

[45] I. Ahmad, S. Shahabuddin, T. Kumar, J. Okwuibe, A. Gurtov, & M. Ylianttila, Security for 5G and beyond, IEEE Commun. Surv. Tutor., 21 (2019) 3682–3722. Available From: https://doi.org/10.1109/COMST.2019.2916180.

[46] C. Cheng, R. Lu, A. Petzoldt, T. Takagi, Securing the Internet of things in a quantum world, IEEE Commun. Mag. 55 (2) (2017) 116–120.

[47] 5G PPP security WG.

[48] J. Cao, P. Yu, M. Ma, W. Gao, Fast authentication and data transfer scheme for massive NB-IoT devices in 3GPP 5G network, IEEE Internet Things J. 6 (2) (2018) 1561–1575.

[49] C. Lai, H. Zhou, N. Cheng, X.S. Shen, Secure group communications in vehicular networks: a software-defined network-enabled architecture and solution, IEEE Veh. Technol. Mag. 12 (4) (2017) 40–49.

18

Hardware and software platforms for low-power wide-area networks

Anjali Askhedkar[1], Bharat Chaudhari[1], Marco Zennaro[2]

[1]SCHOOL OF ELECTRONICS AND COMMUNICATION ENGINEERING, MIT WORLD PEACE UNIVERSITY, PUNE, INDIA [2]T/ICT4D LABORATORY, THE ABDUS SALAM INTERNATIONAL CENTRE FOR THEORETICAL PHYSICS, TRIESTE, ITALY

18.1 Introduction

Low-power wide-area networks (LPWANs) are gaining attention as it offers long-range and wide-area communication at low power and low cost for both the devices and infrastructure, not provided by legacy wireless technologies. This emerging paradigm of Internet of things (IoT) overcomes the range limits and scalability issues of traditional short-range wireless sensor networks (WSNs). LPWAN technologies support a variety of applications including smart city, smart grid and smart metering, home automation and safety, logistics, industrial assets monitoring, critical infrastructure monitoring, wildlife monitoring and tracking, agriculture, health care, and many others. Recently, there are several LPWAN [1] technologies proposed, namely LoRa, SigFox, IQRF, random phase multiple access, DASH7, Weightless, SNOW, long-term evolution category M1, extended coverage - global system for mobile - internet of things, narrowband-IoT (NB-IoT), and 5G-based. These technologies have the potential to wirelessly connect a large number of devices that are geographically spread. Although the performance of some of these technologies is yet to be fully explored, many variants are already available [2].

The purpose of any IoT device is to sense some parameter and connect with other IoT devices and applications to communicate information using the Internet. An IoT platform connects the data network to the sensor arrangement and uses back-end applications to provide understanding about the vast amount of data generated by hundreds of sensors. Many such platforms are now available, which facilitate the deployment of IoT applications [3]. Similarly, LPWAN platform consists of a large number of connected objects, spread over a large area, connecting them to gateways and data networks to cloud services and applications. A centralized management or processing unit controls the connected devices. The generic block diagram of LPWAN device (node) is shown in Fig. 18−1.

LPWAN platform is basically made up of (1) a sensing and actuating component that include sensors, actuators, and devices; (2) a communication and identification component

LPWAN Technologies for IoT and M2M Applications. DOI: https://doi.org/10.1016/B978-0-12-818880-4.00019-3

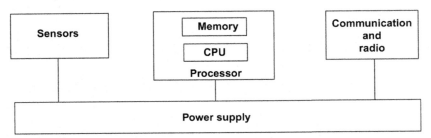

FIGURE 18–1 Generic block diagram of LPWAN node.

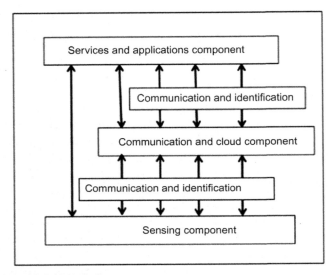

FIGURE 18–2 Components of LPWAN platform.

including the communication protocols and a gateway; (3) a computation and cloud component representing the tasks of processing unit; and (4) the services and applications component that represents the provided services and features offered to the user to connect and control. Fig. 18–2 illustrates the components of a typical LPWAN platform. Although the functionality provided by various existing platforms is similar, the implementation and base technology may differ. Choosing an appropriate platform that is scalable, flexible, available, and less expensive for a given field of application is a challenge [4].

The number of connected devices is rapidly increasing, and as a result, using simulators to study and prepare a project of installing new networks before real deployment is inevitable [5]. Appropriate tools to carry out what-if analysis and predeployment studies are required to understand the consequences of choices that are made at the time of design. Various platforms for the research and development of LPWAN, either open-source or commercial, are available today. In the following sections, currently available hardware and software platforms for LPWAN are discussed.

18.2 Hardware platforms

We present here a brief study and discussion about some of the significant LPWAN hardware platforms available.

18.2.1 Pycom platform

Pycom [6] offers powerful and economical MicroPython-enabled, multinetwork microcontroller development boards such as WiPy, LoPy, and SiPy for IoT. The firmware is open source enabling developers to download the source code and make contributions. The hardware includes a complete variety of development boards, expansion boards, sensor shields, and original equipment manufacturer hardware with a combination of different networks. A few steps can make an IoT out of the box possible: program any Pycom module with MicroPython, connect over Wi-Fi, LoRa, or SigFox, onboard the device to the cloud in seconds, send data with one command, prototype online with devices in the field, update firmware and application over-the-air (OTA), visualize data with charts and tables, create dashboards for applications, and integrate with third-party applications. The table given below lists various development boards and their communication capabilities. Some of the prominent Pycom boards and their features are listed in Table 18−1.

LoPy is a MicroPython triple network (LoRa/Wi-Fi/bluetooth low energy device) development platform that can also be configured as a LoRa Nano gateway. It is small in size, light in weight and low-power with years of battery life. It is compatible with Arduino integrated development environment (IDE) and supports new Pymakr IDE. LoRa specifications include a Semtech LoRa transceiver SX1272, long-range wide-area network (LoRaWAN) stack, Class A and C devices, and operating frequencies of 868 and 915 MHz. The node range is up to 40 km (when in line of sight), Nano gateway capacity is up to 100 nodes and range is up to 5 km (when in line of sight). It supports Wi-Fi networking with 802.11b/g/n. Pycom continues to fund the development of core MicroPython features to ensure that they stay open source. PyMakr is also open source. The Things Network [7] can be used on the LoPy as a node. The Things Network is a crowd-sourced, open IoT data network that allows for things to connect to the Internet using small power and low data rates.

Table 18–1 Pycom development boards.

Module	Wi-Fi	Bluetooth	LoRa	SigFox	LTE Cat-M1, NB-IoT
SiFy	✓	✓		✓	
GPy	✓	✓			✓
LoPy	✓	✓	✓		
LoPy4	✓	✓	✓	✓	
FiPy	✓	✓	✓	✓	✓

18.2.2 Lite gateways

LoRa lite gateway [8], a long-range radio is used as an evaluation platform. It consists of a precertified concentrator iC880A, a Raspberry Pi, and a preconfigured SD card. This device enables LoRa connection to existing servers (e.g., iot.semtech.com, Loriot, The Things Network). The open-source GitHub project LoRa-net (https://github.com/Lora-net/lora_gateway) provides the source code of the firmware.

18.2.3 iM880B-L

This is a compact, low-cost, long-range radio module that operates in the unlicensed 868 MHz band and combines a powerful Cortex-M3 controller with the new Semtech LoRa transceiver SX1272 [8]. A preprogrammed embedded LoRaWAN protocol stack is also available with iM880B-L. It enables LPWANs for IoT and machine-to-machine (M2M) applications and supports direct or OTA activation. It is easy to configure and comes with the WiMOD LoRaWAN host controller (HCI) interface.

18.2.4 Remote eye platform

Remote eye platform [9] helps to connect existing sensors to the Internet for remote monitoring via LPWAN networks such as SigFox, NB-IoT, and LoRa. It is an end-to-end, modular IoT platform (hardware plus software)—sensors, adaptor, low-power wireless connectivity, Cloud, and applications, that enables quick and easy deployment of IoT solutions for smart cities, utilities, buildings, factories, and agriculture. The main components are hardware device (rEye LPWAN adaptor), wireless connectivity (LPWAN), rEye IoT Cloud software, and rEye IoT mobile app.

18.2.5 Arm Cordio-N Internet protocol for narrowband-Internet of things

This is an ARM Cortex-M processor-based solution [10] path to integrate LPWAN IoT connectivity into their devices. It provides an Internet protocol (IP) block with layer control, digital front end, radio-frequency (RF) interface, and software. This protocol stack solution is designed for low-footprint memory, low-power, low-bandwidth (BW) IoT platforms, conforming to the latest Third Generation Partnership Project standard.

18.2.6 CableLabs LoRa server

LoRa server is a community-sourced open-source LoRaWAN network server for setting up and managing LoRaWAN networks [11]. The LPWAN server can be used to simply migrate or operate between two LoRaWAN network servers, such as the LoRa server and The Things Network. It can also be used to enable the routing of multiple LPWAN technologies such as LoRaWAN and SigFox or LoRaWAN and NB-IoT.

18.2.7 Libelium

Libelium [12] is a WSN platform provider that delivers open-source, low-power consumption devices that are easy to program and implement for smart cities solutions and a wide range of M2M and sensor applications. Libelium's Waspmote, the open-source wireless sensor platform, works with different protocols (ZigBee, Bluetooth, 3G/GPRS) and frequencies (2.4 GHz, 868 MHz, 900 MHz) and is capable of communicating over long distances with low-power consumption. The Libelium LoRaWAN module is integrated into the main sensor lines Waspmote and Plug & Sense. It has also been ported to Arduino and Raspberry Pi, which enables creating own LPWAN networks. To visualize the information, Actility, Orbiwise, and Loriot Cloud platforms could be used with these LoRaWAN radios.

18.2.8 The Things Uno and Nodes

The Things Uno board is based on the Arduino Leonardo with an additional Microchip LoRaWAN module and is fully compatible with the Arduino IDE and existing shields. It is suitable for prototyping IoT or wireless projects. It uses a long-range and low-power radio-frequency protocol called LoRaWAN and for short-range Bluetooth 4.2 and connects with The Things Network [7]. The Things Node is a LoRa node based on SparkFun Pro Micro with added Microchip LoRaWAN module and several sensors.

18.2.9 Mainflux

Mainflux [13] offers end-to-end, secure, scalable, open-source patent-free IoT cloud platform and IoT edge gateway with plug and play components for the development of complex IoT solutions and smart, connected products. It accepts user and thing connections over various network protocols such as Hypertext Transfer Protocol (HTTP), Message Queuing Telemetry Transport (MQTT), WebSocket, and constrained application protocol (CoAP). The platform serves as software infrastructure and middleware that support device management, data aggregation and management, connectivity and message routing, event management, core analytics, user interface, and application enablement services for IoT.

18.2.10 Silabs STK3400 Happy Gecko board

The module hosts an EFM32 Happy Gecko developer board [14] that combines the powerful but low-power, ARM Cortex-M0 +, with real-time power monitoring. Together with a customized extension board with LoRa transceiver, that is, Semtech SX1272 chip and different types of sensors, it creates a versatile IoT module for LPWAN IoT networks. The software as well as the hardware is open source.

18.2.11 OpenMote

OpenMote [15] is a modular open hardware ecosystem designed within Berkeley's OpenWSN open-source project, apt for industrial IoT standards such as IEEE802.15.4e TSCH and IETF 6TiSCH. An advanced computing and communication device, the OpenMote-CC2538, interfaces with other accessories or skins that include boards to provide power, boards that enable a developer to easily debug the platform, and boards to allow smooth integration of an OpenMote network into the Internet. It also offers a set of software tools and ports to common open-source IoT implementations. It is an open platform, giving users basic access to up-to-date hardware. Currently, work is being carried out to use it within several additional open-source IoT communities such as Contiki, RIOT, and FreeRTOS.

18.2.12 BigClown

BigClown [16] is an open-source ecosystem of hardware modules and software tools, specially designed for IoT. BigClown supports not only generic communication but can also communicate by other LPWANs such as LoRa, SigFox, and NB-IoT. With BigClown, a secure radio network for various devices can be built for home automation applications, security alarms, smart metering, etc.

18.2.13 Arduino-based platforms

18.2.13.1 Arduino MKR WAN 1300

The Arduino MKR WAN 1300 [17] is a learning and development board with the ATMEL SAMD21 microcontroller that has the advantages of the core's low-power consumption and high performance and user friendliness of Arduino. With a Murata CMWX1ZZABZ LoRa module, it adds LoRa connectivity to the MKR ZERO Arduino platform. The Arduino MKR WAN 1300 is programmed using the Arduino software IDE, common to all the boards and running both online and offline.

18.2.13.2 WiMOD Shield for Arduino

The WiMOD Shield [18] is a small, low-cost expansion board that enables users of Arduino/Arduino-compatible boards to use WiMOD radio modules based on LoRa technology. Along with the hardware board, an example software library compatible with the Arduino IDE and providing easy access to the WiMOD HCI interface is also available. Hence a prototype of a LoRaWAN end-node can be easily created.

18.2.13.3 Seeeduino LoRaWAN

Seeeduino LoRaWAN [19] is an Arduino development board with embedded LoRaWAN protocol. It is compatible with LoRaWAN Class A or C and supports various communication frequencies, using the communication module RHF76-052AM. The onboard standard Grove connectors allow it to connect with hundreds of Grove sensors and actuators from Seeedstudio easily. The board has embedded an integrated lithium battery management chip that allows charging via universal serial bus (USB) interface.

18.2.14 KRATOS

It is a low-cost, wireless, multisensor, dual-radio LoRa platform running Contiki OS [20] that allows researchers to design and explore energy-efficient protocols for LoRa networks. Both hardware and software designs are released as an open-source to the research community that can download the code, fabricate LoRa mote, and write applications. Being an open-source hardware platform, KRATOS is useful for rapid prototyping and testing of large-scale LoRa networks using commercial off-the-shelf components.

18.2.15 Low-power wide-area network universal serial bus dongle

There are various USB dongles available that support LPWAN capabilities. A few prominent ones are discussed here.

18.2.15.1 Long-range wide-area network universal serial bus dongle

This LoRaWAN USB dongle [21] supports the AT command set and plug and play with Raspberry Pi, Arduino, LinKit, and other popular IoT platforms and is designed to provide LoRa modem functionality. This enables developers to do rapid prototyping of LoRa-enabled sensors. It can be connected to a Raspberry Pi that is connected to the desired sensors and programmed to read the sensors and send data via LoRa up to a network server in few hours.

18.2.15.2 LoStik universal serial bus dongle

Previously known as LoRa Stick, LoStik [22] is an open-source hardware USB dongle that plugs into any computer or device and provides LoRa network connectivity. Using the Microchip RN2903/RN2483 system on module, it can be employed in packet and LoRaWAN modes, making it compatible with The Things Network. LoStik also works on Linux or with boards such as Raspberry Pi or BeagleBone. It is offered on Crowd Supply for reasonable rates, while the hardware and software sources are available on the project's GitHub repository. The LoStik enables you to connect to a LoRa network faster, diagnose network issues more quickly, and build new and exciting connected devices.

18.2.15.3 LD-20 LoRa universal serial bus dongle

GlobalSat LD-20 [23] is a low-power, half-duplex dongle that can wirelessly transmit data to long distance. It has a built-in high-speed and low-power MCU and SX1276 modulation chipset. It allows any personal computer (PC) or Raspberry Pi to monitor sensor data and function as a low-cost LoRa private network gateway.

18.2.16 Universal software radio peripheral

The universal software radio peripheral device is a tunable transceiver for designing, prototyping, and deploying wireless communication systems. Paired with the LabVIEW development and GNU Radio [24] environment, National Instruments universal software radio peripherals (USRPs) [25] and Ettus USRPs [26] provide an affordable solution to validate wireless algorithms with OTA signals. Ettus USRPs come with the necessary Ettus Research's software defined radio (SDR) software platform. All platforms support USRP hardware driver (UHD), which ensures cross-platform code portability. UHD also supports Linux, Windows, and mac OS. All USRP SDRs support GNU Radio, a free open-source software development framework. Most USRP SDRs also have support for the following:

- RFNoC, an open-source software package from Ettus Research that integrates into GNU Radio, enabling field-programmable gate array (FPGA) development without having to write VHDL or Verilog.
- LabVIEW, a graphical programming tool for managing complex system configurations, multirate digital signal processing design of the FPGA and float-to-fixed point conversion.
- MATLAB and Simulink, which connect to the USRP family of SDRs to provide an environment for SISO and MIMO wireless system design, prototyping, and verification.

18.3 Software platforms

The number of connected devices is rapidly increasing. Simulators play an important role in the study and analysis of a network prior to deployment. A network simulator is a relatively inexpensive tool for the design evaluation that provides an early insight into different aspects that can affect a network design's performance and predict the feasibility of the deployment in terms of location, interferences, communication, and cost [5]. Some prominent software platforms and network simulators for LPWAN are briefly discussed here.

18.3.1 CupCarbon

CupCarbon [5,27] is a smart city and IoT WSN (SCI-WSN) simulator developed to design, visualize, debug, test, and validate distributed algorithms for monitoring, environmental data collection, and to create various environmental and mobility scenarios generally within educational and scientific projects. Networks can be designed and prototyped using the OpenStreetMap framework to deploy sensors directly on the map. SenScript allows to program and configure each sensor node individually. It is also possible to generate codes for hardware platforms such as Arduino/XBee. CupCarbon permits dynamic configuration of the nodes so as to split nodes into separate networks or to join different networks, including ZigBee, LoRa, and Wi-Fi protocols.

CupCarbon represents the main kernel of the ANR project PERSEPTEUR that aims to develop algorithms for an accurate simulation of the propagation and interference of signals in a 3D urban environment.

18.3.2 LoRaSim and extended LoRaSim

LoRaSim [28], a LoRa network simulator, is a discrete-event simulator developed using the Python programming language's SimPy package. The simulator implements a radio propagation model based on the well-known log-distance path loss model. The sensitivity of a radio transceiver at room temperature with respect to different LoRa spreading factors and BWs settings can be calculated. It is useful for simulating collisions in LoRa networks and to analyze scalability. LoRaSim allows to simulate LoRaWAN networks with a single application. Practical deployments of multiple IoT applications such as utility meters, intelligent transport, parking systems, and many such applications are deployed on a single LoRaWAN network. Extended LoRaSim [29] allows simulating a LoRaWAN running multiple applications. LoRaWANSim is a simulator which extends the LoRaSim tool to add support for the LoRaWAN MAC protocol, which employs bidirectional communication. This is a prominent feature not available in any other LoRa simulator. Consequently, it is suitable to predict the performance of LoRaWAN-based networks such as achievable network capacity and energy consumption versus reliability trade-offs associated with the choice of number of retransmission attempts through extensive simulations.

LoRaWAN packet generator [30] is a command-line tool for generation of user datagram protocol (UDP) packets that can be sent from the PC host to the LoRa network server. It simulates LoRaWAN gateway and sends the UDP packages defined by the "Gateway to Server Interface Protocol" defined in Semtech document ANNWS.01.2.1.W.SYS [31]. Basically, it acts as a LoRa node and a gateway and useful to test the LoRa network server deployments and integrations in the absence of expensive network hardware setup.

18.3.3 Other simulators

Simple IoT simulator [32] is an easy to use IoT device simulator that creates test environments made up of multiple sensors and gateways, all on just one computer. It supports many of the common IoT protocols including MQTT, CoAP, HTTP, and LoRa and also IPv4 and IPv6. Scripted error scenarios can also be created. Example scripts to work with popular platforms such as Azure IoT, Amazon AWS, IBM Bluemix, and others are available to facilitate quick setup. It can be used to simulate LoRaWAN networks to demonstrate and test user application software and LoRa network servers.

FLoRa (framework for LoRa) [33] is a simulation framework that allows the creation of LoRa networks with modules for LoRa nodes, gateways, and a network server, for carrying out end-to-end simulations for LoRa networks. It is based on the OMNeT++ network simulator and uses components from the INET framework as well. It supports dynamic management of configuration parameters, and the energy consumption statistics can be collected for every node.

myDevices Cayenne [34] is a drag and drop IoT platform to create IoT prototypes. It consists of Cayenne Mobile Apps to remotely monitor and control IoT devices from the Android or iOS Apps and Cayenne Online Dashboard that uses customizable widgets to visualize data, set up rules, schedule events, and more. It requires a Raspberry Pi or Arduino device connected to the Internet, or a LoRa device connected to a public or private gateway. It is designed to work from iOS and Android smartphones and popular browsers.

THiNX [35] is an open-source platform that supports the latest ATMega168/ATMega328, ESP32 and ESP8266, ILI7697 MCUs together with Arduino, and all legacy platforms. THiNX Javascript library supports basically any platform running on Linux—such as Onion, Raspberry Pi, and even PCs. It also supports SigFox and LoRaWAN.

NS-3 [36] is a discrete-event network simulator for Internet systems, primarily for research and educational use. It is free software, licensed under the GNU GPLv2 license, and is publicly available for research, development, and use. LoRaWAN networks can be modeled in NS-3.

18.3.4 The Things Network

The Things Network, also known as TTN [7], is a crowd-sourced open infrastructure aiming to provide free LoRaWAN network platform. This project is developed by a growing community across the world and is based on voluntary contributions. It provides a set of open tools and a global network to build a secure and scalable IoT application at low cost. The TTN supports LoRaWAN for long-range, low-power, and low-BW communication. It supports the Things Nodes, Uno, and any certified LoRaWAN devices. To connect a device, TTN needs to have a LoRaWAN module either on board as a shield or wired. Most modules communicate via a serial interface. Users can connect their devices via TTN LoRaWAN gateways and access their application data on TTN GUI globally.

18.3.5 GNU Radio

GNU Radio [37] is a free and open-source software development toolkit that provides signal processing blocks to implement software radios. It can be used with readily available low-cost external RF hardware to create SDRs, or without hardware in a simulation-like environment. GNU Radio is widely used in research, industry, government, and academia to support both wireless communications research and real-world radio systems. The GNU Radio applications themselves are generally known as flowgraphs, which are a series of signal processing blocks connected together, thus describing a data flow. The flowgraphs can be written in either C++ or the Python programming language. GNU Radio can be used to develop implementations of basically any band-limited communication standard, including LPWAN for the different applications.

References

[1] D. Ismail, M. Rahman, A. Saifullah, Low-power wide-area networks: opportunities, challenges, and directions, Proceedings of the Workshop Program of the 19th International Conference on Distributed Computing and Networking, ACM, 2018, p. 8.

[2] J. Petajajarvi, K. Mikhaylov, A. Roivainen, T. Hanninen, M. Pettissalo, On the coverage of LPWANs: range evaluation and channel attenuation model for LoRa technology, 2015 14th International Conference on ITS Telecommunications (ITST), IEEE, 2015, pp. 55–59.

[3] <https://internetofthingswiki.com/top-20-iot-platforms/634/>.

[4] H. Hamdan, H. Rajab, Tr Cinkler, L. Lengyel, Survey of platforms for massive IoT, 2018 IEEE International Conference on Future IoT Technologies (Future IoT), IEEE, 2018, pp. 1–8.

[5] A. Bounceur, L. Clavier, P. Combeau, O. Marc, R. Vauzelle, A. Masserann, et al., CupCarbon: a new platform for the design, simulation and 2D/3D visualization of radio propagation and interferences in IoT networks, 2018 15th IEEE Annual Consumer Communications & Networking Conference (CCNC), IEEE, 2018, pp. 1−4.

[6] <https://pycom.io/>.

[7] <https://www.thethingsnetwork.org>.

[8] <https://www.wireless-solutions.de>.

[9] <https://www.spaceagelabs.com.sg>.

[10] <https://www.arm.com>.

[11] <https://www.cablelabs.com>.

[12] <www.libelium.com>.

[13] <https://www.mainflux.com>.

[14] G. Callebaut, et al., Remote IoT devices: sleepy strategies and signal processing to the rescue for a long battery life, arXiv:1901.06836v1 [eess.SP], 21 January 2019.

[15] <www.openwsn.org>.

[16] <https://developers.bigclown.com/basics/about-bigclown>.

[17] <https://www.arduino.cc/en/Guide/MKRWAN1300>.

[18] <https://www.thethingsnetwork.org/marketplace/product/wimod-shield-for-arduinotm>.

[19] <https://www.seeedstudio.com>.

[20] <https://contikios4lora.github.io/contikios-lora>.

[21] <https://www.giotnetwork.com/end_node/node_dongle.jsp>.

[22] <https://www.cnx-software.com/2018/06/29/lora-stick-lora-usb-dongle>.

[23] <http://www.vinduino.com/portfolio-view/lora-usb-dongle>.

[24] <https://www.gnuradio.org>.

[25] <http://www.ni.com>.

[26] <https://www.ettus.com>.

[27] <http://www.cupcarbon.com>.

[28] <https://www.lancaster.ac.uk/scc/sites/lora>.

[29] M.O. Farooq, D. Pesch, Poster: Extended LoRaSim to simulate multiple IoT applications in a LoRaWAN, in: International Conference on Embedded Wireless Systems and Networks (EWSN) 2018, 14−16 February, Madrid, Spain.

[30] <https://github.com/donadonny/lora-pktgen>.

[31] <https://www.semtech.com>.

[32] <https://www.simplesoft.com>.

[33] <https://flora.aalto.fi>.

[34] <http://mydevices.com/cayenne/docs/intro>.

[35] <https://thinx.cloud>.

[36] <https://www.nsnam.org>.

[37] <https://www.gnuradio.org>.

Index

Note: Page numbers followed by "*f*" and "*t*" refer to figures and tables, respectively.